43 iwe 530
esp 529
- 10. Expl. -

IET ELECTROMAGNETIC WAVES SERIES 48

Series Editors: Professor P.J.B. Clarricoats
Professor E.V. Jull

Theory and Design of Microwave Filters

Other volumes in this series:

Volume 1	**Geometrical theory of diffraction for electromagnetic waves, 3rd edition** G.L. James
Volume 10	**Aperture antennas and diffraction theory** E.V. Jull
Volume 11	**Adaptive array principles** J.E. Hudson
Volume 12	**Microstrip antenna theory and design** J.R. James, P.S. Hall and C. Wood
Volume 15	**The handbook of antenna design, volume 1** A.W. Rudge, K. Milne, A.D. Oliver and P. Knight (Editors)
Volume 16	**The handbook of antenna design, volume 2** A.W. Rudge, K. Milne, A.D. Oliver and P. Knight (Editors)
Volume 18	**Corrugated horns for microwave antennas** P.J.B. Clarricoats and A.D. Oliver
Volume 19	**Microwave antenna theory and design** S. Silver (Editor)
Volume 21	**Waveguide handbook** N. Marcuvitz
Volume 23	**Ferrites at microwave frequencies** A.J. Baden Fuller
Volume 24	**Propagation of short radio waves** D.E. Kerr (Editor)
Volume 25	**Principles of microwave circuits** C.G. Montgomery, R.H. Dicke and E.M. Purcell (Editors)
Volume 26	**Spherical near-field antenna measurements** J.E. Hansen (Editor)
Volume 28	**Handbook of microstrip antennas, 2 volumes** J.R. James and P.S. Hall (Editors)
Volume 31	**Ionospheric radio** K. Davies
Volume 32	**Electromagnetic waveguides: theory and application** S.F. Mahmoud
Volume 33	**Radio direction finding and superresolution, 2nd edition** P.J.D. Gething
Volume 34	**Electrodynamic theory of superconductors** S.A. Zhou
Volume 35	**VHF and UHF antennas** R.A. Burberry
Volume 36	**Propagation, scattering and diffraction of electromagnetic waves** A.S. Ilyinski, G. Ya.Slepyan and A. Ya.Slepyan
Volume 37	**Geometrical theory of diffraction** V.A. Borovikov and B.Ye. Kinber
Volume 38	**Analysis of metallic antenna and scatterers** B.D. Popovic and B.M. Kolundzija
Volume 39	**Microwave horns and feeds** A.D. Olver, P.J.B. Clarricoats, A.A. Kishk and L. Shafai
Volume 41	**Approximate boundary conditions in electromagnetics** T.B.A. Senior and J.L. Volakis
Volume 42	**Spectral theory and excitation of open structures** V.P. Shestopalov and Y. Shestopalov
Volume 43	**Open electromagnetic waveguides** T. Rozzi and M. Mongiardo
Volume 44	**Theory of nonuniform waveguides: the cross-section method** B.Z. Katsenelenbaum, L. Mercader Del Rio, M. Pereyaslavets, M. Sorella Ayza and M.K.A. Thumm
Volume 45	**Parabolic equation methods for electromagnetic wave propagation** M. Levy
Volume 46	**Advanced electromagnetic analysis of passive and active planar structures** T. Rozzi and M. Farinai
Volume 47	**Electromagnetic mixing formulae and applications** A. Sihvola
Volume 48	**Theory and design of microwave filters** I.C. Hunter
Volume 49	**Handbook of ridge waveguides and passive components** J. Helszajn
Volume 50	**Channels, propagation and antennas for mobile communications** R. Vaughan and J. Bach-Anderson
Volume 51	**Asymptotic and hybrid methods in electromagnetics** F. Molinet, I. Andronov and D. Bouche
Volume 52	**Thermal microwave radiation: applications for remote sensing** C. Matzler (Editor)
Volume 502	**Propagation of radiowaves, 2nd edition** L.W. Barclay (Editor)

Theory and Design of Microwave Filters

Ian Hunter

The Institution of Engineering and Technology

Published by The Institution of Engineering and Technology, London, United Kingdom

First edition © 2001 The Institution of Electrical Engineers
Reprint with new cover © 2006 The Institution of Engineering and Technology

First published 2001
Reprinted 2006

This publication is copyright under the Berne Convention and the Universal Copyright Convention. All rights reserved. Apart from any fair dealing for the purposes of research or private study, or criticism or review, as permitted under the Copyright, Designs and Patents Act, 1988, this publication may be reproduced, stored or transmitted, in any form or by any means, only with the prior permission in writing of the publishers, or in the case of reprographic reproduction in accordance with the terms of licences issued by the Copyright Licensing Agency. Inquiries concerning reproduction outside those terms should be sent to the publishers at the undermentioned address:

The Institution of Engineering and Technology
Michael Faraday House
Six Hills Way, Stevenage
Herts, SG1 2AY, United Kingdom

www.theiet.org

While the author and the publishers believe that the information and guidance given in this work are correct, all parties must rely upon their own skill and judgement when making use of them. Neither the author nor the publishers assume any liability to anyone for any loss or damage caused by any error or omission in the work, whether such error or omission is the result of negligence or any other cause. Any and all such liability is disclaimed.

The moral rights of the author to be identified as author of this work have been asserted by him in accordance with the Copyright, Designs and Patents Act 1988.

British Library Cataloguing in Publication Data
Hunter, I.C. (Ian C.)
  Theory and design of microwave filters.
  – (IEE electromagnetic waves series; no. 48)
  1. Microwave filters
  I. Title
  621.3'81331

ISBN (10 digit) 0 85296 777 2
ISBN (13 digit) 978-0-85296-777-5

Printed in the UK by University Press, Cambridge
Reprinted in the UK by Lightning Source UK Ltd, Milton Keynes

To Anne, for understanding that a blank stare is not a sign of madness.

Contents

Foreword	xi
Preface	xiii

1	**Introduction**	**1**
	1.1 Applications of RF and microwave filters	1
	1.2 Ideal lowpass filters	5
	1.3 Minimum phase networks	7
	1.4 Amplitude approximation	10
	1.5 Practical realisations of microwave filters	12
	1.6 Summary	12
	1.7 References	13
2	**Basic network theory**	**15**
	2.1 Linear passive time-invariant networks	15
	2.1.1 Linearity	16
	2.1.2 Time invariance	16
	2.1.3 Passivity	17
	2.1.4 The bounded real condition	17
	2.2 Lossless networks	18
	2.3 Ladder networks	21
	2.4 Synthesis of two-port networks – Darlington synthesis	22
	2.4.1 Cascade synthesis	25
	2.4.2 All-pole networks	29
	2.5 Analysis of two-port networks – the $ABCD$ matrix	29
	2.6 Analysis of two-port networks – the scattering matrix	34
	2.6.1 Relationships between $ABCD$ parameters and S parameters	40
	2.7 Even- and odd-mode analysis of symmetrical networks	41
	2.8 Analysis by image parameters	44
	2.9 Analysis of distributed circuits	46
	2.10 Summary	47
	2.11 References	48

viii *Contents*

3 Designs of lumped lowpass prototype networks — 49
- 3.1 Introduction — 49
- 3.2 The maximally flat prototype — 49
  - 3.2.1 Impedance inverters — 55
- 3.3 The Chebyshev prototype — 56
- 3.4 The elliptic function prototype — 64
- 3.5 The generalised Chebyshev prototype — 68
- 3.6 Filters with specified phase and group delay characteristics — 71
  - 3.6.1 The maximally flat group delay lowpass prototype — 73
  - 3.6.2 The equidistant linear phase approximation — 75
  - 3.6.3 Combined phase and amplitude approximation — 77
- 3.7 Filters with specified time domain characteristics — 78
- 3.8 Synthesis of generalised Chebyshev filters — 81
  - 3.8.1 Synthesis of generalised Chebyshev prototypes with symmetrically located transmission zeros — 81
  - 3.8.2 Synthesis of generalised Chebyshev prototypes with ladder-type networks — 86
  - 3.8.3 Asymmetrically located transmission zeros — 90
- 3.9 Summary — 98
- 3.10 References — 99

4 Circuit transformations on lumped prototype networks — 101
- 4.1 Introduction — 101
- 4.2 Impedance transformations — 101
  - 4.2.1 Example — 103
- 4.3 Lowpass to arbitrary cut-off frequency lowpass transformation — 103
  - 4.3.1 Example — 104
- 4.4 Lowpass to highpass transformation — 105
  - 4.4.1 Example — 106
- 4.5 Lowpass to bandpass transformation — 107
  - 4.5.1 Example — 114
  - 4.5.2 Nodal admittance matrix scaling — 116
- 4.6 Lowpass to bandstop transformation — 118
  - 4.6.1 Design example — 121
- 4.7 Effects of losses on bandpass filters — 125
- 4.8 Practical procedures — 131
  - 4.8.1 Measurement of input coupling — 131
  - 4.8.2 Measurement of inter-resonator coupling — 133
  - 4.8.3 Measurement of resonator Q factor — 134
- 4.9 Summary — 135
- 4.10 References — 136

5 TEM transmission line filters — 137
- 5.1 Commensurate distributed circuits — 137
- 5.2 Stepped impedance unit element prototypes — 144
  - 5.2.1 Physical realisation of stepped impedance lowpass filters — 149

5.3		Broadband TEM filters with generalised Chebyshev characteristics	151
	5.3.1	Generalised Chebyshev highpass filters	162
5.4		Parallel coupled transmission lines	165
5.5		The interdigital filter	167
	5.5.1	Design example	173
	5.5.2	Narrowband interdigital filters	174
	5.5.3	Design example	176
	5.5.4	Physical design of the interdigital filter	177
5.6		The combline filter	182
	5.6.1	Design example	192
	5.6.2	Tunable combline filters	194
5.7		The parallel coupled-line filter	194
5.8		Narrowband coaxial resonator filters	197
5.9		Summary	198
5.10		References	199

6 Waveguide filters — 201

6.1		Introduction	201
6.2		Basic theory of waveguides	202
	6.2.1	TE modes	203
	6.2.2	TM modes	208
	6.2.3	Relative cut-off frequencies of modes	209
	6.2.4	Rectangular waveguide resonators	209
	6.2.5	Numerical example	210
	6.2.6	Spurious resonances	212
	6.2.7	Circular waveguides	212
	6.2.8	TE modes	213
	6.2.9	TM modes	215
	6.2.10	Circular waveguide resonators	217
	6.2.11	Numerical example	217
6.3		Design of waveguide bandpass filters	220
	6.3.1	Design example	228
6.4		The generalised direct-coupled cavity waveguide filter	230
6.5		Extracted pole waveguide filters	239
	6.5.1	Realisation in waveguide	249
	6.5.2	Design example	252
	6.5.3	Realisation in TE_{011} mode cavities	254
6.6		Dual-mode waveguide filters	255
	6.6.1	Numerical example	263
	6.6.2	Asymmetric realisations for dual-mode filters	265
6.7		Summary	267
6.8		References	268

x *Contents*

7 Dielectric resonator filters 271
7.1 Introduction 271
7.2 Dielectric rod waveguides and the $TE_{01\delta}$ mode 272
7.3 Dual-mode dielectric resonator filters 283
7.3.1 Dual-mode conductor-loaded dielectric resonator filters 285
7.4 Triple-mode dielectric resonator filters 290
7.4.1 Spherical dielectric resonators 290
7.4.2 Cubic dielectric resonators 298
7.4.3 Design of triple-mode dielectric resonator reflection filters 301
7.4.4 Design example 305
7.5 Dielectric-loaded filters 309
7.5.1 Dielectric-loaded waveguide filters 314
7.6 Summary 317
7.7 References 318

8 Miniaturisation techniques for microwave filters 321
8.1 Introduction 321
8.2 Dielectric resonator filters 322
8.3 Superconducting filters 322
8.4 Surface acoustic wave filters 323
8.5 Active microwave filters 325
8.6 Lossy filters 327
8.6.1 Design of lossy filters – classical predistortion 331
8.6.2 Design of lossy filters – reflection mode type 335
8.6.3 Design example 339
8.7 Summary 343
8.8 References 344

Index 345

Foreword

The microwave region of the electromagnetic spectrum has certain unique properties. These enable microwave signals to propagate over long distances through the atmosphere under all but the most severe weather conditions. Both civilian and military applications abound, including radar, navigation, and the latest "hot application", wireless communications. However the microwave spectrum is a finite resource which must be divided, cared for, and treated with respect. And this is where microwave filters come in. Although the now classic book on microwave filters and couplers by Matthaei, Young and Jones, published 36 years ago, was never revised, it is still widely used as a handbook. However it needs updating. Dr Ian Hunter's book is therefore a significant event as it includes filter types and design theories which simply did not exist (either in concept or practice) 36 years ago. Dr Hunter has himself been active and enthusiastic in developing and enlarging some of these new technologies. He has also taught University courses on Microwave filters and his ability to elucidate and communicate the subject is evident in these pages. This book will be most useful to serious students of the subject, as well as to practitioners of the art and science of microwave filters.

Dr Leo Young

Preface

Microwave filters are vital components in a huge variety of electronic systems, including cellular radio, satellite communications and radar. The specifications on these devices are usually severe, often approaching the limit of what is theoretically achievable in terms of frequency selectivity and phase linearity. Consequently an enormous amount of published material on this topic is available and anyone new to the subject is in danger of being overwhelmed with information. The design of filters is unusual in that it uses network synthesis, with which it is possible to apply systematic procedures to work forward from a specification to a final theoretical design. This is the converse of most engineering disciplines which tend to use design rules based on analysis. A pre-requisite to skills in network synthesis is a thorough grounding in the circuit theory of passive networks, a subject often treated superficially in modern electrical engineering degree courses. However, a knowledge of network synthesis is not the only tool needed in order to design filters. Synthesis provides the designer with a prototype network which can then be transformed into a variety of microwave networks including TEM transmission lines, waveguides and dielectric resonator realisations. Thus the designer also has to have a reasonable knowledge of the properties of the electromagnetics of these devices. This book evolved from a series of lectures on filter design which the author gave to engineers at Filtronic plc and MSc students at Leeds and Bradford Universities. The purpose of the book is to provide a single source for filter design which includes basic circuit theory, network synthesis and the design of a variety of microwave filter structures. The philosophy throughout the book is to present design theories, followed by specific examples with numerical simulations of the designs. Where possible pictures of real devices have been used to illustrate the theory.

It is expected that the book will be useful to final year undergraduate, MSc and PhD students. It should also form a useful reference for research workers and engineers who are designing and/or specifying filters for commercial systems.

I would like to thank Filtronic plc and the University of Leeds, Institute of Microwaves and Photonics, for allowing me time to write this book. I would also like to thank the following for providing practical and moral support: Duncan Austin, Christine Blair, Stephen Chandler, Peter Clarricoats, Vanessa Dassonville, John Dean, Wael Fathelbab, Keith Ferguson, Dharshika Fernando, Patrick Geraghty, Peter Hardcastle, Eric Hawthorn, Kimmo Koskiniemni, Neil McEwan, Chris Mobbs, Marco Morelli, Richard Parry, Sharon Pickles, Richard Ranson, David Rhodes, Richard Rushton, Philip Sleigh, Chris Snowden, and Stewart Walker.

Ian Hunter, April 2000
Filtronic plc
The Waterfront
Salts Mill Road
Saltaire
West Yorkshire
BD18 3TT
England

Chapter 1
Introduction

1.1 Applications of RF and microwave filters

Microwave systems have an enormous impact on modern society. Applications are diverse, from entertainment via satellite television, to civil and military radar systems. In the field of communications, cellular radio is becoming as widespread as conventional telephony. Microwave and RF filters are widely used in all these systems in order to discriminate between wanted and unwanted signal frequencies. Cellular radio provides particularly stringent filter requirements both in the base stations and in mobile handsets. A typical filtering application is shown in Figure 1.1 which is a block diagram of the RF front end of a cellular radio base station.

The GSM system uses a time division multiple access technique (TDMA) [1]. Here the base station is transmitting and receiving simultaneously. Mobile

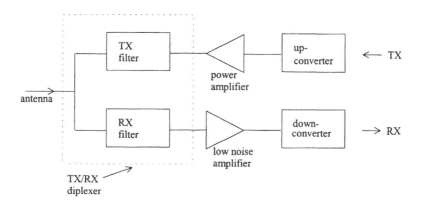

Figure 1.1   RF front end of a cellular base station

propagation effects require a system dynamic range in excess of 100 dB. The transmit power amplifier produces out-of-band intermodulation products and harmonics. These must be filtered to prevent leakage into the receiver and to satisfy regulatory requirements on out-of-band radiation. Therefore the transmit filter must have a high level of attenuation in the receive band. Furthermore, the transmit filter must have low passband insertion loss, to satisfy power amplifier linearity and efficiency requirements. Similarly the receiver must be protected by a filter with high attenuation in the transmit band to isolate the high power (30 W) transmitter. This filter must have low passband insertion loss to preserve system sensitivity. A typical specification for a GSM transmit filter is given in Table 1.1. Similar specifications are required for the receive band and for GSM in the 1800 MHz band.

In summary the base station filters must achieve a remarkable performance. Very low loss in the passband with high rejection at frequencies close to the passband is required. This high selectivity is illustrated in Figure 1.2.

We shall see in later chapters that the selectivity of a filter increases with the number of resonant sections. Furthermore, the insertion loss in the passband is inversely proportional to the filter bandwidth and the resonator Q factor and is proportional to the number of resonators used. The above specifications require at least eight resonators with unloaded Q factors of at least 5000. The Q requirement dictates a certain physical size, resulting in typical sizes for commercial coaxial resonator filters of 15 cm × 30 cm × 5 cm. Considerable research is under way in order to achieve smaller filters with improved performance. Some of this research will be described in later chapters although obviously we first need to understand the basic principles of filter design.

A second example of a base station filter is a requirement for a notch filter for the US AMPS system. In this case the two operators A and B have been assigned interleaving spectra as shown in Figure 1.3. In this situation a mobile which is far from base station A and is thus transmitting maximum power can cause interference to base station B if it is physically close to B. The requirement is for a notch filter with the specification given in Table 1.2. These extremely

Table 1.1 Specification of a GSM base station filter

Passband	925–960 MHz
Insertion loss	0.8 dB (max)
Input and output return loss	20 dB (min)
Stopband	
Frequency/MHz	Attenuation/dB (min)
d.c.–880	50
880–915	80
970–980	20
980–12750	50
Temperature range	$-10\,°C$ to $+70\,°C$
System impedance	50 Ω

Introduction 3

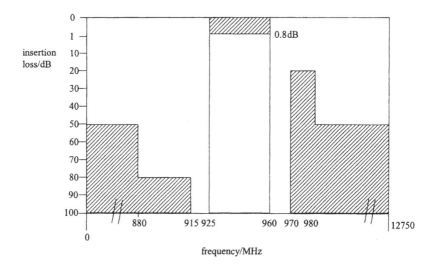

Figure 1.2 Frequency response of a GSM transmit filter

narrowband filters require resonators with unloaded Q factors in excess of 25 000. This requires the use of dielectric resonators with physical size of the cavities $9 \times 9 \times 9$ cm per resonator. At least four resonators are required per filter, resulting in physically large devices.

A third application of filters in cellular systems is in microwave links for communicating between base stations. These links operate at much higher frequencies; a typical specification for a 38 GHz filter is given in Table 1.3. These high frequency filters are normally constructed using waveguide technology.

A completely different filter technology is required in mobile handsets. The handset is only handling one call at a given time and in GSM does not transmit and receive simultaneously. In some handsets the transmitter and receiver are

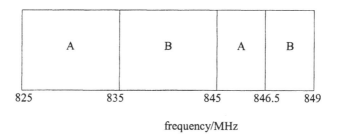

Figure 1.3 Advanced Mobile Phone System (AMPS) spectrum allocations

4 *Theory and design of microwave filters*

Table 1.2 Specification of an AMPS base station notch filter

Passband edges	845 and 846.5 MHz
Insertion loss	1 dB (max)
Stopband	845.2–846.3 MHz
Attenuation	20 dB (min)

Table 1.3 Specification of a filter for a microwave link

Passband	38–38.3 GHz
Insertion loss	1.5 dB (max)
Stopband	39.26–39.56 MHz
Attenuation	70 dB (min)
Local oscillator harmonic rejection	10 dB (min) at 74–74.6 GHz

isolated by a PIN diode switch and there is a requirement for only a receive filter (Figure 1.4).

The main purpose of the front end filter is to protect the LNA and mixer in the down-converter from being over-driven by extraneous signals. For example, this situation may occur if two mobiles are being operated simultaneously within a vehicle. Typical specifications for a 900 MHz GSM receive filter are given in Table 1.4.

Although the electrical specifications on these filters are much less severe than for base station filters, the miniaturisation required means that they are still an extremely challenging design problem.

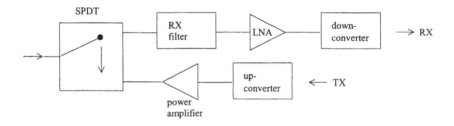

Figure 1.4 Typical GSM mobile handset RF front end

Table 1.4 Typical specification for a GSM handset receive filter

Passband	925–960 MHz	
Insertion loss	3.5 dB (max)	
Stopband	850–905 MHz	20 dB (min)
	905–915 MHz	12 dB (min)
Physical size	10 × 7 × 2 mm (typical)	

1.2 Ideal lowpass filters

As we have already seen, filters must achieve a specified selectivity. In other words the transition from passband to stopband must be achieved within a certain bandwidth. It is interesting to consider the fundamental limits on achievable selectivity by examining an 'ideal' lowpass filter. This is defined as a two-port device with infinite selectivity as in Figure 1.5.

The magnitude of the gain of the lowpass filter is unity in the passband and immediately drops to zero in the stopband with no transition region. Hence

$$|H(j\omega)| = 1 \qquad |\omega| < \omega_c \qquad (1.1)$$

$$|H(j\omega)| = 0 \qquad |\omega| > \omega_c \qquad (1.2)$$

The phase response of the filter is assumed to be linear in the passband. Hence

$$\psi(\omega) - k\omega \qquad (1.3)$$

and the group delay through the device is

$$T_g = \frac{-\mathrm{d}\psi(\omega)}{\mathrm{d}\omega} \qquad (1.4)$$

Linear phase implies constant group delay, which ensures zero phase distortion for finite bandwidth signals.

It is instructive to examine the impulse response of the filter. The impulse response is the time domain output from an infinitely short (delta function) excitation at the input. It is the inverse Fourier transform of the transfer function [2]. Now

$$H(j\omega) = \exp(-jk\omega) \qquad |\omega| < \omega_c \qquad (1.5)$$

$$H(j\omega) = 0 \qquad |\omega| > \omega_c \qquad (1.6)$$

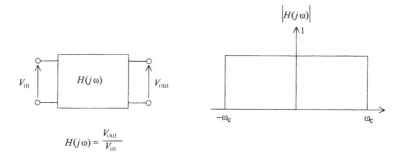

Figure 1.5   Ideal lowpass filter and its frequency response

Hence the impulse response is given by

$$h(t) = \frac{1}{2\pi} \int_{-\infty}^{\infty} H(j\omega) \exp(j\omega t)\, d\omega$$

$$= \int_{-\omega_c}^{\omega_c} \frac{\exp[j\omega(t-k)]\, d\omega}{2\pi}$$

$$= \frac{1}{\pi} \frac{\sin[(t-k)\omega_c]}{t-k} \tag{1.7}$$

If $\omega_c = 1$

$$h(t) = \frac{1}{\pi} \operatorname{sinc}(t-k) \tag{1.8}$$

This function has zeros when

$$t - k = m\pi \qquad m = \pm 1, \pm 2, \ldots \tag{1.9}$$

That is,

$$t = m\pi + k \tag{1.10}$$

This is the familiar sinc function with the main peak occurring at k s, which is the passband group delay of the filter. This is shown in Figure 1.6. Here we can see that an output occurs for $t < 0$. In other words the ideal lowpass filter is non-causal because an output occurs before the input is applied! The only way to stop the response existing for $t < 0$ is to let k increase to infinity. This infinite group delay is also physically unrealisable. This is equivalent to telling us something we know intuitively, that an infinitely selective filter has an infinite group delay, or an infinite number of filter elements. In order to make the filter

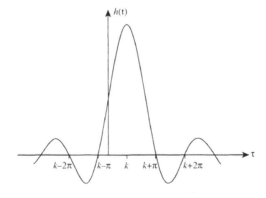

Figure 1.6 Impulse response of an ideal lowpass filter

realisable we can truncate the impulse response by removing the part which occurs for negative time to give a causal response which is physically realisable. In other words $h(t)$ is zero for t less than zero.

Obviously if the delay of the filter is very low then simply removing a part of the impulse response results in a considerable distortion of the frequency response. In fact, it may be shown that this corresponds to reducing the frequency selectivity of the filter. In reality filter design involves a compromise between removing too much of the impulse response curve and having too much delay. Practical filters use transfer functions which approximate to the ideal response with the minimum amount of delay. This will be dealt with in more detail in Chapter 3.

1.3 Minimum phase networks

Consider a lumped element filter with the transfer function

$$S_{12}(p) = \frac{N(p)}{D(p)} \tag{1.11}$$

where p is the complex frequency variable. This is defined as a minimum phase network if there are no poles or zeros in the right half p plane, i.e.

$$N(p) \neq 0 \qquad D(p) \neq 0 \qquad \text{Re } p > 0 \tag{1.12}$$

which defines $N(p)$ and $D(p)$ as Hurwitz polynomials.

Physical systems where the transmission of energy between input and output can only take one path are minimum phase. Examples include ladder networks (Figure 1.7) and coaxial cables.

It may be shown that if a minimum phase network has a transfer function

$$H(j\omega) = \exp[-\alpha(\omega) - j\psi(\omega)] \tag{1.13}$$

then the amplitude $\alpha(\omega)$ and phase $\psi(\omega)$ characteristics of a minimum phase

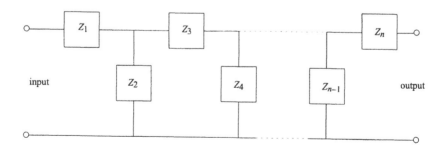

Figure 1.7 Ladder network

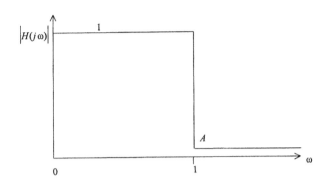

Figure 1.8 Transfer function of a rectangular filter

network are related by a pair of Hilbert transforms [3]:

$$\psi(\omega) = \frac{\omega}{\pi} \int_{-\infty}^{\infty} \frac{\alpha(y)}{y^2 - \omega^2} dy \tag{1.14}$$

$$\alpha(\omega) = \alpha(0) + \frac{\omega^2}{\pi} \int_{-\infty}^{\infty} \frac{\psi(y)}{y(y^2 - \omega^2)} dy \tag{1.15}$$

In other words, if the amplitude characteristic of a minimum phase network is known, then the phase characteristic is uniquely determined and vice versa within a constant gain $\alpha(0)$.

As an example consider the transfer function in Figure 1.8. Here

$$|H(j\omega)| = 1 \quad |\omega| < 1$$
$$= A \quad |\omega| > 1 \tag{1.16}$$

and

$$A \ll 1 \tag{1.17}$$

Now

$$|H(j\omega)| = \exp[-\alpha(\omega)] \tag{1.18}$$

and

$$\alpha(\omega) = 0 \quad |\omega| < 1 \tag{1.19}$$

$$\alpha(\omega) = -L_n A \quad |\omega| > 1 \tag{1.20}$$

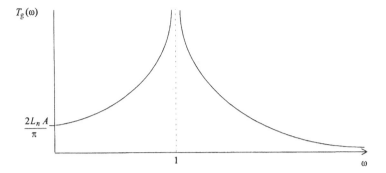

Figure 1.9 Group delay of ideal minimum phase filter

Now

$$\psi(\omega) = \frac{\omega}{\pi} \int_{-\infty}^{\infty} \frac{\alpha(y)}{y^2 - \omega^2} \, dy$$

$$= \frac{2\omega}{\pi} \int_{0}^{\infty} \frac{\alpha(y)}{y^2 - \omega^2} \, dy$$

$$= \frac{-2L_n A}{\pi} \omega \int_{1}^{\infty} \frac{\alpha(y)}{y^2 - \omega^2} \, dy$$

$$= \frac{-L_n A}{\pi} \left[L_n \left| \frac{y - \omega}{y + \omega} \right| \right]_{1}^{\infty}$$

$$= \frac{-L_n A}{\pi} L_n \left| \frac{1 + \omega}{1 - \omega} \right| \tag{1.21}$$

Hence

$$T_g(\omega) = \frac{-d\psi(\omega)}{d\omega} = \frac{2L_n A}{\pi} \left(\frac{1}{|1 - \omega^2|} \right) \tag{1.22}$$

The group delay is shown in Figure 1.9.

Thus we see that the group delay of the ideal minimum phase filter is inverse parabolic in the passband rising to infinity at the band-edge. In reality, filters are not infinitely selective and the amplitude response and group delay response are as shown in Figure 1.10.

Considerable research in the area of non-minimum phase filters has resulted in so-called selective linear phase filters where a similar amplitude response to

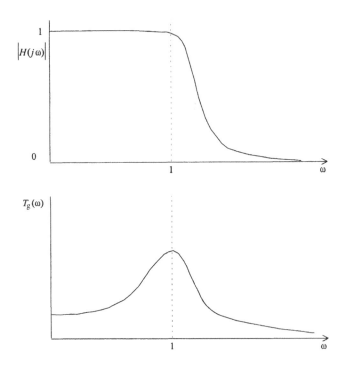

Figure 1.10 Amplitude and group delay response of a real minimum phase filter

Figure 1.10 is obtained but with much reduced group delay variation. These devices require multiple paths between the input and output of the filter.

1.4 Amplitude approximation

We will now briefly consider theoretical approximations to ideal lowpass filters. Consider a lowpass ladder filter operating between resistive terminations as shown in Figure 1.11. Scrutiny of this circuit shows that it has zero gain at infinite frequency. As the frequency is increased from d.c. the series inductors

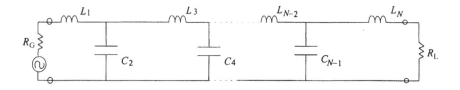

Figure 1.11 Lowpass ladder network

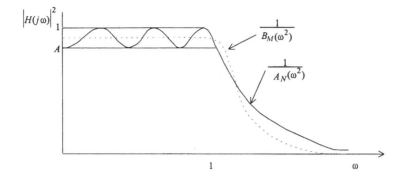

Figure 1.12   Equiripple filter characteristics ($N = 6$)

become open circuited and the shunt capacitors become short circuited. Each of these open or short circuits contributes a 'transmission zero' to the response of the network. Thus a ladder network with N circuit elements has N transmission zeros at infinity. Since all the transmission zeros are at infinity the gain response must be a rational function with constant numerator. The zeros of transmission are the zeros of the numerator or the infinities of the denominator.

Hence the power gain is given by

$$|H(j\omega)|^2 = \frac{1}{A_N(\omega^2)} \qquad (1.23)$$

where $A_N(\omega^2)$ is a polynomial of degree N in ω^2.

Now let us assume that $|H(j\omega)|^2$ is equiripple in the filter passband and rolls off monotonically to zero in the filter stopband as in Figure 1.12. Here we see that $|H(j\omega)|^2$ ripples the maximum number of times between unity and A.

Now consider a second transfer function which is more selective than the equiripple transfer function and is at least as flat in the passband. Again we assume this transfer function arises from a ladder network, so all its transmission zeros are at infinity. This function is shown as the dotted curve in Figure 1.12. The two curves intercept at least $N + 1$ times so we can say that

$$\frac{1}{A_N(\omega^2)} - \frac{1}{B_M(\omega^2)} = 0 \text{ at least } N + 1 \text{ times} \qquad (1.24)$$

and $B_M(\omega^2) - A_N(\omega^2)$ is at least of degree $N + 1$. Thus either $B_M(\omega^2)$ is identical to $A_N(\omega^2)$ or M is of degree $N + 1$ or higher.

Hence for this class of ladder networks or 'all-pole' transfer function the equiripple characteristics will always provide the optimum selectivity for a given degree (number of elements) of filter. Other classes of filter have equiripple response in the passband and stopband [4] but these are not realisable by ladder networks.

1.5 Practical realisations of microwave filters

As we have seen, filters with equiripple amplitude characteristics achieve optimum selectivity for a given number of circuit elements. The use of network synthesis enables lumped element lowpass prototype networks to be designed as described in Chapter 3. These lumped prototype networks may be converted into bandpass filters using the transformations described in Chapter 4. However, lumped element realisations of microwave filters are not often used because the wavelength is so short compared with the dimensions of circuit elements. For this reason a variety of 'distributed' element realisations are used. A distributed circuit element has one or more dimensions which are comparable with wavelength, and connections of distributed elements may be described by distributed network theory, which is an extension of the theory of lumped element networks. One example of a distributed element is the TEM transmission line and various types of microwave filters may be designed using interconnections of these elements. One of the most common TEM filters is the interdigital filter which consists of an array of coupled TEM lines with coupling constrained between nearest neighbours. Such devices enable practical realisations of microwave filters with relatively high resonator Q factors (typically 1–5000) enabling quite severe specifications to be achieved. The design theory for interdigital and other TEM devices is described in Chapter 5.

One of the fundamental problems of filter design is that the passband loss is inversely proportional to the filter bandwidth. Thus for very narrow band applications it is often the case that very high resonator Q factors must be used in order to achieve low passband loss. Air-filled waveguide resonators enable Q factors from 5 to 20 000 to be realised. Further increases in Q, up to 50 000, may be achieved by using dielectric resonators as the resonant elements within filters. The electromagnetic properties of these devices and the design theories for waveguide and dielectric resonator filters are described in Chapters 6 and 7.

1.6 Summary

There are numerous applications for microwave and RF filters in the communications industry requiring many different design approaches. In addition there are fundamental limits on the achievable performance of electrical filters, regardless of the physical construction. No finite device can produce an 'ideal' or infinitely selective amplitude response. Furthermore, there are strict relationships between the phase and amplitude characteristics of minimum phase networks. The remainder of the book is devoted to developing design techniques which enable filters to approach these theoretical limits as closely as possible.

Chapter 2 presents the basic linear passive network theory which is required for a theoretical understanding of filter design. This includes network parameters, network analysis and network synthesis. Chapter 3 concentrates on the design of lumped lowpass prototypes which can be considered as building blocks for many classes of lumped and distributed filters. Both approximation theory and network synthesis of prototypes are included for both amplitude and phase responses. Chapter 4 includes material on frequency transformations from lumped lowpass prototypes to highpass, bandpass and bandstop filters. It also includes the effects of dissipation loss in filters and methods for practical filter development. In Chapter 5 the design of distributed filters using TEM transmission lines is covered including the Richards transformation, stepped impedance and coupled-line filters. Chapter 6 concentrates on the design of waveguide filters. The basic theory of waveguides is followed by design techniques for iris-coupled bandpass filters, generalised cross-coupled filters, extracted pole filters and dual-mode filters. In Chapter 7 the principles of dielectric resonator filters are presented. Starting with the basic theory of dielectric resonators, design techniques for single, multi-mode and dielectric-loaded structures are described. Finally in Chapter 8 techniques for miniaturisation are described. These include SAW filters, superconducting filters, active filters and new system architectures using lossy filters.

1.7 References

1 REDL, S.M., WEBER, M.K., and OLIPHANT, M.W.: 'An introduction to GSM' (Artech House, Norwood, MA, 1995) pp. 71–75
2 GLOVER, I.A., and GRANT, P.M.: 'Digital communications' (Prentice Hall, Englewood Cliffs, NJ, 1997) pp. 146–49
3 PAPOULIS, A: 'The Fourier integral and its applications' (McGraw-Hill, New York, 1962)
4 RHODES, J.D.: 'Theory of electrical filters' (Wiley, New York, 1976) pp. 12–17

Chapter 2
Basic network theory

2.1 Linear passive time-invariant networks

This book is concerned with the design of passive RF and microwave filters. These devices are manufactured using a variety of technologies, e.g. coaxial resonators, microstrip, waveguide etc. However, they are normally designed using lowpass prototype networks as a starting point, regardless of the eventual physical realisation. Lowpass prototype networks are two-port lumped element networks with an angular cut-off frequency of $\omega = 1$, operating from a $1\,\Omega$ generator into a $1\,\Omega$ load. A typical lowpass prototype network is shown in Figure 2.1.

The design of lowpass prototype networks is dealt with in detail in Chapter 3. In this chapter we develop useful techniques for the analysis and synthesis of such networks. These network methods assume a basic understanding of Laplace transform theory and of the operation of inductors, capacitors and resistors. We will restrict ourselves to linear, time-invariant, passive networks, which are defined as follows.

First we consider a one-port network (Figure 2.2). This one-port network is excited by a voltage $v(t)$ producing a current flow $i(t)$. The Laplace transform of the voltage is $V(p)$ and the resultant current is $I(p)$.

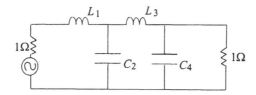

Figure 2.1 A typical lowpass prototype network

16 Theory and design of microwave filters

Figure 2.2 A one-port network and its Laplace transform equivalent

2.1.1 Linearity

If a voltage $v_1(t)$ across the terminals of N produces a current $i_1(t)$ then

$$v_1(t) \Rightarrow i_1(t) \tag{2.1}$$

Similarly

$$v_2(t) \Rightarrow i_2(t) \tag{2.2}$$

Now if the network is linear then the principle of superposition holds and we have

$$\alpha v_1(t) + \beta v_2(t) \Rightarrow \alpha i_1(t) + \beta i_2(t) \tag{2.3}$$

where α and β are constants.

2.1.2 Time invariance

If the network is invariant with time and if

$$v(t) \Rightarrow i(t) \tag{2.4}$$

then

$$v(t - \tau) \Rightarrow i(t - \tau) \tag{2.5}$$

where τ is an arbitrary time delay.

If a linear time-invariant network is excited by a voltage $v(t)$ where

$$v(t) = 0 \text{ for } t < 0 \tag{2.6}$$

then the relationship between voltage and current may be expressed as follows:

$$V(p) = Z(p)I(p) \tag{2.7}$$

where $V(p)$ is the Laplace transform of $v(t)$ and $I(p)$ is the Laplace transform of $i(t)$. Here p is the complex frequency variable (sometimes denoted s); $Z(p)$ is the input impedance of the network, which is independent of $v(t)$ and for a finite lumped network is a rational function of p. In this case $Z(p)$ may be expressed as the ratio of two polynomials:

$$Z(p) = \frac{N(p)}{D(p)} \tag{2.8}$$

2.1.3 Passivity

If the network is also passive then the amount of energy which may be extracted from the network up to any point in time may not exceed the energy supplied to the network up to that point in time. Combining this property with the property that all physical networks give rise to real responses to real stimuli yields [1] that $Z(p)$ is a 'positive real function', i.e.

$Z(p)$ is real for p real (2.9)

Re $Z(p) > 0$ for Re $p > 0$ (2.10)

Relation (2.9) implies that the coefficients of $N(p)$ and $D(p)$ are all real. Relation (2.10) implies that $Z(p)$ has no poles or zeros in the right half-plane, i.e. both $N(p)$ and $D(p)$ are Hurwitz polynomials [2].

2.1.4 The bounded real condition

The input impedance of passive linear time-invariant networks is a positive real function. In microwave filter design it is often desirable to work with reflection coefficients rather than input impedances. The reflection coefficient $\Gamma(p)$ of a network with an input impedance $Z(p)$ is related to $Z(p)$ by

$$\Gamma(p) = \pm \frac{Z(p) - 1}{Z(p) + 1} \qquad (2.11)$$

$\Gamma(p)$ may be shown to be a bounded real function, i.e.

$\Gamma(p)$ is real for p real (2.12)

$0 \leq |\Gamma(p)| \leq 1 \qquad$ for Re $p > 0$ (2.13)

Relation (2.13) may be demonstrated as follows: for Re $p > 0$, let

$$Z(p) = R + jX \qquad (2.14)$$

Now since $Z(p)$ is positive real then

$$R > 0 \qquad (2.15)$$

(i.e. the real part of the input impedance is always positive). Hence

$$\Gamma = \pm \frac{R + jX - 1}{R + jX + 1} \qquad (2.16)$$

$$|\Gamma|^2 = \frac{(R-1)^2 + X^2}{(R+1)^2 + X^2}$$

$$= 1 - \frac{4R}{(R+1)^2 + X^2} \qquad (2.17)$$

Thus if $R \geq 0$ then $|\Gamma| \leq 1$.

2.2 Lossless networks

Lossless networks consist entirely of reactive elements, i.e. they contain no resistors. In reality all real microwave filters contain resistive elements but it is useful in the initial design process to simplify things by working with lossless networks.

Now for our positive real $Z(p)$

$$Z(p)|_{p=j\omega} = Z(j\omega) = R(\omega) + jX(\omega) \tag{2.18}$$

where

$$R(\omega) = \operatorname{Re} Z(j\omega) \tag{2.19}$$

$$X(\omega) = \operatorname{Im} Z(j\omega) \tag{2.20}$$

By definition for a lossless network

$$R(\omega) = 0 \tag{2.21}$$

Now

$$Z(p) = \frac{m_1 + n_1}{m_2 + n_2} \tag{2.22}$$

where m_1 and n_1 are the even and odd parts of $N(p)$ and m_2 and n_2 are the even and odd parts of $D(p)$.

$Z(p)$ may be split into an even polynomial plus an odd polynomial, i.e.

$$Z(p) = \operatorname{Ev} Z(p) + \operatorname{Odd} Z(p) \tag{2.23}$$

Now even polynomials contain only even powers of p and odd polynomials contain only odd powers of p. Hence $\operatorname{Ev} Z(j\omega)$ is purely real and $\operatorname{Odd} Z(j\omega)$ is purely imaginary. Thus for a lossless network $R(\omega) = 0$ implies $\operatorname{Ev} Z(p) = 0$ and

$$\operatorname{Ev} Z(p) = \frac{Z(p) + Z(-p)}{2} = 0 \tag{2.24}$$

Therefore

$$\frac{m_1 + n_1}{m_2 + n_2} + \frac{m_1 - n_1}{m_2 - n_2} = 0 \tag{2.25}$$

and

$$\frac{m_1 m_2 - n_1 n_2}{m_2^2 - n_2^2} = 0 \tag{2.26}$$

Hence

$$\frac{m_1}{n_1} = \frac{n_2}{m_2} \tag{2.27}$$

Thus for example

$$Z(p) = \frac{n_1}{m_2} \frac{m_1/n_1 + 1}{1 + n_2/m_2} = \frac{n_1}{m_2} \tag{2.28}$$

or

$$Z(p) = \frac{m_1}{n_2} \frac{1 + n_1/m_1}{1 + m_2/n_2} = \frac{m_1}{n_2} \quad (2.29)$$

$Z(p)$ is thus the ratio of an even polynomial to an odd polynomial or an odd polynomial to an even polynomial. $Z(p)$ is then known as a 'reactance function'.

Now since $Z(p)$ is positive real it cannot have any right half-plane poles or zeros and $Z(-p)$ cannot have any left half-plane poles or zeros. However

$$Z(p) = -Z(-p) \quad (2.30)$$

Thus $Z(p)$ can have neither right half-plane nor left half-plane zeros. The poles and zeros of a reactance function must thus lie on the imaginary axis. Furthermore, the poles of $Z(p)$ may be shown to have positive real residues [3], yielding a general solution for a reactance function of the form

$$Z(p) = A_\infty p + \frac{A_0}{p} + \sum_{i=1}^{m} \frac{2A_i p}{p^2 + \omega_i^2} \quad (2.31)$$

Furthermore, for $p = j\omega$,

$$Z(j\omega) = jX(\omega) \quad (2.32)$$

where

$$X(\omega) = A_\infty \omega - \frac{A_0}{\omega} + \sum_{i=1}^{m} \frac{2A_i \omega}{\omega_i^2 - \omega^2} \quad (2.33)$$

and

$$\frac{dX(\omega)}{d\omega} = A_\infty + \frac{A_0}{\omega^2} + \sum_{i=1}^{m} 2A_i \frac{\omega_i^2 + \omega^2}{(\omega_i^2 - \omega^2)^2} \quad (2.34)$$

Therefore

$$\frac{dX(\omega)}{d\omega} > 0 \quad (2.35)$$

The fact that the differential of $X(\omega)$ is always positive implies that the poles and zeros of $X(\omega)$ must be interlaced. Hence a typical plot of a reactance function is as shown in Figure 2.3.

Now consider the parallel tuned circuit shown in Figure 2.4. The impedance of this circuit is

$$Z(p) = \frac{1}{Cp + (1/Lp)} = \frac{P/C}{p^2 + (1/LC)} \quad (2.36)$$

From (2.31) and (2.36) we can see that the general equivalent circuit for a reactance function is the network shown in Figure 2.5. This process of working backward from an impedance function to the actual physical circuit is known as 'synthesis'.

20 Theory and design of microwave filters

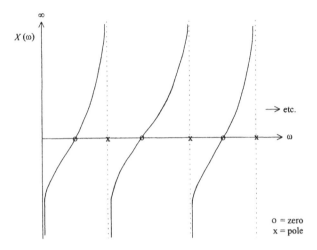

Figure 2.3   A typical reactance function

The particular method shown here is known as Foster synthesis [4], where the circuit is derived by a partial fraction expansion of the impedance function.

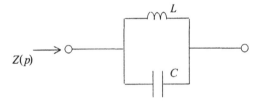

Figure 2.4   Parallel tuned circuit

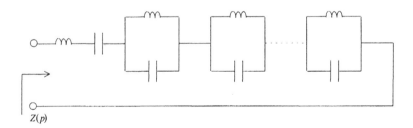

Figure 2.5   Foster realisation of $Z(p)$

2.3 Ladder networks

A common realisation of impedance functions used in filter design is the ladder network shown in Figure 2.6.

As an example consider the impedance function given in partial fraction form as

$$Z(p) = 2p + \frac{1}{p} + \frac{2p}{p^2+1}$$

$$= \frac{2p^4 + 5p^2 + 1}{p^3 + p} \qquad (2.37)$$

This may be synthesised using a continued fraction expansion. From (2.37) we see that $Z(p)$ tends to $2p$ as p tends to infinity. Thus we first evaluate the residue at $p = \infty$, i.e.

$$\left.\frac{Z(p)}{p}\right|_{p=\infty} = 2 \qquad (2.38)$$

Now we remove a series inductor of value $L = 2$, leaving a remaining impedance $Z_1(p)$ where

$$Z_1(p) = Z(p) - 2p$$

$$= \frac{2p^4 + 5p^2 + 1}{p^3 + p} - 2p$$

$$= \frac{3p^2 + 1}{p^3 + p} \qquad (2.39)$$

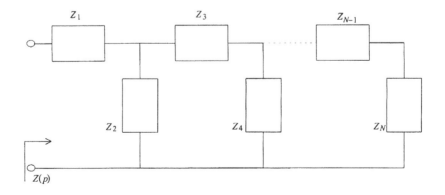

Figure 2.6 Ladder network

Again we evaluate the residue at $p = \infty$:

$$\left.\frac{Y_1(p)}{p}\right|_{p=\infty} = \frac{1}{3} \tag{2.40}$$

So we extract a shunt capacitor of value $C_2 = 1/3$, leaving a remaining admittance $Y_2(p)$ where

$$Y_2(p) = Y_1(p) - p/3$$

$$= \frac{2p/3}{3p^2 + 1} \tag{2.41}$$

Now we invert $Y_2(p)$ to form an impedance $Z_2(p)$ where

$$Z_2(p) = \frac{3p^2 + 1}{2p/3} \tag{2.42}$$

Again we evaluate the residue at $p = \infty$, i.e.

$$\left.\frac{Z_2(p)}{p}\right|_{p=\infty} = \frac{9}{2} \tag{2.43}$$

Now we extract a series inductor of value $L_3 = 9/2$ leaving an impedance $Z_3(p)$ where

$$Z_3(p) = Z_2(p) - \frac{9p}{2}$$

$$= \frac{3}{2p} \tag{2.44}$$

Now invert $Z_3(p)$ to form an admittance

$$Y_3(p) = \frac{2p}{3} \tag{2.45}$$

which is a capacitor of value

$$C_3 = 2/3 \tag{2.46}$$

The complete synthesis process is shown in Figure 2.7.

2.4 Synthesis of two-port networks – Darlington synthesis

Historically it was first proven by Brune that any positive real function can be synthesised using a network composed of resistors, capacitors, inductors and mutual inductances [5]. However, in the practical world of filter design we are more concerned with two-port networks with a pair of terminals at the input and a pair of terminals at the output. Darlington [6] proved that any positive real function can by synthesised as the input impedance of a lossless passive reciprocal two-port network which is terminated in a (load) resistor (Figure 2.8).

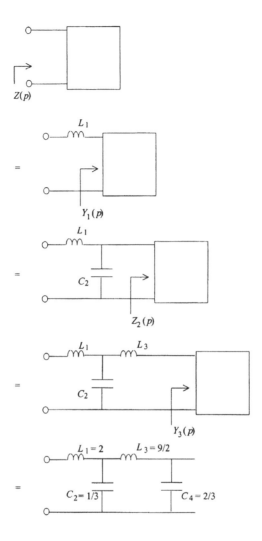

Figure 2.7 Synthesis of a ladder network

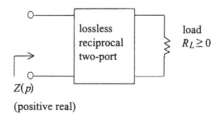

Figure 2.8 Darlington synthesis

The lossless two-port network may be decomposed into a cascade of first-, second- and fourth-degree networks depending on the locations of the zeros of the even part of $Z(p)$. These zeros are called the transmission zeros of the network. In other words they are the values of p for which there is zero transmission of power to the load. A zero on the $j\omega$ axis would correspond to zero transmission at a 'real' sinusoidal frequency (a real frequency transmission zero). This corresponds to a measured zero in the swept frequency response of the network.

Consider a lossless network driven from a $1\,\Omega$ generator and terminated in a $1\,\Omega$ load, as shown in Figure 2.9. The input impedance $Z_{in}(p)$ is

$$Z_{in}(p) = \frac{m_1 + n_1}{m_2 + n_2} \quad \text{(with } m_1, m_2 \text{ even, } n_1, n_2 \text{ odd)} \tag{2.47}$$

The even part of the input impedance which is the real part of $Z(j\omega)$ is given by

$$\text{Ev } Z_{in}(p) = \frac{Z(p) + Z(-p)}{2}$$

$$= \frac{m_1 m_2 - n_1 n_2}{m_2^2 - n_2^2} \tag{2.48}$$

The input power to the two-port network is given by

$$P_{in} = |I_{in}(j\omega)|^2 \operatorname{Re} Z_{in}(j\omega)$$

$$= \tfrac{1}{2} I_{in}(j\omega) I_{in}(-j\omega)[Z_{in}(j\omega) + Z_{in}(-j\omega)]$$

$$= \frac{V_g V_g^*[Z_{in}(j\omega) + Z_{in}(-j\omega)]}{[1 + Z_{in}(j\omega)][1 + Z_{in}(-j\omega)]} \tag{2.49}$$

Thus in the complex frequency plane the transmission zeros are the zeros of

$$\frac{Z_{in}(p) + Z_{in}(-p)}{[1 + Z_{in}(p)][1 + Z_{in}(-p)]} \tag{2.50}$$

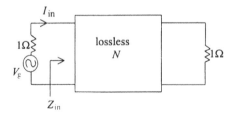

Figure 2.9   *Doubly terminated lossless two-port network*

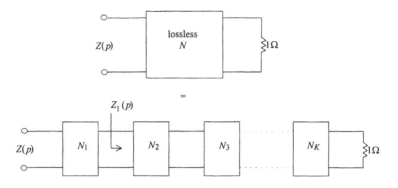

Figure 2.10 Realisation of a network N as a cascade of subnetworks

The transmission zeros are thus the zeros of the even part of the input impedance. In addition, transmission zeros occur at values of p which are simultaneously poles of $Z_{in}(p)$ and $Z_{in}(-p)$. These poles must occur at d.c., infinity or on the imaginary axis. They may be removed as elements of a reactance function by Foster synthesis. The remaining transmission zeros are not poles of $Z_{in}(p)$ and $Z_{in}(-p)$ and may be removed by second-order or fourth-order networks. Finite real frequency transmission zeros are extracted in complex conjugate pairs by a second-order network known as a Brune section. Transmission zeros on the real axis are removed by extraction of a second-order Darlington C section. Complex transmission zeros are removed by a fourth-order Darlington D section.

2.4.1 Cascade synthesis

The purpose of cascade synthesis is to synthesise an input impedance as a cascade of Brune, C and D sections terminated in a resistor. It is assumed that any transmission zeros which are simultaneously poles of $Z(p)$ and $Z(-p)$ have already been removed. Thus each of the sections contains transmission zeros of the entire network and progressive removal of basic sections lowers the degree of the network until only the positive load resistor remains [7].

The objective is shown in Figure 2.10 where

$$Z(p) = \frac{m_1 + n_1}{m_2 + n_2} \qquad (2.51)$$

(where m is even and n is odd) and the even part of $Z(p)$ is

$$\text{Ev } Z(p) = \frac{m_1 m_2 - n_1 n_2}{m_2^2 - n_2^2} \qquad (2.52)$$

The zeros of transmission are the zeros of $m_1 m_2 - n_1 n_2$ which we shall assume is

26 *Theory and design of microwave filters*

a perfect square. Thus

$$m_1 m_2 - n_1 n_2 = \left\{ \prod_{i=1}^{q}\left(1+\frac{p^2}{\omega_i^2}\right) \prod_{i=1}^{r}\left(1-\frac{p^2}{\sigma_i^2}\right) \prod_{i=1}^{s}[p^4 + 2(\omega_i^2 - \sigma_i^2)p^2 + (\omega_i^2 + \sigma_i^2)^2] \right\}^2 \tag{2.53}$$

which gives three types of transmission zeros, an imaginary axis pair, a real axis pair or a complex quadruplet. Asymmetrically located zeros are dealt with in Chapter 3.

Now in Figure 2.10 assume a transmission zero at $p = p_1$ is assigned to N_1. Then this transmission zero must not be a transmission zero of the remaining network, N_2 etc. with impedance $Z_1(p)$. The transfer matrix of N_1 is

$$[T_1] = \frac{1}{F(p)} \begin{bmatrix} A_1 & B_1 \\ C_1 & D_1 \end{bmatrix} \tag{2.54}$$

where $F(p)$ is even and

$$F(p)|_{p=p_1} = 0 \tag{2.55}$$

Then

$$Z_1(p) = \frac{B_1 - Z(p)D_1}{Z(p)C_1 - A_1} \tag{2.56}$$

and

$$Z_1(p) + Z_1(-p) = \frac{(A_1 D_1 - B_1 C_1)[Z(p) + Z(-p)]}{[A_1 - C_1 Z(p)][A_1 + C_1 Z(-p)]} \tag{2.57}$$

where since N_1 is lossless A_1, D_1 are even, B_1, C_1 are odd, and from reciprocity

$$A_1 D_1 - B_1 C_1 = F^2(p) \tag{2.58}$$

Thus

$$Z_1(p) + Z_1(-p) = \frac{F^2(p)[Z(p) + Z(-p)]}{[A_1 - C_1 Z(p)][A_1 + C_1 Z(-p)]}$$

$$= \frac{F^2(p)[Z(p) + Z(-p)]}{[A_1 - C_1 Z(p)]\{A_1 - C_1 Z(p) + C_1 [Z(p) + Z(-p)]\}} \tag{2.59}$$

Now p_1 should not be a factor of $Z_1(p) + Z_1(-p)$. However, since $Z_1(p) + Z_1(-p)$ contains the factor $F^2(p)$, then the numerator of $Z_1(p) + Z_1(-p)$ contains a factor $F^4(p)$. This factor must be cancelled by a similar factor in the denominator. Thus $A_1 - C_1 Z(p)$ must contain a factor $F^2(p)$. This is the condition on the network N_1 such that the transmission zero has been successfully extracted. Hence

$$A_1 - C_1 Z(p) = 0 \quad \text{for} \quad F(p) = 0 \tag{2.60}$$

$$\frac{d}{dp}[A_1 - C_1 Z(p)] = 0 \quad \text{for} \quad F(p) = 0 \tag{2.61}$$

or

$$A_1'(p_1) - C_1'(p_1)Z(p_1) - C_1(p_1)Z'(p_1) = 0 \tag{2.62}$$

Equations (2.60)–(2.62) determine the transfer matrix of N_1 such that the transmission zeros have been successfully extracted. $Z_1(p)$ can be found from (2.56) and the process repeated until we are left with the load resistor.

Three types of section are required for the three types of transmission zero. The Brune section realises imaginary axis (real frequency) transmission zeros with the transfer matrix

$$[T] = \frac{1}{1 + (p^2/\omega_i^2)} \begin{bmatrix} 1 + ap^2 & bp \\ cp & 1 + dp^2 \end{bmatrix} \tag{2.63}$$

with

$$(1 + ap^2)(1 + dp^2) - bcp^2 = [1 + (p^2/\omega_i^2)]^2 \tag{2.64}$$

The Darlington C section realises real axis transmission zeros with

$$[T] = \frac{1}{1 - (p^2/\sigma_i^2)} \begin{bmatrix} 1 + ap^2 & bp \\ cp & 1 + dp^2 \end{bmatrix} \tag{2.65}$$

with

$$(1 + ap^2)(1 + dp^2) - bcp^2 = [1 - (p^2/\sigma_i^2)]^2 \tag{2.66}$$

The Darlington D section realises a complex quadruplet of transmission zeros with

$$[T] = \frac{1}{(\sigma_i^2 + \omega_i^2)^2 + 2(\omega_i^2 - \sigma_i^2)p^2 + p^4} \begin{bmatrix} p^4 + a_1 p^2 + a_2 & b_1 p + b_2 p^3 \\ c_1 p + c_2 p^3 & p^4 + d_1 p^2 + d_2 \end{bmatrix} \tag{2.67}$$

with

$$(p^4 + a_1 p^2 + a_2)(p^4 + d_1 p^2 + d_2) - (b_1 p + b_2 p^3)(c_1 p + c_2 p^3)$$
$$= [(\sigma_i^2 + \omega_i^2)^2 + 2(\omega_i^2 - \sigma_i^2)p^2 + p^4]^2 \tag{2.68}$$

As an example consider the maximally flat amplitude and linear phase filter described in Chapter 3 (see Section 2.6 for a description of scattering parameters). Here

$$S_{12}(p) = \frac{9 - 2p^2}{9 + 18p + 16p^2 + 8p^3 + 2p^4} \tag{2.69}$$

This has two transmission zeros at infinity and a pair on the real axis at

$p = \pm 3/\sqrt{2}$ which requires a C section. Now

$$S_{11}(p)S_{11}(-p) = 1 - S_{12}(p)S_{12}(-p) \tag{2.70}$$

Hence

$$S_{11}(p) = \frac{2p^4}{9 + 18p + 16p^2 + 8p^3 + 2p^4} \tag{2.71}$$

and

$$Z(p) = \frac{1 + S_{11}(p)}{1 - S_{11}(p)}$$

$$= \frac{4p^4 + 8p^3 + 16p^2 + 18p + 9}{8p^3 + 16p^2 + 18p + 9} \tag{2.72}$$

First we extract all the transmission zeros which are poles of $Z(p)$ and $Z(-p)$. Thus we extract a series inductor of value 0.5 and then a shunt capacitor of value 8/7 from the admittance. The remaining impedance is

$$Z(p) = \frac{49p^2 + 94.5p + 63}{4p^2 + 54p + 63} \tag{2.73}$$

Now the remaining transmission zeros are on the real axis requiring a C section. From (2.60), (2.62) and (2.65) we obtain

$$1 + ap_1^2 - cZ(p_1) = 0 \tag{2.74}$$

$$2ap_1^2 - cZ(p_1) - cp_2Z'(p_1) = 0 \tag{2.75}$$

Thus

$$a = \frac{Z(p_1) + p_1 Z'(p_1)}{p^2[Z(p_1) - p_1 Z'(p_1)]} \tag{2.76}$$

$$c = \frac{2}{p^2[Z(p_1) - p_1 Z'(p_1)]} \tag{2.77}$$

(with $p_1 = \pm 3\sqrt{2}$, $Z(3/\sqrt{2}) = 2.477$ and $Z'(3/\sqrt{2}) = 0.648$).
b and d can be obtained from (2.66) giving

$$d = \frac{1}{a\sigma_i^4} \tag{2.78}$$

$$b = \frac{a + d + 2/\sigma_i^2}{c} \tag{2.79}$$

Thus $a = 0.7764$, $b = 1.5019$, $c = 0.8552$ and $d = 0.0636$.

The remaining network is a resistor of value $1\,\Omega$. The complete synthesis has produced the network shown in Figure 2.11.

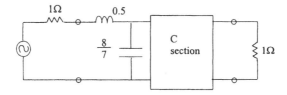

Figure 2.11 Synthesis of a linear phase filter

Various physical realisations are available for C sections including resonant circuits, coupled coils and cross-coupled resonant circuits. These will be discussed later.

2.4.2 All-pole networks

Certain filtering functions can be met using networks where all the transmission zeros are at $p = \infty$ (see Chapter 3), i.e.

$$\operatorname{Ev} Z(p)|_{p=\infty} = 0 \tag{2.80}$$

and

$$|S_{12}(j\omega)|^2 = \frac{1}{D_N(j\omega)D_N(-j\omega)} \tag{2.81}$$

In this case the Darlington synthesis yields a lowpass ladder network terminated in a positive resistor. The synthesis is identical to the continued fraction expansion technique previously described except that the final element is a resistor (Figure 2.12).

2.5 Analysis of two-port networks – the *ABCD* matrix

To progress further in the understanding of filter theory it is appropriate to review some basic network analysis techniques for two-port networks. One of the most useful tools is the *ABCD* or transfer matrix.

Consider the two-port network shown in Figure 2.13. The network may be

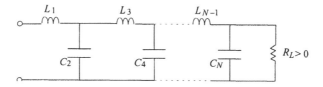

Figure 2.12 A typical lowpass ladder network

30 Theory and design of microwave filters

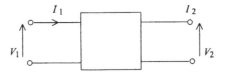

Figure 2.13 A two-port network

described using the matrix equation

$$\begin{bmatrix} V_1 \\ I_1 \end{bmatrix} = \begin{bmatrix} A & B \\ C & D \end{bmatrix} \begin{bmatrix} V_2 \\ I_2 \end{bmatrix} \tag{2.82}$$

where

$$\begin{bmatrix} A & B \\ C & D \end{bmatrix} = [T] \tag{2.83}$$

The ABCD or transfer matrix [T] relates the input voltage and current vector to the output voltage and current vector. One of the main uses of the transfer matrix is for the analysis of cascaded networks as shown in Figure 2.14.

Now

$$\begin{bmatrix} V_1 \\ I_1 \end{bmatrix} = [T_1] \begin{bmatrix} V_2 \\ I_2 \end{bmatrix}$$

$$= [T_1][T_2] \begin{bmatrix} V_3 \\ I_3 \end{bmatrix}$$

$$= [T_1][T_2][T_3] \begin{bmatrix} V_4 \\ I_4 \end{bmatrix}$$

$$= [T] \begin{bmatrix} V_4 \\ I_4 \end{bmatrix} \tag{2.84}$$

where

$$[T] = [T_1][T_2][T_3] \tag{2.85}$$

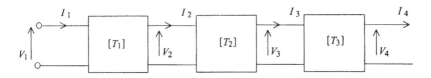

Figure 2.14 Cascaded two-port networks

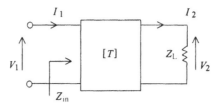

Figure 2.15 Two-port network with termination

Thus $[T]$, the transfer matrix of a cascade of two-port networks, is equal to the product of the individual matrices of each of the circuits.

Furthermore the transfer matrix has some interesting properties, depending on the network. If the network is symmetrical then

$$A = D \tag{2.86}$$

If the network is reciprocal then

$$AD - BC = 1 \tag{2.87}$$

If the network is lossless then, for $p = j\omega$, A and D are purely real and B and C are purely imaginary.

The input impedance of a two-port network terminated in a load impedance can be readily calculated (Figure 2.15).

$$Z_{in} = \frac{V_1}{I_1} \tag{2.88}$$

Now

$$\begin{bmatrix} V_1 \\ I_1 \end{bmatrix} = \begin{bmatrix} A & B \\ C & D \end{bmatrix} \begin{bmatrix} V_2 \\ I_2 \end{bmatrix} \tag{2.89}$$

Thus

$$\frac{V_1}{I_1} = \frac{AV_2 + BI_2}{CV_2 + DI_2}$$

$$= \frac{AV_2/I_2 + B}{CV_2/I_2 + D} \tag{2.90}$$

Now

$$V_2/I_2 = Z_L \tag{2.91}$$

Therefore

$$Z_{in} = \frac{AZ_L + B}{CZ_L + D} \tag{2.92}$$

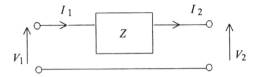

Figure 2.16 Series circuit element

ABCD matrices can be defined for series and shunt connected elements. Consider the series connected impedance Z in Figure 2.16. Now

$$\begin{bmatrix} V_1 \\ I_1 \end{bmatrix} = \begin{bmatrix} A & B \\ C & D \end{bmatrix} \begin{bmatrix} V_2 \\ I_2 \end{bmatrix} \qquad (2.93)$$

where

$$A = \left.\frac{V_1}{V_2}\right|_{I_2=0} = 1 \qquad (2.94)$$

$$B = \left.\frac{V_1}{I_2}\right|_{V_2=0} = Z \qquad (2.95)$$

$$C = \left.\frac{I_1}{V_2}\right|_{I_2=0} = 0 \qquad (2.96)$$

$$D = \left.\frac{I_1}{I_2}\right|_{V_2=0} = 1 \qquad (2.97)$$

Therefore

$$[T] = \begin{bmatrix} 1 & Z \\ 0 & 1 \end{bmatrix} \qquad (2.98)$$

For a shunt element, consider the shunt connected admittance Y in Figure 2.17.

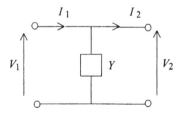

Figure 2.17 Shunt connected circuit element

Now

$$A = \left.\frac{V_1}{V_2}\right|_{I_2=0} = 1 \tag{2.99}$$

$$B = \left.\frac{V_1}{I_2}\right|_{V_2=0} = 0 \tag{2.100}$$

$$C = \left.\frac{I_1}{V_2}\right|_{I_2=0} = Y \tag{2.101}$$

$$D = \left.\frac{I_1}{I_2}\right|_{V_2=0} = 1 \tag{2.102}$$

Hence

$$[T] = \begin{bmatrix} 1 & 0 \\ Y & 1 \end{bmatrix} \tag{2.103}$$

As an example let us now use transfer matrices to compute the input impedance of the network shown in Figure 2.18. The transfer matrix of the total network is the product of the transfer matrices of the individual series inductors and shunt capacitors, i.e.

$$\begin{aligned}
[T] &= \begin{bmatrix} 1 & L_1 p \\ 0 & 1 \end{bmatrix} \begin{bmatrix} 1 & 0 \\ C_2 p & 1 \end{bmatrix} \begin{bmatrix} 1 & L_3 p \\ 0 & 1 \end{bmatrix} \begin{bmatrix} 1 & 0 \\ C_4 p & 1 \end{bmatrix} \\
&= \begin{bmatrix} 1 & 2p \\ 0 & 1 \end{bmatrix} \begin{bmatrix} 1 & 0 \\ p/3 & 1 \end{bmatrix} \begin{bmatrix} 1 & 9p/2 \\ 0 & 1 \end{bmatrix} \begin{bmatrix} 1 & 0 \\ 2p/3 & 1 \end{bmatrix} \\
&= \begin{bmatrix} 1 + 2p^2/3 & 2p \\ p/3 & 1 \end{bmatrix} \begin{bmatrix} 1 + 3p^2 & 9p/2 \\ 2p/3 & 1 \end{bmatrix} \\
&= \begin{bmatrix} (1 + 2p^2/3)(1 + 3p^2) + 4p^2/3 & (9p/2)(1 + 2p^2/3) + 2p \\ (p/3)(1 + 3p^2) + 2p/3 & 1 + 3p^2/2 \end{bmatrix}
\end{aligned} \tag{2.104}$$

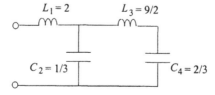

Figure 2.18 Example network

Now
$$Z_{in} = \frac{AZ_L + B}{CZ_L + D} \tag{2.105}$$

In this case there is no termination so $Z_L = \infty$. Therefore

$$Z_{in} = \frac{A}{C}$$

$$= \frac{1 + 5p^2 + 2p^4}{p + p^3} \tag{2.106}$$

which agrees with the impedance given in (2.37).

2.6 Analysis of two-port networks – the scattering matrix

We shall now introduce a new set of two-port parameters known as scattering parameters or S parameters.

Consider the two-port network shown in Figure 2.19. This may be described by the well-known impedance matrix equation

$$[V] = [Z][I] \tag{2.107}$$

where

$$[V] = \begin{bmatrix} V_1 \\ V_2 \end{bmatrix} \qquad [I] = \begin{bmatrix} I_1 \\ I_2 \end{bmatrix} \tag{2.108}$$

and

$$[Z] = \begin{bmatrix} Z_{11} & Z_{12} \\ Z_{21} & Z_{22} \end{bmatrix} \tag{2.109}$$

Now let

$$[a] = \begin{bmatrix} a_1 \\ a_2 \end{bmatrix} = \frac{[V] + [I]}{2} \tag{2.110}$$

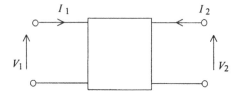

Figure 2.19   A two-port network

Basic network theory 35

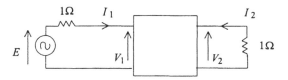

Figure 2.20 A resistively terminated two-port network

and

$$[b] = \begin{bmatrix} b_1 \\ b_2 \end{bmatrix} = \frac{[V] - [I]}{2} \tag{2.111}$$

Also let
$$[b] = [S][a] \tag{2.112}$$
where
$$[S] = \begin{bmatrix} S_{11} & S_{12} \\ S_{21} & S_{22} \end{bmatrix} \tag{2.113}$$

$[S]$ is called the scattering matrix of the two-port network.

Now from (2.110) and (2.111)

$$[V] = [a] + [b] \tag{2.114}$$
$$[I] = [a] - [b] \tag{2.115}$$

Now since
$$[V] = [Z][I] \tag{2.116}$$
then
$$[a] + [b] = [Z][[a] - [b]] \tag{2.117}$$
and substituting (2.112) in (2.117)
$$(1 + [S])[a] = [Z](1 - [S])[a] \tag{2.118}$$
and
$$[Z] = \frac{1 + [S]}{1 - [S]} \tag{2.119}$$

Now let the two-port network be terminated in a $1\,\Omega$ source and load as shown in Figure 2.20. From basic circuit analysis the transducer power gain of this network is given by

$$G(\omega) = 4\frac{|I_2|^2}{|E|^2} \tag{2.120}$$

36 Theory and design of microwave filters

and from (2.114) and (2.115)

$$V_1 = a_1 + b_1 \tag{2.121}$$

$$V_2 = a_2 + b_2 \tag{2.122}$$

$$I_1 = a_1 - b_1 \tag{2.123}$$

$$I_2 = a_2 - b_2 \tag{2.124}$$

and from Figure 2.20

$$V_2 = -I_2 \tag{2.125}$$

Hence

$$a_2 = 0 \tag{2.126}$$

and

$$E = I_1 + V_1 = 2a_1 \tag{2.127}$$

and

$$I_2 = -b_2 \tag{2.128}$$

Hence substituting (2.127) and (2.128) in (2.120)

$$G(j\omega) = \frac{|b_2|^2}{|a_1|^2} \tag{2.129}$$

Now since $a_2 = 0$, $b_2 = S_{21}a_1$, so

$$G(\omega) = |S_{21}(j\omega)|^2 \tag{2.130}$$

For a reciprocal network, since $Z_{21} = Z_{12}$, then

$$S_{12} = S_{21} \tag{2.131}$$

Furthermore, we can compute the input impedance at port (1). Since

$$Z_{in} = \frac{V_1}{I_1} = \frac{a_1 + b_1}{a_1 - b_1} \tag{2.132}$$

and

$$b_1 = S_{11}a_1 \tag{2.133}$$

so

$$Z_{in} = \frac{1 + S_{11}}{1 - S_{11}} \tag{2.134}$$

Thus the scattering parameters relate to measured transmission and reflection through the network. Since

$$\begin{bmatrix} b_1 \\ b_2 \end{bmatrix} = \begin{bmatrix} S_{11} & S_{12} \\ S_{21} & S_{22} \end{bmatrix} \begin{bmatrix} a_1 \\ a_2 \end{bmatrix} \tag{2.135}$$

then

$$S_{11} = \left. \frac{b_1}{a_1} \right|_{a_2 = 0} \tag{2.136}$$

$$S_{12} = \left. \frac{b_1}{a_2} \right|_{a_1 = 0} \tag{2.137}$$

$$S_{21} = \left. \frac{b_2}{a_1} \right|_{a_2 = 0} \tag{2.138}$$

$$S_{22} = \left. \frac{b_2}{a_2} \right|_{a_1 = 0} \tag{2.139}$$

S_{11} and S_{12} are the input and output reflection coefficients. S_{12} is the reverse transmission coefficient and S_{21} is the forward transmission coefficient; b_1 and b_2 can be considered as reflected signals at ports (1) and (2), while a_1 and a_2 are incident signals at ports (1) and (2).

$Z(p)$ is a positive real function, thus $S_{11}(p)$ is a bounded real function:

$$|S_{11}(p)| < 1 \text{ for Re } p > 0 \tag{2.140}$$

or

$$S_{11}(p) \text{ is analytic for Re } p \geq 0 \tag{2.141}$$

i.e. $S_{11}(p)$ contains no poles or zeros in the right half p plane.

Furthermore, from (2.140), using the maximum modulus theorem [8], we obtain

$$|S_{11}(j\omega)| \leq 1 \tag{2.142}$$

This is another way of stating conservation of energy, i.e. a passive device cannot reflect more energy than is incident upon it.

If the network is unterminated and lossless, i.e. $Z(p)$ is a reactance function, then all incident energy must be reflected from the network, i.e.

$$|S_{11}(j\omega)| = 1 \tag{2.143}$$

Furthermore, since $|S_{12}(j\omega)|^2$ is the transducer power gain, then

$$0 \leq |S_{12}(j\omega)|^2 \leq 1 \tag{2.144}$$

Again this is a statement of conservation of energy for a passive network.

Now consider power entering the two-port network at real frequencies.

$$P = \text{Re}\,(V_1 I_1^* + V_2 I_2^*)$$
$$= \text{Re}\,[(a_1 + b_1)(a_1^* - b_1^*) + (a_2 + b_2)(a_2^* - b_2^*)] \tag{2.145}$$

Now we know that $a_2 = 0$. Hence

$$P = a_1 a_1^* - b_1 b_1^* - b_2 b_2^* \tag{2.146}$$

and

$$b_1 = S_{11} a_1 \qquad b_2 = S_{12} a_1 \tag{2.147}$$

Hence

$$P = a_1 a_1^* (1 - S_{11} S_{11}^* - S_{12} S_{12}^*) \tag{2.148}$$

For a lossless network $P = 0$ and

$$S_{11} S_{11}^* + S_{12} S_{12}^* = 1 \tag{2.149}$$

This is known as the unitary condition and states that for a lossless network

$$|S_{11}(j\omega)|^2 + |S_{12}(j\omega)|^2 = 1 \tag{2.150}$$

This condition is extremely important as it relates the input reflection coefficient and power gain for lossless networks and enables us to synthesise networks.

In general we can state for two-port networks [9]

$$[S(j\omega)][S^*(j\omega)] = [1] \tag{2.151}$$

yielding

$$|S_{11}(j\omega)|^2 + |S_{12}(j\omega)|^2 = 1 \tag{2.152}$$

$$|S_{22}(j\omega)|^2 + |S_{12}(j\omega)|^2 = 1 \tag{2.153}$$

$$S_{11}(j\omega) S_{12}^*(j\omega) + S_{12}(j\omega) S_{22}^*(j\omega) = 0 \tag{2.154}$$

Now if

$$S_{11}(p) = \frac{N(p)}{D(p)} \tag{2.155}$$

then solutions of (2.152)–(2.154) yield

$$S_{22}(p) = -\frac{N(-p)}{D(p)} \tag{2.156}$$

and

$$S_{12}(p) = \frac{F(p)}{D(p)} \tag{2.157}$$

where

$$F(p) F(-p) = D(p) D(-p) - N(p) N(-p) \tag{2.158}$$

The transfer function $S_{12}(p)$ is often expressed in decibels and is called the insertion loss L_A of the network, i.e.

$$L_A = -20 \log_{10} |S_{12}(j\omega)| \, \text{dB} \tag{2.159}$$

The reflection coefficient of $S_{11}(p)$ expressed in decibels is known as the return loss L_R of the network

$$L_R = -20 \log_{10} |S_{11}(j\omega)| \, \text{dB} \tag{2.160}$$

The insertion loss is a measure of attenuation through the network. The return loss is a measure of how well matched the network is. This is because it is a measure of reflected signal attenuation. A perfectly matched lossless network would have zero insertion loss and infinite return loss. A typical 'good value' for return loss in a well-matched system is between 15 and 25 dB depending on the application.

As an example of the application of S parameters, consider the following insertion loss function which is for a degree 3 Butterworth filter.

$$L_A = 10 \log(1 + \omega^6) \, \text{dB} \tag{2.161}$$

That is,

$$|S_{12}(j\omega)|^2 = \frac{1}{1 + \omega^6} \tag{2.162}$$

Hence from the unitary condition

$$|S_{11}(j\omega)|^2 = 1 - |S_{12}(j\omega)|^2 = \frac{\omega^6}{1 + \omega^6} \tag{2.163}$$

Let

$$S_{11}(p) = \frac{N(p)}{D(p)} \tag{2.164}$$

Then

$$N(p)N(-p)|_{p=j\omega} = \omega^6 \tag{2.165}$$

Therefore

$$N(p) = \pm p^3 \tag{2.166}$$

Also

$$D(p)D(-p)|_{p=j\omega} = 1 + \omega^6 \tag{2.167}$$

Thus

$$\begin{aligned} D(p)D(-p) &= 1 - p^6 \\ &= (1 + 2p + 2p^2 + p^3)(1 - 2p + 2p^2 - p^3) \end{aligned} \tag{2.168}$$

40 Theory and design of microwave filters

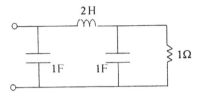

Figure 2.21 Lowpass ladder network

Therefore taking the left half-plane roots we have

$$D(p) = 1 + 2p + 2p^2 + p^3 \tag{2.169}$$

and

$$S_{11}(p) = \frac{\pm p^3}{1 + 2p + 2p^2 + p^3} \tag{2.170}$$

Now

$$Z(p) = \frac{1 + S_{11}(p)}{1 - S_{11}(p)}$$

$$= \frac{D(p) + N(p)}{D(p) - N(p)}$$

and

$$Z(p) \text{ or } Y(p) = \frac{1 + 2p + 2p^2 + 2p^3}{1 + 2p + 2p^2} \tag{2.171}$$

Since all the zeros of $|S_{12}(j\omega)|^2$ occur at $\omega = \infty$ then $Z(p)$ can be synthesised using the Darlington technique as a ladder network terminated in a resistor, as shown in Figure 2.21.

2.6.1 Relationships between ABCD parameters and S parameters

It may be shown by analysis that

$$S_{11} = \frac{A - D + B - C}{A + B + C + D} \tag{2.172}$$

$$S_{22} = \frac{D - A + B - C}{A + B + C + D} \tag{2.173}$$

$$S_{12} = \frac{2}{A + B + C + D} \tag{2.174}$$

Hence

$$|S_{12}(j\omega)|^2 = \frac{4}{|A + B + C + D|^2} \tag{2.175}$$

and the insertion loss L_A is given by

$$L_A = 10 \log_{10} \left| \tfrac{1}{4} |A + B + C + D|^2 \right| \text{ dB} \tag{2.176}$$

Now for a lossless network B and C are purely imaginary, i.e.

$$B = jB' \tag{2.177}$$

$$C = jC' \tag{2.178}$$

and for a reciprocal network

$$AD + B'C' = 1 \tag{2.179}$$

Thus for a lossless reciprocal network

$$L_A = 10 \log_{10} \left\{ 1 + \tfrac{1}{4}[(A - D)^2 + (B' - C')^2] \right\} \text{ dB} \tag{2.180}$$

and for a symmetrical network $A = D$.
Similarly the return loss is given by

$$L_R = 10 \log_{10} \left[1 + \frac{4}{(A - D)^2 + (B' - C')^2} \right] \text{ dB} \tag{2.181}$$

These formulae apply in a $1\,\Omega$ system.

As an example we can analyse the lowpass ladder network given in Figure 2.21. By multiplication of transfer matrices the overall transfer matrix of the lossless part of the circuit is

$$[T] = \begin{bmatrix} 1 - 2\omega^2 & j2\omega \\ j\omega(2 - 2\omega^2) & 1 - 2\omega^2 \end{bmatrix} \tag{2.182}$$

Hence

$$A = D \qquad B' = 2\omega \qquad C' = 2\omega - 2\omega^3 \tag{2.183}$$

Hence

$$L_A = 10 \log_{10}[1 + \tfrac{1}{4}(2\omega - 2\omega + 2\omega^3)^2]$$

$$= 10 \log_{10}(1 + \omega^6) \tag{2.184}$$

2.7 Even- and odd-mode analysis of symmetrical networks

Given the symmetrical circuit shown in Figure 2.22, it is possible to simplify the analysis by the use of even- and odd-mode networks.

An even-mode excitation implies that equal potentials are applied at each end of the circuit; hence there is an open circuit along the line of symmetry. In

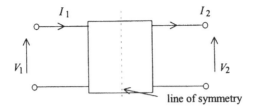

Figure 2.22 Symmetrical two-port network

this case
$$V_2 = V_1 \tag{2.185}$$
and
$$I_2 = -I_1 \tag{2.186}$$
Now
$$\begin{bmatrix} V_1 \\ I_1 \end{bmatrix} = \begin{bmatrix} A & B \\ C & D \end{bmatrix} \begin{bmatrix} V_2 \\ I_2 \end{bmatrix} \tag{2.187}$$
Hence
$$V_1 = AV_1 - BI_1 \tag{2.188}$$
Rearranging
$$Y_e = \frac{A-1}{B} \tag{2.189}$$
An odd-mode excitation implies opposite potentials at each end of the circuit. Hence there is a short circuit along the line of symmetry. In this case
$$V_2 = -V_1 \tag{2.190}$$
$$I_2 = I_1 \tag{2.191}$$
and the odd-mode admittance is given by
$$Y_o = \frac{1+A}{B} \tag{2.192}$$
From (2.189) and (2.192) we obtain
$$A = \frac{Y_e + Y_o}{Y_o - Y_e} = D \tag{2.193}$$
$$B = \frac{2}{Y_o - Y_e} \tag{2.194}$$

and from reciprocity and symmetry

$$A^2 - BC = 1 \tag{2.195}$$

Hence

$$C = \frac{2Y_e Y_o}{Y_o - Y_e} \tag{2.196}$$

Thus the transfer matrix of a symmetrical network may be given in terms of the even- and odd-mode admittances:

$$[T] = \begin{bmatrix} \dfrac{Y_e + Y_o}{Y_o - Y_e} & \dfrac{2}{Y_o - Y_e} \\ \dfrac{2Y_e Y_o}{Y_o - Y_e} & \dfrac{Y_e + Y_o}{Y_o - Y_e} \end{bmatrix} \tag{2.197}$$

Now by combining (2.172)–(2.174) with (2.197) we can obtain expressions for the S parameters in terms of Y_e and Y_o, i.e.

$$S_{11} = \frac{1 - Y_o Y_e}{(1 + Y_o)(1 + Y_e)} = S_{22} \tag{2.198}$$

$$S_{12} = S_{21} = \frac{Y_o - Y_e}{(1 + Y_o)(1 + Y_e)} \tag{2.199}$$

As an example consider the third-order Butterworth filter shown in Figure 2.21. Here

$$Y_e = p + \frac{1}{p} \tag{2.200}$$

and

$$Y_o = p \tag{2.201}$$

Hence

$$S_{12}(p) = \frac{1/p}{(1 + p)(1 + 1/p + p)}$$

$$= \frac{1}{1 + 2p + 2p^2 + 2p^3} \tag{2.202}$$

and

$$|S_{12}(j\omega)|^2 = \frac{1}{1 + \omega^6} \tag{2.203}$$

2.8 Analysis by image parameters

The method of image parameters is often overlooked these days. However, it is a useful technique for analysing simple network structures consisting of cascades of identical elements.

Thus given a symmetrical two-port network with transfer matrix

$$[T] = \begin{bmatrix} \dfrac{Y_e + Y_o}{Y_o - Y_e} & \dfrac{2}{Y_o - Y_e} \\ \dfrac{2Y_e Y_o}{Y_o - Y_e} & \dfrac{Y_e + Y_o}{Y_o - Y_e} \end{bmatrix} \qquad (2.204)$$

$[T]$ can be expressed as [10]

$$[T] = \begin{bmatrix} \cosh\gamma & Z_1 \sinh\gamma \\ Y_1 \sinh\gamma & \cosh\gamma \end{bmatrix} \qquad (2.205)$$

where

$$\gamma = \cosh^{-1}\left(\frac{Y_e + Y_o}{Y_o - Y_e}\right) \qquad (2.206)$$

and

$$Y_1 = \frac{1}{Z_1} = (Y_e Y_o)^{1/2} \qquad (2.207)$$

γ is known as the image propagation function, Z_1 is the image impedance.

Now consider a cascade of identical sections each with transfer matrix $[T]$. Then

$$[T_N] = [T]^N$$

$$= \begin{bmatrix} \cosh(N\gamma) & Z_1 \sinh(N\gamma) \\ Y_1 \sinh(N\gamma) & \cosh(N\gamma) \end{bmatrix} \qquad (2.208)$$

Here we see the power of the technique in that the problem of computing the transfer matrix of a cascade of identical sections is reduced to a relatively trivial result.

The S parameters of the cascade are thus given by

$$S_{11} = S_{22} = \frac{(Z_1 - Y_1)\sinh(N\gamma)}{2\cosh(N\gamma) + (Z_1 + Y_1)\sinh(N\gamma)} \qquad (2.209)$$

$$S_{12} = \frac{2}{2\cosh(N\gamma) + (Z_1 + Y_1)\sinh(N\gamma)} \qquad (2.210)$$

As an example consider the ladder network shown in Figure 2.23. This

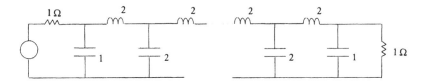

Figure 2.23 Ladder network

network is of degree $2N+1$ but it consists of a cascade of N identical sections (Figure 2.24). The even- and odd-mode admittances of the basic section are

$$Y_e = p \tag{2.211}$$

and

$$Y_o = p + \frac{1}{p} = \frac{1+p^2}{p} \tag{2.212}$$

Hence

$$Y_1 = (Y_e Y_o)^{1/2} = (1+p^2)^{1/2} \tag{2.213}$$

and

$$\cosh \gamma = \frac{Y_e + Y_o}{Y_o - Y_e}$$

$$= 1 + 2p^2 \tag{2.214}$$

Now

$$\frac{S_{11}}{S_{12}} = \frac{Z_1 - Y_1}{2} \sinh(N\gamma)$$

$$= \frac{\frac{1}{(1+p^2)^{1/2}} - (1+p^2)^{1/2}}{2} \sinh[N \cosh^{-1}(1+2p^2)]$$

$$= \frac{-p^2}{2(1+p^2)^{1/2}} \sinh[2N \sinh^{-1} p)] \tag{2.215}$$

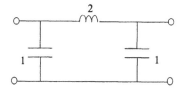

Figure 2.24 Basic section of the ladder network

Hence

$$\frac{|S_{11}|^2}{|S_{12}|} = \frac{\omega^4 \sin^2[2N\sin^{-1}(\omega)]}{4(1-\omega^2)} \qquad (2.216)$$

2.9 Analysis of distributed circuits

At microwave frequencies the use of distributed circuit elements is widespread. These differ from zero-dimensional lumped circuits by the fact that one or more dimensions are a significant fraction of the operating wavelength. There are many texts on this subject and all we are concerned with here is the ability to analyse circuits containing transmission lines.

For a one-dimensional line (Figure 2.25) it can be shown [11] that the transfer matrix is given by

$$\begin{bmatrix} V_1 \\ I_1 \end{bmatrix} = \begin{bmatrix} \cosh(\gamma\ell) & Z_o \sinh(\gamma\ell) \\ Y_o \sinh(\gamma\ell) & \cosh(\gamma\ell) \end{bmatrix} \qquad (2.217)$$

where Z_o is the characteristic impedance of the line. The value of Z_o depends on the physical construction of the line and is the ratio of voltage to current at any point p along the line. γ is the propagation constant of the line and

$$\gamma = \alpha + j\beta \qquad (2.218)$$

where α is the attenuation constant and β is the phase constant.

For a lossless line

$$\gamma = j\beta \qquad (2.219)$$

and the transfer matrix reduces to

$$[T] = \begin{bmatrix} \cos(\beta\ell) & jZ_o \sin(\beta\ell) \\ j\sin(\beta\ell)/Z_o & \cos(\beta\ell) \end{bmatrix} \qquad (2.220)$$

Now

$$\beta = \frac{2\pi}{\lambda} \qquad (2.221)$$

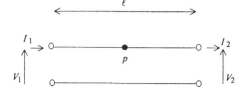

Figure 2.25   One-dimensional transmission line

Basic network theory 47

and for any wave propagating at velocity v with wavelength λ

$$v = f\lambda \qquad (2.222)$$

Hence

$$\beta = \frac{\omega}{v} \qquad (2.223)$$

and

$$\beta\ell = \frac{\omega\ell}{v} = a\omega \qquad (2.224)$$

where

$$a = \frac{\omega}{v} \qquad (2.225)$$

Thus the section of lossless transmission line has a transfer matrix which is a function of frequency as follows.

$$[T] = \begin{bmatrix} \cos(a\omega) & jZ_o \sin(a\omega) \\ jY_o \sin(a\omega) & \cos(a\omega) \end{bmatrix} \qquad (2.226)$$

where

$$Y_o = \frac{1}{Z_o} \qquad (2.227)$$

If a transmission line is terminated in a short circuit then the input impedance is

$$Z_{in}(j\omega) = jZ_o \tan(a\omega) \qquad (2.228)$$

and if a transmission line is terminated in an open circuit then the input impedance is

$$Z_{in}(j\omega) = -jZ_o/\tan(a\omega) \qquad (2.229)$$

It will be seen in later chapters that sections of transmission line in cascade or shunt or series connection have useful properties as circuit elements in microwave filters.

2.10 Summary

In this chapter an attempt has been made to summarise various network theoretical concepts which are relevant to modern filter design. The book is concerned entirely with linear passive time-invariant networks and so these properties have been precisely defined. The concept of the input impedance of a network in terms of the complex frequency variable has led to the properties of positive real and bounded real functions. The synthesis of lossless one-port networks has led on to Darlington synthesis of terminated two-port networks, with lowpass ladder networks being one particular case.

Various analysis techniques in terms of *ABCD* matrices, *S* parameters, even- and odd-mode networks and image parameters have been discussed. The extension of these techniques to distributed circuits is introduced at the end of the chapter. The material in this chapter gives sufficient background for the more advanced material in Chapter 3.

2.11 References

1 BAHER, H.: 'Synthesis of electrical networks' (Wiley, New York, 1984) pp. 23–29
2 KUO, F.F.: 'Network analysis and synthesis' (Wiley, New York, 1966) pp. 294–99
3 VAN VALKENBURG, M.E.: 'Introduction to network synthesis' (Wiley, New York, 1960) pp. 121–137
4 FOSTER, R.M.: 'A reactance theorem', *Bell System Technical Journal*, 1924, (3), pp. 259–67
5 BRUNE, O.: 'Synthesis of finite two terminal network whose driving point impedance is a prescribed function of frequency', *Journal of Mathematics and Physics*, 1931 (10), pp. 191–236
6 DARLINGTON, S.: 'Synthesis of reactance 4-poles with prescribed insertion loss characteristics', *Journal of Mathematics and Physics*, 1939 (18), pp. 257–353
7 SCANLAN, J.O., and RHODES, J.D.: 'Unified theory of cascade synthesis', *Proceedings of the IEE*, 1970, **17**, pp. 665–70
8 KREYSIG, E.: 'Advanced engineering mathematics' (Wiley, New York, 1988) p. 934
9 HELSAJN, J.: 'Synthesis of lumped element, distributed and planar filters' (McGraw-Hill, New York, 1990) pp. 7–9
10 SCANLAN, J.O., and LEVY, R.: 'Circuit theory, vol. 1' (Oliver and Boyd, Edinburgh, 1970) pp. 205–17
11 SANDER, K.F., and REED, G.A.L.: 'Transmission and propagation of electromagnetic waves' (Cambridge University Press, Cambridge, 1986) pp. 170–71

Chapter 3
Design of lumped lowpass prototype networks

3.1 Introduction

A lowpass prototype is a passive, reciprocal, normally lossless two-port network which is designed to operate from a 1 Ω generator into a 1 Ω load. The network response has a lowpass characteristic with its band-edge frequency at $\omega = 1$. The amplitude response of the network is designed to at least meet a minimum specification on passband return loss L_R and stopband insertion loss L_A. For example,

$$L_R \geq 20\,\text{dB} \quad 0 \leq \omega \leq 1 \quad \text{(passband)} \tag{3.1}$$

$$L_A \geq 50\,\text{dB} \quad 1.2 \leq \omega \leq \infty \quad \text{(stopband)} \tag{3.2}$$

Since the network is normally lossless there is no need to specify the passband insertion loss since this is related to the return loss by the unitary condition.

There may also be a phase linearity or group delay specification on the passband characteristic of the filter.

The lowpass prototype which may be of lumped or distributed realisation is a 'building block' from which real filters may be constructed. Various transformations may be used to convert it into a bandpass or other filter of arbitrary centre frequency and bandwidth.

3.2 The maximally flat prototype

The maximally flat or Butterworth approximation is the simplest meaningful approximation to an ideal lowpass filter. The approximation is defined by [1]

$$|S_{12}(j\omega)|^2 = \frac{1}{1 + \omega^{2N}} \tag{3.3}$$

Hence the insertion loss is given by

$$L_A = 10\log_{10}(1+\omega^{2N}) \tag{3.4}$$

Now for $\omega < 1$, ω^{2N} rapidly becomes very small. For $\omega > 1$, ω^{2N} rapidly becomes very large. The 3 dB frequency is at $\omega = 1$ and marks the transition between passband and stopband.

N is the degree of the network. The larger the value of N the more rapid is the transition from passband to stopband.

More exactly, the behaviour of $|S_{12}(j\omega)|^2$ is maximally flat at $\omega = 0$ and $\omega = \infty$. Given

$$F(\omega) = 1 + \omega^{2N} \tag{3.5}$$

then

$$\frac{dF(\omega)}{d\omega} = 2N\omega^{2N-1} = 0 \quad \text{at} \quad \omega = 0 \tag{3.6}$$

and

$$\frac{d^{2N-1}F(\omega)}{d\omega^{2N-1}} = 2N!\omega = 0 \quad \text{at} \quad \omega = 0 \tag{3.7}$$

In other words the first $2N - 1$ derivatives of the insertion loss characteristic are zero at $\omega = 0$. This implies a very flat response across the passband. Now

$$|S_{12}(j\omega)|^2 = \frac{\omega^{-2N}}{1+\omega^{-2N}} \tag{3.8}$$

and the first $2N - 1$ derivatives of the insertion loss function are zero at $\omega = \infty$. The maximally flat behaviour gives rise to an S-shaped frequency response, shown in Figure 3.1 for various values of N.

It is important to be able to calculate the degree N of the filter in order to meet a given specification. We require

$$\text{return loss} \geq L_R \quad \text{for} \quad \omega \leq \omega_p \tag{3.9}$$

$$\text{insertion loss} \geq L_A \quad \text{for} \quad \omega \geq \omega_s \tag{3.10}$$

Hence in the stopband

$$10\log_{10}(1+\omega_s^{2N}) \geq L_A \tag{3.11}$$

If $L_A \gg 1$ then $\omega_s^{2N} \gg 1$ and

$$N \geq \frac{L_A}{20\log_{10}(\omega_s)} \tag{3.12}$$

In the passband

$$10\log_{10}\left[\frac{1}{|S_{11}(j\omega_p)|^2}\right] \geq L_R \tag{3.13}$$

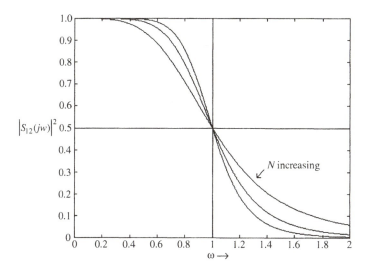

Figure 3.1 Maximally flat filter response

Now

$$|S_{11}(j\omega)|^2 = 1 - |S_{12}(j\omega)|^2 \qquad (3.14)$$

Therefore

$$10 \log_{10}\left(\frac{1 + \omega_p^{2N}}{\omega_p^{2N}}\right) \geq L_R \qquad (3.15)$$

If we define the selectivity S of the filter as being the ratio of stopband to passband frequencies, i.e.

$$S = \frac{\omega_s}{\omega_p} \geq 1 \qquad (3.16)$$

then

$$10 \log_{10}\left[1 + \left(\frac{S}{\omega_s}\right)^{2N}\right] \geq L_R \qquad (3.17)$$

and if $L_R \gg 1$ then $(S/\omega_s)^{2N} \gg 1$

So $20N[\log_{10}(S) - \log_{10}(\omega_s)] \geq L_R \qquad (3.18)$

and from (3.12) and (3.18)

$$20N \log_{10}(S) \geq L_A + L_R \qquad (3.19)$$

52 Theory and design of microwave filters

That is,

$$N \geq \frac{L_A + L_R}{20 \log_{10}(S)} \tag{3.20}$$

For example, if $L_A = 50\,\text{dB}$, and $L_R = 20\,\text{dB}$ and $S = 2$, then $N \geq 11.7$, i.e. we need a twelfth-degree filter to meet the specification.

Synthesis of the maximally flat filter proceeds as follows. Given

$$|S_{12}(j\omega)|^2 = \frac{1}{1+\omega^{2N}} \tag{3.21}$$

then

$$|S_{11}(j\omega)|^2 = \frac{\omega^{2N}}{1+\omega^{2N}} \tag{3.22}$$

Therefore

$$S_{11}(j\omega)S_{11}(-j\omega) = \frac{\omega^{2N}}{1+\omega^{2N}} \tag{3.23}$$

Hence

$$S_{11}(p)S_{11}(-p) = \frac{\pm p^{2N}}{1+(-jp)^{2N}} \tag{3.24}$$

The numerator of $S_{11}(p)$ can be formed by selecting any combination of zeros. Thus, for example, if

$$S_{11}(p) = \frac{N(p)}{D(p)} \tag{3.25}$$

then

$$N(p) = p^N \tag{3.26}$$

However, $S_{11}(p)$ contains left half-plane poles and $S_{11}(-p)$ contains right half-plane poles. We must take care to form $D(p)$ from the left half-plane poles, i.e. we need to find the left half-plane zeros of

$$1 + (-p^2)^N = 0 \tag{3.27}$$

That is,

$$(-p^2)^N = -1 = \exp[j\pi(2r-1)] \tag{3.28}$$

and

$$p_r = j\exp(j\theta_r) \tag{3.29}$$

where

$$\theta_r = \frac{(2r-1)\pi}{2N} \tag{3.30}$$

Lumped lowpass prototype networks 53

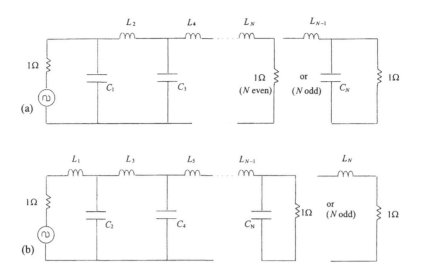

Figure 3.2 Ladder realisation of a maximally flat prototype network

Therefore

$$p_r = -\sin(\theta_r) + j\cos(\theta_r) \qquad r = 1, \ldots, 2n \tag{3.31}$$

These poles lie on a unit circle in the complex plane and the first n roots lie in the left half-plane. Thus

$$S_{11}(p) = \frac{\pm p^N}{\prod_{r=1}^{N}[p - j\exp(j\theta_r)]} \tag{3.32}$$

This prototype may be synthesised by forming the input impedance

$$Z(p) = \frac{1 + S_{11}(p)}{1 - S_{11}(p)} \tag{3.33}$$

and since all the transmission zeros occur at infinity the network can be synthesised as a lowpass ladder network as in Figure 3.2.

The two realisations in Figure 3.2 are subtly different. Type (a) starts with a shunt capacitor and type (b) with a series inductor. The particular type (a) or (b) depends on the choice of sign used for $S_{11}(p)$. Reversing the sign has the effect of inverting the impedance so the input impedance is either short or open circuited at $\omega = \infty$, depending on the sign chosen.

As an example we will synthesise a second-order Butterworth filter. Hence

$$S_{11}(p) = \frac{\pm p^2}{[p - j\exp(j\pi/4)][p - j\exp(j3\pi/4)]}$$

$$= \frac{\pm p^2}{p^2 + \sqrt{2}p + 1} \tag{3.34}$$

Now

$$Z(p) = \frac{1 + S_{11}(p)}{1 - S_{11}(p)} \tag{3.35}$$

and taking the positive sign numerator

$$Z(p) = \frac{2p^2 + \sqrt{2}p + 1}{\sqrt{2}p + 1}$$

$$= \sqrt{2}p + \frac{1}{\sqrt{2}p + 1} \tag{3.36}$$

or

$$Y(p) = \frac{2p^2 + \sqrt{2}p + 1}{\sqrt{2}p + 1} \tag{3.37}$$

giving the two realisations shown in Figure 3.3.

In the case of the maximally flat lowpass prototype, explicit design formulae have been developed for the element values shown in Figure 3.4 [2, 3]. The element values are given by

$$g_r = 2\sin\left[\frac{(2r - 1)\pi}{2N}\right] \quad (r = 1, \ldots, N) \tag{3.38}$$

where

$$\begin{aligned} g_r &= L_r \quad (r \text{ odd}) \\ g &= C_r \quad (r \text{ even}) \end{aligned} \tag{3.39}$$

Figure 3.3 Network realisations of an $N = 2$ maximally flat lowpass prototype filter

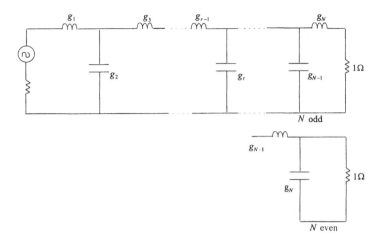

Figure 3.4   Maximally flat lowpass prototype ladder network

3.2.1 Impedance inverters

There is an alternative realisation of ladder networks using elements called impedance or admittance inverters. An impedance inverter is a lossless, reciprocal, frequency-independent, two-port network, defined by its transfer matrix

$$[T] = \begin{bmatrix} 0 & jK \\ j/K & 0 \end{bmatrix} \qquad (3.40)$$

where K is the characteristic impedance or admittance of the inverter.

The main property of an inverter is that of impedance inversion. Consider the circuit of Figure 3.5 consisting of an impedance inverter which is terminated in a load Z_L. Now

$$Z_{in}(p) = \frac{AZ_L + B}{CZ_L + D}$$

$$= \frac{jK}{jZ_L/K} = \frac{K^2}{Z_L} \qquad (3.41)$$

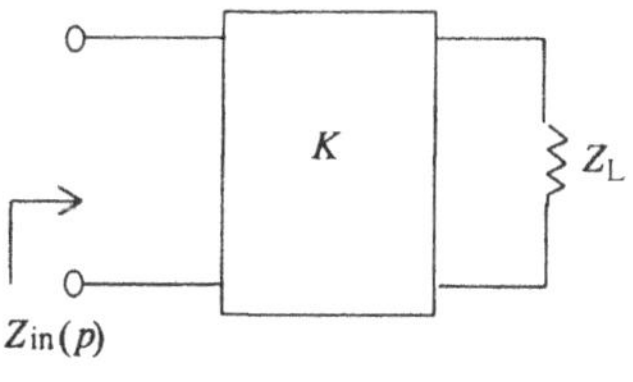

Figure 3.5   Impedance inverter terminated in a load

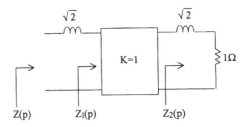

Figure 3.6   Second-order maximally flat filter with inverters

Thus the input impedance is proportional to the inverse of the load impedance. The use of the inverter may be illustrated by the synthesis of the second-order maximally flat filter.

From (3.36)

$$Z(p) = \frac{2p^2 + \sqrt{2}p + 1}{\sqrt{2}p + 1} \tag{3.42}$$

Extracting a series inductor of value $\sqrt{2}$ we obtain

$$Z_1(p) = \frac{2p^2 + \sqrt{2}p + 1}{\sqrt{2}p + 1} - \sqrt{2}p$$

$$= \frac{1}{\sqrt{2}p + 1} \tag{3.43}$$

Normally at this stage we invert $Z_1(p)$ to form $Y_1(p)$. However if we extract a unity impedance inverter the inverse of $Z_1(p)$ is still an impedance. Hence

$$Z_2(p) = \sqrt{2}p + 1 \tag{3.44}$$

Thus the network is synthesised entirely with inductors and inverters as shown in Figure 3.6.

Explicit formulae for the general Nth-degree maximally flat inverter coupled lowpass prototype shown in Figure 3.7 are given below [3].

$$G_r = L_r \text{ or } C_r$$

$$= 2 \sin\left[\frac{(2r - 1)\pi}{2N}\right] \quad (r = 1, \ldots, N) \tag{3.45}$$

$$K_{r,r+1} = 1 \quad (r = 1, \ldots, N - 1) \tag{3.46}$$

3.3 The Chebyshev prototype

The maximally flat approximation is the simplest meaningful approximation to the ideal lowpass filter. It is maximally flat at d.c. and infinity, but rolls off to

Lumped lowpass prototype networks 57

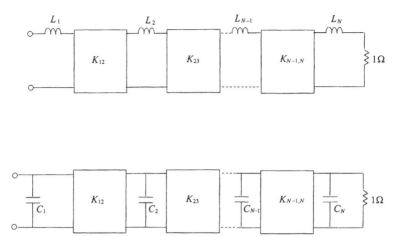

Figure 3.7 Inverter-coupled lowpass prototype maximally flat filter

3 dB at $\omega = 1$. It is thus sometimes called a zero-bandwidth approximation. As discussed in Chapter 1 a better approximation is one which ripples between two values in the passband up to the band-edge at $\omega = 1$, before rolling off rapidly in the stopband. This type of approximation is shown in Figure 3.8 for degrees 5 and 6.

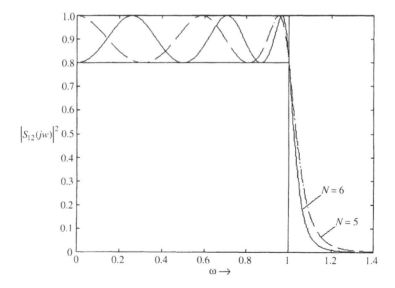

Figure 3.8 Chebyshev lowpass approximation

The insertion loss at ripple level is normally expressed as

$$IL = 10\log(1+\varepsilon^2) \tag{3.47}$$

Thus the ripple in the passband can be controlled by the level of ε.

To achieve this type of behaviour we let

$$|S_{12}(j\omega)|^2 = \frac{1}{1+\varepsilon^2 T_N^2(\omega)} \tag{3.48}$$

Thus

$$IL = 10\log_{10}[1+\varepsilon^2 T_N^2(\omega)] \tag{3.49}$$

$T_N(\omega)$ is a function which must then obtain the maximum value of 1 at the maximum number of points in the region $|\omega| < 1$. $T_N(\omega)$ is thus of the form shown in Figure 3.9.

We need to work out the formula for $T_N(\omega)$ so that we can calculate $|S_{12}(j\omega)|^2$. First we see that all points in the region $|\omega| < 1$ (except $\omega = \pm 1$) where $|T_N(\omega)| = 1$ must be turning points. Thus

$$\frac{dT_N(\omega)}{d\omega} = 0 \text{ when } |T_N(\omega)| = 1 \tag{3.50}$$

except when $|\omega| = 1$. Hence

$$\frac{dT_N(\omega)}{d\omega} = C_N \frac{[1-T_N^2(\omega)]^{1/2}}{(1-\omega^2)^{1/2}} \tag{3.51}$$

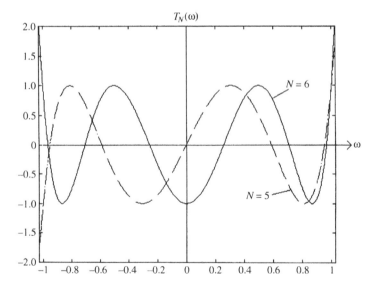

Figure 3.9 Equiripple response

From Figure 3.9 we see that

$$|T_N(\pm 1)| = 1 \tag{3.52}$$

Thus $1 - T_N^2(\pm 1) = 0$ when $(1 - \omega^2)^{1/2} = 0$ and $dT_N(\omega)/d\omega$ is finite at $\omega = \pm 1$.
Rewriting (3.51)

$$\frac{dT_N(\omega)}{[1 - T_N^2(\omega)]^{1/2}} = C_N \frac{d\omega}{(1 - \omega^2)^{1/2}} \tag{3.53}$$

Integrating both sides of (3.53) gives

$$\cos^{-1}[T_N(\omega)] = C_N \cos^{-1}(\omega) \tag{3.54}$$

C_N must be determined so that $T_N(\omega)$ is an nth-degree polynomial in ω.
Let $\cos^{-1}(\omega)$ be written as $\omega = \cos(\theta)$. Then

$$T_N(\omega) = \cos(C_N \theta) \tag{3.55}$$

and $T_N(\omega) = 0$ when

$$C_N \theta = \frac{(2r - 1)\pi}{2} \quad (r = 1, 2 \text{ etc.}) \tag{3.56}$$

or

$$\theta = \frac{(2r - 1)\pi}{2C_N} \tag{3.57}$$

For $T_N(\omega)$ to have N zeros then $C_N = N$, and

$$T_N(\omega) = \cos[N \cos^{-1}(\omega)] \tag{3.58}$$

Thus

$$|S_{12}(j\omega)|^2 = \frac{1}{1 + \varepsilon^2 \cos^2[N \cos^{-1}(\omega)]} \tag{3.59}$$

Now (3.59) must be a polynomial in ω; otherwise it could not represent the response of a real network. In fact $T_N(\omega)$ is known as the Chebyshev polynomial [4] and is given by the formula [3]

$$T_{N+1}(\omega) = 2\omega T_N(\omega) - T_{N-1}(\omega) \tag{3.60}$$

with initial conditions

$$T_0(\omega) = 1 \quad \text{and} \quad T_1(\omega) = \omega \tag{3.61}$$

Thus

$$T_2(\omega) = 2\omega\omega - 1 = 2\omega^2 - 1 \tag{3.62}$$

$$T_3(\omega) = 2\omega(2\omega^2 - 1) - \omega = 4\omega^3 - 3\omega \tag{3.63}$$

Let us now do an evaluation of the response of a third-order Chebyshev filter.

60 *Theory and design of microwave filters*

Say we want 20 dB minimum passband return loss. Then (in the worst case)

$$\text{insertion loss} = 10\log_{10}(1+\varepsilon^2) = 10\log\left(\frac{1}{|S_{12}|^2}\right) \quad (3.64)$$

and

$$\text{return loss} = 10\log\left(\frac{1}{|S_{11}|^2}\right) = L_R \quad (3.65)$$

Then

$$|S_{11}|^2 = 10^{-L_R/10} = 0.01 \quad (3.66)$$

and

$$|S_{12}|^2 = 1 - |S_{11}|^2 = 0.99 \quad (3.67)$$

So

$$1 + \varepsilon^2 = \frac{1}{0.99} \quad (3.68)$$

and

$$\varepsilon = 0.1005 \approx 0.1 \quad (3.69)$$

Thus

$$\text{insertion loss} = 10\log[1 + 0.01(4\omega^3 - 3\omega)^2] \quad (3.70)$$

The insertion loss ripples between zero and 0.043 dB in the region $|\omega| < 1$ and then rolls off, reaching 9 dB at $\omega = 2$. It would appear at first that the function is not as selective as the third-order maximally flat filter. In fact the maximally flat filter had 3 dB insertion loss at $\omega = 1$ so it is not a fair comparison.

The passband insertion loss ripple in the Chebyshev filter is 0.043 dB. The third-degree maximally flat filter achieved this at $\omega = 0.463$. As a comparison the ratio of stopband to passband frequency is 4.64 for the maximally flat filter and 3 for the Chebyshev filter. Thus we see that the Chebyshev response is considerably more selective than the maximally flat response.

A formula to calculate the degree of a Chebyshev filter to meet a specified response is now given. The proof of this formula is similar to the proof for the maximally flat response [3]. The formula is

$$N \geq \frac{L_A + L_R + 6}{20\log_{10}[S + (S^2 - 1)^{1/2}]} \quad (3.71)$$

where L_A is the stopband insertion loss, L_R is the passband return loss and S is the ratio of stopband to passband frequencies.

As an example, for $L_A = 50$ dB, $L_R = 20$ dB, $S = 2$ then N must be greater than 6.64, i.e. $N = 7$. The maximally flat filter required $N = 12$ to meet this specification. Synthesis of the Chebyshev filter proceeds as follows. Given

$$|S_{12}(j\omega)|^2 = \frac{1}{1 + \varepsilon^2 T_N^2(\omega)} \quad (3.72)$$

the poles occur when
$$T_N^2(\omega) = -1/\varepsilon^2 \tag{3.73}$$
That is,
$$\cos^2[N\cos^{-1}(\omega)] = -1/\varepsilon^2 \tag{3.74}$$
To solve this equation we introduce a new parameter η, where
$$\eta = \sinh\left[\frac{1}{N}\sinh^{-1}\left(\frac{1}{\varepsilon}\right)\right] \tag{3.75}$$
or
$$\frac{1}{\varepsilon} = \sinh[N\sinh^{-1}(\eta)] \tag{3.76}$$
Hence from (3.74) and (3.76)
$$\cos^2[N\cos^{-1}(\omega)] = -\sinh^2[N\sinh^{-1}(\eta)]$$
$$= \sin^2[N\sinh^{-1}(j\eta)] \tag{3.77}$$
Therefore
$$\cos^{-1}(\omega) = \sin^{-1}(j\eta) + \theta_r \tag{3.78}$$
where
$$\theta_r = \frac{(2r-1)\pi}{2N} \tag{3.79}$$
and
$$p_r = -j\cos[\sin^{-1}(j\eta) + \theta_r] = +\eta\sin(\theta_r) + j(1+\eta^2)^{1/2}\cos(\theta_r) \tag{3.80}$$
The left half-plane poles occur when $\sin(\theta_r)$ is positive, i.e. $r = 1, \ldots, N$. Therefore
$$p_r = \sigma_r + j\omega_r = \eta\sin(\theta_r) + j(1+\eta^2)^{1/2}\cos(\theta_r) \tag{3.81}$$
Thus
$$\frac{\sigma r^2}{\eta^2} + \frac{\omega r^2}{1+\eta^2} = 1 \tag{3.82}$$
The poles thus lie on an ellipse.

It may be shown that
$$S_{12}(p) = \prod_{r=1}^{N}\left\{\frac{[\eta^2 + \sin^2(r\pi/N)]^{1/2}}{p + j\cos[\sin^{-1}(j\eta) + \theta_r]}\right\} \tag{3.83}$$
Now
$$|S_{11}(j\omega)|^2 = 1 - |S_{12}(j\omega)|^2$$
$$= \frac{\varepsilon^2 T_N^2(\omega)}{1 + \varepsilon^2 T_N^2(\omega)} \tag{3.84}$$

62 Theory and design of microwave filters

The zeros occur when
$$\cos^2[N\cos^{-1}(\omega)] = 0 \tag{3.85}$$
That is,
$$p = -j\cos(\theta_r) \tag{3.86}$$
Now
$$S_{11}(\infty) = 1 \tag{3.87}$$
Then
$$S_{11}(p) = \prod_{r=1}^{N}\left\{\frac{p+j\cos(\theta_r)}{p+j\cos[\sin^{-1}(j\eta)+\theta_r]}\right\} \tag{3.88}$$

The network can then be synthesised as a lowpass ladder network by formulating $Z_{in}(p)$.

As an example we will synthesise a degree 3 Chebyshev filter.

For $r = 1$ $\theta_1 = \pi/6$ $\cos(\theta_1) = \sqrt{3}/2$ (3.89)

For $r = 2$ $\theta_2 = \pi/2$ $\cos(\theta_2) = 0$ (3.90)

For $r = 3$ $\theta_3 = 5\pi/6$ $\cos(\theta_3) = -\sqrt{3}/2$ (3.91)

Therefore
$$S_{11}(p) = \frac{p(p+j\sqrt{3}/2)(p-j\sqrt{3}/2)}{(p+\eta)[p+\eta/2+j(\sqrt{3}/2)(1+\eta^2)^{1/2}][p+\eta/2-j(\sqrt{3}/2)(1+\eta^2)^{1/2}]}$$
$$= \frac{p^3 + 3p/4}{p^3 + 2\eta p^2 + (2\eta^2 + 3/4)p + \eta(\eta^2 + 3/4)} \tag{3.92}$$

Now
$$Z_{in}(p) = \frac{2p^3 + 2\eta p^2 + 2(\eta^2 + 3/4)p + \eta(\eta^2 + 3/4)}{2\eta p^2 + 2\eta^2 p + \eta(\eta^2 + 3/4)} \tag{3.93}$$

Removing a series inductor of value $1/\eta$ we are left with
$$Z_1(p) = Z(p) - p/\eta$$
$$= \frac{(\eta^2 + 3/4)(p+\eta)}{2\eta p^2 + 2\eta^2 p + \eta(\eta^2 + 3/4)} \tag{3.94}$$

Extracting an inverter of characteristic admittance
$$K_{12} = \frac{(\eta^2 + 3/4)^{1/2}}{\eta} \tag{3.95}$$

the remaining impedance is
$$Z_2(p) = \frac{2p^2 + 2\eta p + (\eta^2 + 3/4)}{\eta(p+\eta)} \tag{3.96}$$

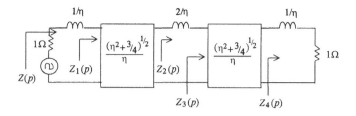

Figure 3.10 Synthesis of an $N = 3$ Chebyshev filter

Now extracting a second inductor of value $2/\eta$ we are left with

$$Z_3(p) = Z_2(p) - 2p/\eta$$

$$= \frac{\eta^2 + 3/4}{\eta(p+\eta)} \tag{3.97}$$

Extracting a second admittance inverter of characteristic admittance

$$K_{23} = \frac{(\eta^2 + 3/4)^{1/2}}{\eta} \tag{3.98}$$

the remaining impedance is

$$Z_4(p) = 1 + p/\eta \tag{3.99}$$

i.e. an inductor of value $1/\eta$ followed by a load resistor of value unity. The complete synthesis cycle is shown in Figure 3.10.

The synthesised element values given here are actually formulae in terms of η. These formulae may be generalised to the Nth-degree prototype shown in Figure 3.11 [3, 5, 6].

$$K_{R,R+1} = \frac{[\eta^2 + \sin^2(r\pi/N)]^{1/2}}{\eta} \qquad r = 1, \ldots, N-1 \tag{3.100}$$

where $K_{R,R+1}$ is the impedance of the inverters.

$$L_R = \frac{2}{\eta} \sin\left[\frac{(2r-1)\pi}{2N}\right] \qquad R = 1, \ldots, N \tag{3.101}$$

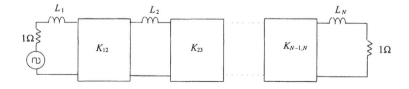

Figure 3.11 General Nth-degree Chebyshev prototype network

where

$$\eta = \sinh\left[\frac{1}{N}\sinh^{-1}\left(\frac{1}{\varepsilon}\right)\right] \tag{3.102}$$

ε is related to the insertion loss ripple and hence the passband return loss. Since

$$|S_{12}|^2 = \frac{1}{1+\varepsilon^2} \tag{3.103}$$

$$|S_{11}|^2 = \frac{\varepsilon^2}{1+\varepsilon^2} \tag{3.104}$$

therefore

$$L_R = 10\log_{10}(1+1/\varepsilon^2) \tag{3.105}$$

Hence

$$\varepsilon = (10^{L_R/10} - 1)^{-1/2} \tag{3.106}$$

Note that the dual of Figure 3.11 would consist of shunt capacitors separated by inverters. Formulae (3.100)–(3.102) still apply but they would then represent the values of the capacitors and the characteristic admittance of the inverters.

3.4 The elliptic function prototype

The elliptic function approximation is equiripple in both the passband and the stopband. It thus has the optimum response in terms of selectivity from passband to stopband. A typical elliptic function filter response is shown in Figure 3.12. The transmission zeros of this network are no longer at infinity and thus the filter cannot be realised with a ladder network. One of the disadvantages of this filter response is that the transmission zeros are prescribed to be at certain frequencies and there is no flexibility in their location, i.e.

$$|S_{12}(j\omega)|^2 = \frac{1}{1+F_N^2(\omega)} \tag{3.107}$$

where

$$F_N(\omega) = \frac{(\omega^2 - \omega_1^2)(\omega^2 - \omega_1^2)\cdots}{(\omega^2 - \omega_A)^2(\omega^2 - \omega_B^2)\cdots} \tag{3.108}$$

i.e. all the values ω_1, ω_A etc. are specified.

The synthesis procedures for the elliptic function filter are simpler if we work with a highpass rather than a lowpass prototype. This is shown in Figure 3.13 for $N=6$. In this case

$$|S_{12}(j\omega)|^2 = \frac{\varepsilon^2 F_N^2(\omega)}{1+\varepsilon^2 F_N^2(\omega)} \tag{3.109}$$

Lumped lowpass prototype networks 65

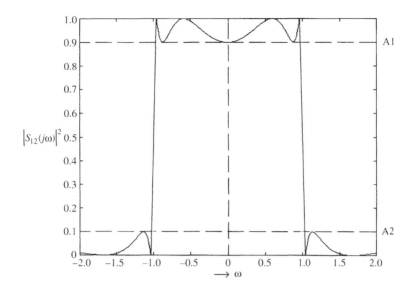

Figure 3.12   Elliptic function lowpass response

$F_N(\omega)$ is a rational function as in (3.108). It oscillates between ± 1 for $|\omega| \leq 1$ and $|F_N(\omega)| \geq m_0^{-1/2}$ for $|\omega| \geq m^{-1/2}$. $F_N(\omega)$ is shown in Figure 3.14.

$F_N(\omega)$ can be determined from a differential equation in a similar way to the Chebyshev filter, as follows. First we define the turning points in the passband and stopband:

$$\left.\frac{\mathrm{d}F_N(\omega)}{\mathrm{d}\omega}\right|_{|F_N(\omega)|=1} = 0 \qquad \text{except when } |\omega| = 1 \qquad (3.110)$$

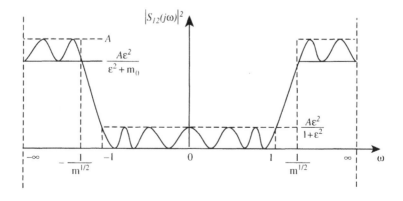

Figure 3.13   Degree 6 highpass elliptic function filter

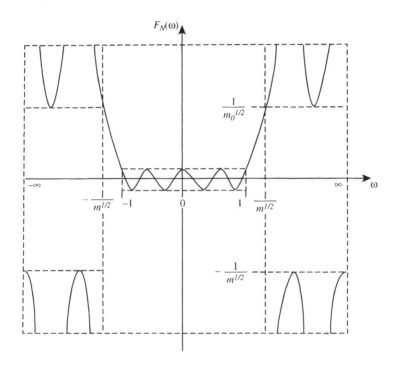

Figure 3.14 $F_N(\omega)$ for the highpass elliptic function filter

$$\left.\frac{dF_N(\omega)}{d\omega}\right|_{|F_N(\omega)|=m_0^{-1/2}} = 0 \qquad \text{except when } |\omega| = m_0^{-1/2} \tag{3.111}$$

Thus

$$\frac{dF_N(\omega)}{d\omega} = C_N \frac{\{[1 - F_N^2(\omega)][1 - m_0 F_N^2(\omega)]\}^{1/2}}{[(1 - \omega^2)(1 - m\omega^2)]^{1/2}} \tag{3.112}$$

Rearranging with $F_N(\omega)$ on the left-hand side we obtain

$$\frac{dF_N(\omega)}{\{[1 - F_N^2(\omega)][1 - m_0 F_N^2(\omega)]\}^{1/2}} = \frac{C_N\, d\omega}{[(1 - \omega^2)(1 - m\omega^2)]^{1/2}} \tag{3.113}$$

and after integrating we obtain

$$cd_0^{-1} F_N(\omega) = C_N cd^{-1} \omega = u \tag{3.114}$$

and

$$F_N(\omega) = cd_0 u \tag{3.115}$$

where the elliptic functions are all dependent on the elliptic parameter m, with the same notation as in Reference 7.

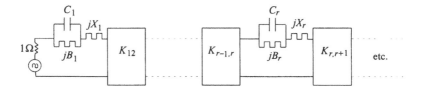

Figure 3.15 Highpass prototype for elliptic function filters

It may be shown [3, 8] that the polynomial form of $F_N(\omega)$ is

$$F_N(\omega) = \frac{B \prod_{r=1}^{N}\{\omega - cd[(2r-1)K/N]\}}{\prod_{r=1}^{N}\{1 - \omega mcd[(2r-1)K/N]\}} \qquad (3.116)$$

where

$$B = \prod_{r=1}^{N} \left\{ \frac{1 - mcd[(2r-1)K/N]}{1 - cd[(2r-1)K/N]} \right\} \qquad (3.117)$$

Synthesis of the elliptic function filter cannot be accomplished using a ladder network as it has finite real frequency transmission zeros. It may be synthesised using the techniques described for generalised Chebyshev filters later in this chapter, or by the techniques described for extracted pole waveguide filters in Chapter 6. It is possible to synthesise the filter using a type of ladder network using series resonators composed of capacitors in parallel with frequency-invariant reactances as shown in Figure 3.15.

Explicit formulae have been developed for the element values of the elliptic function prototype [3, 8]. These are

$$C_r = \frac{ds[(2-1)K/N]dn[(2r-1)K/N]}{2\eta(1-m)} \qquad r = 1, \ldots, N \qquad (3.118)$$

$$B_r = C_r cd\frac{(2r-1)K}{N} \qquad r = 1, \ldots, N \qquad (3.119)$$

$$X_r = -\eta m \left[sn\frac{2(r-1)K}{N} + sn\frac{2rK}{N} \right] cd\frac{K}{N} cd\frac{(2r-1)K}{N} \qquad r = 1, \ldots, N \qquad (3.120)$$

$$K_{r,r+1} = \left(1 + \eta^2 msn^2\frac{2rK}{N}\right)^{1/2} \qquad r = 1, \ldots, N-1 \qquad (3.121)$$

$$\eta = sc\left(\frac{K}{NK_0}sc^{-1}\frac{1}{\varepsilon}\right) \qquad (3.122)$$

An approximate equation for calculating the degree of the filter is

$$N \geq \frac{K(m)}{K'(m)} \frac{L_A + L_R + 12}{13.65} \tag{3.123}$$

where L_A and L_R are the stopband insertion loss and passband return loss, respectively, and K is the quarter period with respect to the elliptic parameter m. As an example, for $L_A = 50$ dB, $L_R = 20$ dB and $\eta = 2$ we obtain $m = 1/\eta^2 = 0.25$. $K(m)$ and $K'(m)$ are obtained from tables of elliptic integrals in Reference 7 giving $K(m) = 1.68575$ and $K'(m) = 2.15651$. Thus $N \geq 4.69$, i.e. $N = 5$ compared with $N = 7$ for the Chebyshev filter.

The elliptic function highpass prototype can also be used for inverse Chebyshev filters. In this case the transfer function is

$$|S_{12}(j\omega)|^2 = \frac{\varepsilon^2 T_N^2(\omega)}{1 + \varepsilon^2 T_N^2(\omega)} \tag{3.124}$$

The filter is maximally flat in the passband and equiripple in the stopband. In this case the element values for the prototype network of Figure 3.15 are

$$C_r = \frac{1}{2\eta \sin[(2r-1)\pi/2N]} \tag{3.125}$$

$$B_r = C_r \cos\left[\frac{(2r-1)\pi}{2N}\right] \tag{3.126}$$

$$X_r = 0 \tag{3.127}$$

$$K_{r,r+1} = 1 \tag{3.128}$$

One of the disadvantages of using the elliptic function filter is that the range of element values required is quite large, up to 10:1. Furthermore, in many applications we wish to specify the locations of the transmission zeros ourselves. This is possible by using the generalised Chebyshev approximation described in the next section.

3.5 The generalised Chebyshev prototype

The generalised Chebyshev approximation provides a filter with equiripple passband amplitude characteristics but with arbitrarily placed attenuation poles (transmission zeros) in the stopband. Because the transmission zeros can be placed arbitrarily then both symmetric and asymmetric frequency responses can be generated. Furthermore, the transmission zeros are not restricted to being at real frequencies but may also be located in the complex plane. A particular example of a generalised Chebyshev filter is shown in Figure 3.16.

Lumped lowpass prototype networks 69

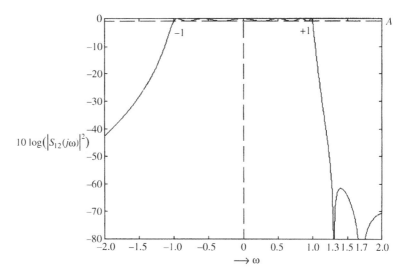

Figure 3.16   Generalised Chebyshev lowpass approximation

The generalised Chebyshev approximation is given by

$$|S_{12}(j\omega)|^2 = \frac{1}{1+\varepsilon^2 F_N^2(\omega)} \tag{3.129}$$

with

$$-1 < F_N(\omega) < +1 \quad \text{for} \quad -1 < \omega < +1 \tag{3.130}$$

and

$$F_N(\omega_r) = \infty \qquad r = 1, \ldots, N \tag{3.131}$$

where ω_r is the frequency of the rth transmission zero.

The normal Chebyshev transfer function can be represented by

$$|S_{12}(j\omega)|^2 = \frac{1}{1+\varepsilon^2 F_N^2(\omega)} \tag{3.132}$$

where

$$F_N(\omega) = \cos[N\cos^{-1}(\omega)]$$

$$= \cos(\theta) \tag{3.133}$$

and

$$\theta = N\cos^{-1}(\omega) \tag{3.134}$$

70 *Theory and design of microwave filters*

It can readily be seen that as ω varies from -1 to $+1$ then θ performs N half-period variations from $-\pi/2$ to $+\pi/2$. Hence $F_N(\omega)$ has N half-period variations from -1 to $+1$. For $|\omega| > 1$, θ is imaginary and $F_N(\omega)$ increases monotonically to infinity at $\omega = \infty$.

In the case of the generalised Chebyshev filter, some of the transmission zeros at infinite frequencies can be brought to finite but not necessarily real frequencies. Let

$$\theta = \sum_{r=1}^{N} \theta_r$$

$$= \sum_{r=1}^{N} \cos^{-1}\left(\frac{1 - \omega \omega_r}{\omega - \omega_r}\right)$$

$$= \sum_{r=1}^{N} \cos^{-1}\left(j\frac{1 + pp_r}{p - p_r}\right) \tag{3.135}$$

where p_r is the position of the rth transmission zero in the complex plane. This preserves the same range of variation of θ across $-1 < \omega < 1$ while producing attenuation poles (transmission zeros) at pr. Hence

$$F_N = \cos \sum_{r=1}^{n} \cos^{-1}\left(j\frac{1 + pp_r}{p - p_r}\right) \tag{3.136}$$

Now given that

$$\cos^{-1}(x) = -\log_e[x + (x^2 - 1)^{1/2}] \tag{3.137}$$

it can be shown that the polynomial form of (3.136) is [9, 10]

$$F_N = \frac{j}{2}\left(\prod_{r=1}^{N}\left\{\frac{1 + pp_r + [(1 + p^2)(1 + p_r^2)]^{1/2}}{p - p_r}\right\}\right.$$

$$\left. + \prod_{r=1}^{N}\left\{\frac{1 + pp_r - [(1 + p^2)(1 + p_r^2)]^{1/2}}{p - p_r}\right\}\right) \tag{3.138}$$

Multiplying out all the products results in

$$F_N = \frac{\sum_{j=0}^{N} a_j p^j}{\sum_{j=0}^{N} b_j p^j} = \frac{A(p)}{B(p)} \tag{3.139}$$

where a_j, b_j are in general complex coefficients.

Now

$$|S_{12}(j\omega)|^2 = \frac{1}{1+\varepsilon^2 A^2/B^2}$$

$$= \frac{1}{(1+j\varepsilon A/B)(1-j\varepsilon A/B)}$$

$$= \frac{B^2}{(B+j\varepsilon A)(B-j\varepsilon A)} \quad (3.140)$$

Furthermore

$$|S_{11}(j\omega)|^2 = 1 - |S_{12}(j\omega)|^2 \quad (3.141)$$

Hence

$$|S_{11}(j\omega)|^2 = \frac{\varepsilon^2 A^2}{(B+j\varepsilon A)(B-j\varepsilon A)} \quad (3.142)$$

The transmission and reflection functions can thus be formed from the A and B polynomials with the denominator being formed from the left half-plane zeros of the factorisation of the denominator of $|S_{12}(j\omega)|^2$.

There is no simple formula for calculating the degree of the generalised Chebyshev filter since the transmission zeros can be placed arbitrarily. The easiest way is to simulate the transfer function on a computer and choose the zero locations as required.

In general the synthesis of such filters can be performed using the cascade synthesis described in Chapter 2. However, certain specific synthesis techniques may be used depending on the degree of the network. These will be dealt with later in this chapter after the group delay and time domain approximations.

3.6 Filters with specified phase and group delay characteristics

The amplitude approximations discussed so far make no attempt to approximate to a prescribed phase or group delay response. However, in some systems applications these are of importance. As an example consider a digitally modulated signal passing through a filter with non-linear phase response. If the bandwidth of the filter is similar to that of the signal then the signal may experience severe phase distortion. This can give rise to inter-symbol interference and hence a degraded bit error rate.

In Chapter 1 we showed that the amplitude and phase responses of ladder networks are related by Hilbert transforms. It is instructive to look at the phase and group delay responses of some simple amplitude approximations. In general, for ladder networks

$$S_{12}(j\omega) = \frac{1}{\alpha(\omega)+jB(\omega)} \quad (3.143)$$

72 Theory and design of microwave filters

(where α is even, B is odd) and the phase response is

$$\psi(\omega) = -\tan^{-1}\left[\frac{B(\omega)}{\alpha(\omega)}\right] \qquad (3.144)$$

The group delay $T_g(\omega)$ is related to the phase by

$$T_g(\omega) = \frac{-d\psi(\omega)}{d\omega} \qquad (3.145)$$

Hence

$$T_g(\omega) = \frac{\alpha B' - B\alpha'}{\alpha^2 + B^2} \qquad (3.146)$$

(α' and B' denote differentiation with respect to ω).
For a Butterworth filter

$$\alpha^2 + B^2 = 1 + \omega^{2N} \qquad (3.147)$$

Hence

$$T_g(\omega) = \frac{\alpha B' - B\alpha'}{1 + \omega^{2N}} \qquad (3.148)$$

For a degree 2 Butterworth filter

$$S_{12}(p) = \frac{1}{p^2 + \sqrt{2}p + 1} \qquad (3.149)$$

That is,

$$S_{12}(j\omega) = \frac{1}{1 - \omega^2 + j\sqrt{2}\omega} \qquad (3.150)$$

$$\alpha = 1 - \omega^2 \qquad (3.151)$$

$$B = \sqrt{2}\omega \qquad (3.152)$$

and

$$T_g(\omega) = \frac{\sqrt{2}(1 + \omega^2)}{1 + \omega^4} \qquad (3.153)$$

$$T_g(0) = \sqrt{2} \qquad T_g(1) = \sqrt{2} \qquad T_g(\infty) = 0 \qquad (3.154)$$

The maximum value of $T_g(\omega)$ occurs when $T_g'(\omega) = 0$, i.e. at $\omega = 0.6435$, with a value $T_{g\,\max}$ of 1.707.
The ratio of $T_{g\,\max}$ to the value at d.c. is given by

$$\frac{T_{g\,\max}}{T_g(0)} = 1.207 \qquad (3.155)$$

3.6.1 The maximally flat group delay lowpass prototype

In the ideal case we would require the group delay to be constant across the passband. If we restrict ourselves to ladder realisations then we will look for a maximally flat approximation to constant group delay, and since all the transmission zeros are at infinity

$$S_{12}(p) = \frac{K}{m(p) + n(p)} \tag{3.156}$$

(m even, n odd). Thus

$$T_g(p) = \frac{m(p)n'(p) - n(p)m'(p)}{m(p)^2 - n(p)^2} \tag{3.157}$$

The objective is to choose m and n such that the group delay is a maximally flat approximation to a constant at $\omega = 0$. Consider

$$S_{12}(p) = \frac{K}{\cosh(ap) + \sinh(ap)} \tag{3.158}$$

(Note that this is physically unrealisable as the polynomial forms of $\cosh(ap)$ and $\sinh(ap)$ require infinite power series in p.) Now

$$m(p) = \cosh(ap) \tag{3.159}$$

$$n(p) = \sinh(ap) \tag{3.160}$$

Hence from (3.157)

$$T_g(p) \equiv a \tag{3.161}$$

Hence, (3.161) states that the group delay is a constant independent of frequency, but this is only true if m and n are of infinite degree. Instead we restrict $n(p)/m(p)$ to be of degree N and to be some approximation to $\tanh(ap)$. Now

$$\tanh(x) = \frac{\sinh(x)}{\cosh(x)} \tag{3.162}$$

and

$$\sinh(x) = x + \frac{x^3}{3!} + \frac{x^5}{5!} \cdots \tag{3.163}$$

$$\cosh(x) = 1 + \frac{x^2}{2!} + \frac{x^4}{4!} \cdots \tag{3.164}$$

Hence the continued fraction of $\tanh(ap)$ is

$$\tanh(ap) = \cfrac{1}{\cfrac{1}{ap} + \cfrac{1}{\cfrac{3}{ap} + \cfrac{1}{\cfrac{5}{ap} + \cdots}}}$$

$$\cdots$$

$$\cfrac{1}{\cfrac{2N-1}{ap} + \cfrac{1}{\cdots}} \quad (3.165)$$

To form n/m we truncate (3.165) after N terms and re-multiply. The numerator is equal to n and the denominator is equal to m.

For example, for $N = 2$

$$\tanh(ap) = \cfrac{1}{\cfrac{1}{ap} + \cfrac{1}{3/ap}}$$

$$= \frac{3ap}{3 + a^2 p^2} = \frac{n(p)}{m(p)} \quad (3.166)$$

Hence

$$S_{12}(p) = \frac{K}{3 + 3ap + a^2 p^2} \quad (3.167)$$

For realisability let $S_{12}(0) = 1$. Hence $K = 3$. Hence

$$S_{12}(p) = \frac{3}{3 + 3ap + a^2 p^2} \quad (3.168)$$

and from (3.157)

$$T_g(p) = \frac{9a - 3a^3 p^2}{9 - 3a^2 p^2 + a^4 p^4} \quad (3.169)$$

and

$$T_g(0) = a \quad (3.170)$$

Thus a is chosen to determine the group delay at d.c. Furthermore, it is readily shown that the group delay is maximally flat around $p = 0$.

Examining the amplitude characteristics of $S_{12}(p)$, without loss of generality we can let $a = 1$ in (3.168) giving

$$S_{12}(p) = \frac{1}{1 + p + p^2/3} \quad (3.171)$$

Hence

$$|S_{12}(j\omega)|^2 = \frac{1}{1+\omega^2/3+\omega^4/9} \tag{3.172}$$

This has considerably less amplitude selectivity that the normal maximally flat filter. This is not surprising as we have restricted ourselves to a minimum phase realisation. However, this type of characteristic can be useful where group delay flatness is more important than absolute selectivity.

It can be shown that a general solution for $S_{12}(p)$ [11] is given by

$$S_{12}(p) = \frac{K}{D_N(ap)} \tag{3.173}$$

where

$$D_N(ap) = (2N-1)D_{N-1}(ap) + a^2p^2 D_{N-2}(ap) \tag{3.174}$$

where

$$D_1 = 1 + ap \tag{3.175}$$

$$D_2 = 3 + 3ap + a^2p^2 \tag{3.176}$$

Hence from (3.173), (3.174) and (3.175)

$$D_3 = 15 + 15ap + 6a^2p^2 + a^3p^3 \tag{3.177}$$

Hence for $S_{12}(0) = 1$

$$S_{12}(p) = \frac{15}{15 + 15ap + 6a^2p^2 + a^3p^3} \tag{3.178}$$

which is an all-pole transfer function, realisable by a ladder network.

3.6.2 The equidistant linear phase approximation

The equiripple approximation to a linear phase response is a much better approximation than the maximally flat solution [3]. In fact the equiripple solution which minimises the maximum deviation from linear phase is the optimum solution. This solution uses a polynomial which approximates the ideal linear phase given by

$$\psi(\omega) = \omega \tag{3.179}$$

The error function $\psi(\omega) = \omega$ is designed to be zero at equal increments of α in ω and is shown in Figure 3.17.

The equidistant linear phase polynomial is given by

$$A_N(j\omega|\alpha) = A(\omega)\exp[j\psi(\omega)] \tag{3.180}$$

where

$$[\omega - \psi(\omega)]|_{\omega=r\alpha} = 0 \quad r = 0, 1, 2, \ldots, N \tag{3.181}$$

The phase is linear at equidistant frequency increments and is called the

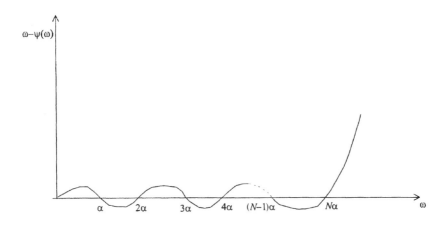

Figure 3.17 Equidistant linear phase approximation

equidistant linear phase polynomial with the recurrence formula [12]

$$A_{N+1}(p|\alpha) = A_N(p|\alpha) + \frac{\tan^2(\alpha)}{\alpha^2} \frac{p^2 + (\alpha N)^2}{4N^2 - 1} A_{N-1}(p|\alpha) \qquad (3.182)$$

where $\alpha < \pi/2$. Initial conditions are

$$A_0 = 1 \qquad A_1 = 1 + \frac{\tan(\alpha)}{\alpha} p \qquad (3.183)$$

and $S_{12}(p)$ is given by

$$S_{12}(p) = \frac{1}{A(p|\alpha)} \qquad (3.184)$$

Note that with $\alpha = 0$ the solution degenerates into the maximally flat case discussed previously.

For example, for $N = 3$, from (3.182)

$$A_2 = A_1 + \frac{\tan^2(\alpha)}{\alpha^2} \frac{p^2 + \alpha^2}{3} A_0$$

$$= 1 + \frac{\tan(\alpha)}{\alpha} p + \frac{\tan^2(\alpha)}{\alpha^2} \frac{p^2 + \alpha^2}{3} \qquad (3.185)$$

$$A_3 = A_2 + \frac{\tan^2(\alpha)}{\alpha^2} \frac{p^2 + 4\alpha^2}{15} A_1$$

$$= 1 + \frac{\tan(\alpha)}{\alpha} p + \frac{\tan^2(\alpha)}{\alpha^2} \frac{p^2 + \alpha^2}{3} + \frac{\tan^2(\alpha)}{\alpha^2} \frac{p^2 + 4\alpha^2}{15} \left[1 + \frac{\tan(\alpha)}{\alpha} \right] p$$

$$(3.186)$$

$S_{12}(p)$ may be synthesised as a ladder network.

3.6.3 Combined phase and amplitude approximation

As already stated, restricting a prototype realisation to an all-pole or ladder network means that the amplitude and phase (or delay) characteristics are related by Hilbert transform pairs. Consequently it is impossible to achieve a combined good approximation to selective amplitude response and linear phase (constant group delay) response with this type of realisation. It is only possible to proceed by using non-minimum phase realisations as follows.

A minimum phase transfer function of the form

$$S_{12}(p) = \frac{N(p)}{D(p)} \qquad (3.187)$$

can be augmented by multiplying by an all-pass function of the form

$$A(p) = \frac{H(-p)}{H(p)} \qquad (3.188)$$

That is,

$$|A(j\omega)|^2 \equiv 1 \qquad (3.189)$$

Hence

$$S'_{12}(p) = \frac{N(p)}{D(p)} \frac{H(-p)}{H(p)} \qquad (3.190)$$

The magnitude response of S'_{12} is unchanged from S_{12} while the phase response is modified to

$$\psi S'_{12}(j\omega) = \psi S_{12}(j\omega) - 2\tan^{-1}\left[\frac{O(j\omega)}{E(j\omega)}\right] \qquad (3.191)$$

(where $H(j\omega) = E(\omega) + jO(\omega)$) and the group delay is modified to

$$T'_g(\omega) = T_g(\omega) - 2\frac{EO' - OE'}{E^2 + O^2} \qquad (3.192)$$

The additional term may thus be used to modify the group delay of the original filter. Note from Chapter 1 that the group delay of an ideal lowpass filter is infinite near band-edge. Thus although modifications to the delay of a selective filter are relatively easy to achieve near $\omega = 0$, corrections close to band-edge require $A(p)$ to be of high degree. Typically correction tends to give flat delay (linear phase) across most of the passband while allowing a peak in delay near band-edge.

One possible solution in this case is the maximally flat amplitude and maximally flat linear phase approximation [3]. In the even-degree case this has $2n - 1$ derivatives of $|S_{12}|^2$ equal to zero at the origin, and $n/2 + 1$ derivatives of group delay equal to zero at the origin. It has two transmission zeros at infinity, and the transfer function is given by

$$S_{12}(p) = \frac{E_{2m-2}(p)}{D_{2m}(p)} \qquad (3.193)$$

where

$$E_{2m-2}(p) = Q_m(p)Q_m(-p) + \frac{p^2}{(2m-1)^2}Q_{m-1}(p)Q_{m-1}(-p) \qquad (3.194)$$

and $Q_m(p)$ is given from the recurrence relation

$$Q_{m+1}(p) = Q_m(p) + \frac{p^2}{4m^2-1}Q_{m-1}(p) \qquad (3.195)$$

and

$$Q_0 = 1 \qquad Q_1 = 1 + p \qquad (3.196)$$

The denominator is given by

$$D_{2m}(p) = Q_m^2 + \frac{p^2}{(2m-1)^2}Q_{m-1}^2 \qquad (3.197)$$

As an example, for $N = 4$ we have $m = 2$ and

$$Q_2(p) = 1 + p + p^2/3 \qquad (3.198)$$

$$S_{12}(p) = \frac{9 - 2p^2}{p + 18p + 16p^2 + 8p^3 + 2p^4} \qquad (3.199)$$

This has two transmission zeros at infinity, realisable as a conventional ladder network. It also has a pair of finite real-axis zeros at $p = \pm 3/\sqrt{2}$ which require a C section. The network can thus be synthesised by the cascade synthesis methods discussed in Chapter 2 or by the methods used for generalised Chebyshev filters, to be discussed later in this chapter. Further examples of filters with combined selective amplitude and linear phase are given in Reference 13 and in the section on dual-mode waveguide filters in Chapter 7.

We have observed that one particular solution to combined selective amplitude and linear phase requires real-axis transmission zeros. However, the generalised Chebyshev approximation discussed previously allows a completely arbitrary placement of transmission zeros. Thus it is possible to generate transfer functions with equiripple passband amplitude characteristics and with transmission zeros placed to increase both amplitude selectivity and phase linearity. This is the approach which is now generally adopted. Synthesis of generalised Chebyshev filters will be discussed later in this chapter.

3.7 Filters with specified time domain characteristics

The impulse response of a filter can be of significance in certain applications. For example in narrowband channelised electronic warfare receivers a large pulse will cause 'ringing' in the filter for a period of time. If the amplitudes of the time domain sidelobes are sufficiently high they can block the ability of the receiver to detect a second pulse. It is thus desirable to have filter characteristics where the time domain sidelobes are damped as rapidly as possible. One particular solution to this problem is the maximally flat impulse response approximation [14] where

$$S_{12}(t) = K_N \exp(-t/2) \sin^{N-1}(\varepsilon t/2) \tag{3.200}$$

with zeros at $t = 2\pi m/\varepsilon$.

For example let $N = 1$. Then

$$S_{12}(t) = K_1 \exp(-t/2) \tag{3.201}$$

and letting

$$S_{12}(p)|_{p=0} = 1 \tag{3.202}$$

then

$$K_1 = 1/2 \tag{3.203}$$

Now $S_{12}(p)$ is the Laplace transform of $S_{12}(t)$. Hence

$$S_{12}(p) = \frac{1}{2p+1} \tag{3.204}$$

For $N = 2$

$$S_{12}(t) = K_2 \exp(-t/2) \sin(\varepsilon t/2) \tag{3.205}$$

Hence

$$S_{12}(p) = \frac{K_2 \varepsilon/2}{(p+1/2)^2 + \varepsilon^2/4} \tag{3.206}$$

Hence for $S_{12}(p)|_{p=0} = 1$

$$K_2 \varepsilon/2 = \varepsilon^2/4 + 1/4 \tag{3.207}$$

Therefore

$$K_2 = \frac{1+\varepsilon^2}{2\varepsilon} \tag{3.208}$$

In general, for N even, it can be shown that $S_{12}(p)$ is an all-pole transfer function (hence is synthesised as a conventional ladder network) given by

$$S_{12}(p) = \frac{1}{\prod_{r=1}^{N/2} \left[\frac{4p(1+p)}{1+(2r-1)^2 \varepsilon^2} + 1 \right]} \tag{3.209}$$

where

$$K_2 = \frac{1+\varepsilon^2}{2\varepsilon} \tag{3.210}$$

and
$$\frac{K_{N+2}}{K_N} = \frac{1+(N+1)^2\varepsilon^2}{N(N+1)\varepsilon^2} \tag{3.211}$$

The value of ε must now be determined. This is done by forcing $|S_{12}(j\omega)|^2$ to have a first-order maximally flat response around $\omega = 0$. Now from (3.209)

$$|S_{12}(j\omega)|^2 = \frac{1}{\prod_{r=1}^{N/2}\left\{1+\frac{8[1-(2r-1)^2\varepsilon^2]\omega^2}{[1+(2r-1)^2\varepsilon^2]^2}+\frac{16\omega^4}{[1+(2r-1)^2\varepsilon^2]^2}\right\}} \tag{3.212}$$

(for N even). For a first-order maximally flat response the coefficients of the ω^2 term must be zero. Hence

$$\sum_{r=1}^{N/2}\frac{(2r-1)^2\varepsilon^2-1}{[1+(2r-1)^2\varepsilon^2]^2} = 0 \tag{3.213}$$

For N odd the coefficient of ω^2 is always positive so the odd-degree solution is of little value.

For $N=2$ $\quad\dfrac{\varepsilon^2-1}{(1+\varepsilon^2)^2} = 0 \quad$ hence $\varepsilon = 1$ $\tag{3.214}$

For $N=4$ $\quad\dfrac{\varepsilon^2-1}{(1+\varepsilon^2)^2}+\dfrac{9\varepsilon^2-1}{(1+9\varepsilon^2)^2} = 0 \quad$ hence $\varepsilon = 0.84336$ $\tag{3.215}$

For $N=6$ $\quad\dfrac{\varepsilon^2-1}{(1+\varepsilon^2)^2}+\dfrac{9\varepsilon^2-1}{(1+9\varepsilon^2)^2}+\dfrac{25\varepsilon^2-1}{(1+25\varepsilon^2)^2} = 0 \quad$ hence $\varepsilon = 0.76591$

$$\tag{3.216}$$

The attenuation of time domain sidelobes can be computed from the time domain response:

$$S_{12}(t) = K_N \exp(-t/2)\sin^{N-1}(\varepsilon t/2) \tag{3.217}$$

Differentiating to obtain the turning points it can be shown that

$$S'_{12}(t) = 0 \quad \text{when } t = \frac{2}{\varepsilon}\{r\pi+\tan^{-1}[\varepsilon(N-1)]\} \tag{3.218}$$

The maximum value of the impulse response occurs for $r=1$ and the sidelobes peak at $r \geq 2$. At the peaks the sine function is unity and

$$S_{12}(t) = K_N \exp(-t/2) \tag{3.219}$$

The attenuation of the mth sidelobe is thus

$$A = \exp(m\pi/\varepsilon)$$

$$= \frac{8.686m\pi}{\varepsilon} \text{ dB} \tag{3.220}$$

Lumped lowpass prototype networks 81

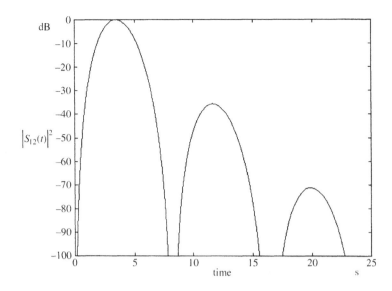

Figure 3.18   Impulse response of a degree 6 maximally flat impulse response filter

For $N = 2$ with $\varepsilon = 1$ $\qquad\qquad A = 27.94\,\text{dB}$ (3.221)

For $N = 4$ with $\varepsilon = 0.84336$ $\qquad A = 32.36\,\text{dB}$ (3.222)

For $N = 6$ with $\varepsilon = 0.76591$ $\qquad A = 35.63\,\text{dB}$ (3.223)

The impulse response of a degree 6 prototype is shown in Figure 3.18.

3.8 Synthesis of generalised Chebyshev filters

Generalised Chebyshev filters have equiripple passband amplitude characteristics and arbitrarily placed transmission zeros. They cannot be synthesised by ladder networks but various synthesis techniques are possible. The type of synthesis depends on the location of the zeros in the complex plane, whether or not the zeros are symmetrically located and whether the degree is even or odd. Three different synthesis techniques will be discussed.

3.8.1 Synthesis of generalised Chebyshev prototypes with symmetrically located transmission zeros

If the transmission zeros are symmetrically located in the complex plane then the synthesis procedure can be simplified by using even- and odd-mode admittances. From (2.198) and (2.199) we have

$$S_{12}(p) = \frac{Y_e - Y_o}{(1 + Y_e)(1 + Y_o)} \tag{3.224}$$

$$S_{11}(p) = \frac{1 - Y_e Y_o}{(1 + Y_e)(1 + Y_o)} \tag{3.225}$$

where Y_e and Y_o are both reactance functions. Now

$$|S_{12}(j\omega)|^2 = \frac{1}{1 + \varepsilon^2 F_N^2(\omega)} = \frac{1}{[1 + j\varepsilon F_N(\omega)][1 - j\varepsilon F_N(\omega)]} \tag{3.226}$$

and

$$|S_{11}(j\omega)|^2 = 1 - |S_{12}(j\omega)|^2 = \frac{\varepsilon^2 F_N^2(\omega)}{1 + \varepsilon^2 F_N^2(\omega)} \tag{3.227}$$

Now from (3.226) and (3.227)

$$\frac{S_{11}(j\omega)}{S_{12}(j\omega)} = j\varepsilon F_N(\omega) \tag{3.228}$$

and from (3.224) and (3.225)

$$j\varepsilon F_N(\omega) = \pm \frac{Y_e Y_o - 1}{Y_e - Y_o} \tag{3.229}$$

Thus

$$\frac{1}{1 - j\varepsilon F_N(p/j)} = \frac{Y_o - Y_e}{(Y_e + 1)(Y_o + 1)} \tag{3.230}$$

and

$$\frac{1}{1 + j\varepsilon F_N(p/j)} = \frac{Y_e - Y_o}{(Y_o + 1)(Y_e + 1)} \tag{3.231}$$

Now since Y_e and Y_o are reactance functions the left half-plane zeros of $1 - j\varepsilon F_N(p/j)$ are the zeros of $Y_e + 1$. Similarly the left half-plane zeros of $1 - j\varepsilon F_N(p/j)$ are the zeros of $Y_o + 1$. These two sets of zeros are the poles of $S_{11}(p)$. Either set can be identified from the poles of $S_{11}(p)$ by taking poles in alternative order from the largest imaginary part. Y_e and Y_o can then be formed from these poles [15].

Two possible network realisations are possible depending on whether the transfer function is even degree or odd degree. First we will examine the even-degree case. As an example we will synthesise a degree 4 transfer function with two transmission zeros at infinity and a pair of transmission zeros at $\omega = \pm 2$, i.e. $p = \pm j2$.

$F_N(\omega)$ can be calculated from (3.138) as follows.

$$F_N(p/j) = \frac{j}{2(p-j2)(p+j2)} \{[p + (1+p^2)^{1/2}]^2[1 + j2p + j\sqrt{3}(1+p^2)^{1/2}]$$

$$\times [1 - j2p - j\sqrt{3}(1+p^2)^{1/2}]$$

$$+ [p - (1+p^2)^{1/2}]^2[1 + j2p - j\sqrt{3}(1+p^2)^{1/2}]$$

$$\times [1 - j2p + j\sqrt{3}(1+p^2)^{1/2}]\}$$

$$= \frac{j}{2(p^2+4)} \{[1 + 2p(1+p^2)^{1/2} + 2p^2]$$

$$\times [1 + 4p^2 + 3(1+p^2) + 4\sqrt{3}(1+p^2)^{1/2}]$$

$$+ [1 - 2p(1+p^2)^{1/2} + 2p^2][1 + 4p^2 + 3(1+p^2) - 4\sqrt{3}p(1+p^2)^{1/2}]\}$$

$$= \frac{j}{p^2+4}[4 + (15 + 8\sqrt{3})p^2 + (14 + 8\sqrt{3})p^4] \tag{3.232}$$

The poles of $S_{11}(p)$ are thus the zeros of

$$(4 + p^2) + \varepsilon^2(4 + 28.88564p^2 + 27.8504p^4)^2 = 0 \tag{3.233}$$

For 20 dB return loss $\varepsilon = 0.1$; hence we find the zeros of

$$p^8 + 2.07179p^6 + 1.48914p^4 + 1.32845p^2 + 2.08253 = 0 \tag{3.234}$$

The left half-plane zeros are easily found numerically using Matlab and are

$$p = -0.80347 \pm j0.58582 \tag{3.235}$$

and

$$p = -0.24621 \pm j1.18275 \tag{3.236}$$

This transfer function may be synthesised as the cross-coupled ladder network shown in Figure 3.19.

The even-mode network is shown in Figure 3.20. The frequency-invariant reactances in Figure 3.20 arise from bisecting the coupling inverters of admittance jK_r with an open circuit.

The odd-mode network is thus the complex conjugate of the even-mode network, i.e.

$$Y_o = Y_e^* \tag{3.237}$$

We can now formulate Y_e by constructing a polynomial $P(p)$ from a pair of the four roots of $S_{11}(p)$ choosing opposite signs for the imaginary part. Thus

$$P(p) = (p + 0.80347 - j0.58582)(p + 0.24621 + j1.18275)$$

$$= p^2 + 1.04968p + j0.59693 + 0.89351 + j0.80606 \tag{3.238}$$

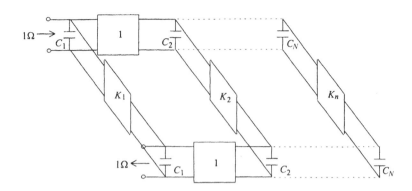

Figure 3.19 Cross-coupled prototype network

Now the zeros of $P(p)$ are the zeros of $1 + Y_e(p)$ and

$$1 + Y_e(p) = 1 + \frac{N(p)}{D(p)} \tag{3.239}$$

That is, the zeros of $1 + Y_e(p)$ are the zeros of $N(p) + D(p)$ where $N(p)$ and $D(p)$ are complex even and odd polynomials.

Thus we formulate

$$N(p) = p^2 + j0.59693p + 0.89351 \tag{3.240}$$

$$D(p) = 1.04968p + j0.80606 \tag{3.241}$$

Hence

$$Y_e(p) = \frac{N(p)}{D(p)}$$

$$= \frac{p^2 + j0.59693p + 0.89351}{1.04968p + j0.80606} \tag{3.242}$$

Now

$$\left.\frac{Y_e(p)}{p}\right|_{p=\infty} = \frac{1}{1.04968} = 0.95267 \tag{3.243}$$

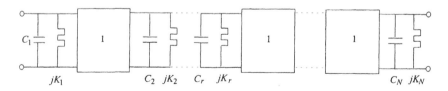

Figure 3.20 Even-mode network of the cross-coupled prototype

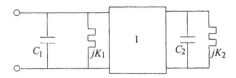

Figure 3.21 Even-mode network

Thus extracting a shunt capacitor of value $C_1 = 0.95267$ we obtain

$$Y_1(p) = Y_e(p) - C_1 p = \frac{-j0.17098p + 0.89351}{1.04968p + j0.80606} \quad (3.244)$$

Now

$$Y_1(p)|_{p=\infty} = -j0.16288 \quad (3.245)$$

Thus we extract a frequency-invariant reactance of value jK_1 where $K_1 = -0.16288$ leaving

$$Y_2(p) = Y_1(p) - jK_1$$

$$= Y_1(p) + j0.16288$$

$$= \frac{0.7622}{1.04968p + j0.80606} \quad (3.246)$$

Inverting we obtain

$$Y_3(p) = 1.3772p + j1.0575$$

$$= C_2 p + jK_2 \quad (3.247)$$

Y_e is shown in Figure 3.21.

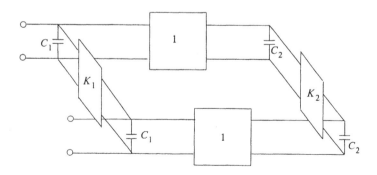

Figure 3.22 Complete fourth-degree cross-coupled filter

86 *Theory and design of microwave filters*

Now $Y_o = Y_e^*$ and the complete network is shown in Figure 3.22. The simulated response is shown in Figure 3.23.

Note that the complete filter is a fourth-degree ladder network with a non-adjacent coupling between the first and last resonators with opposite sign to the main couplings. This is necessary for the real frequency (j axis) transmission zeros. If the zero had been on the real axis the cross-coupling would have been of the same sign.

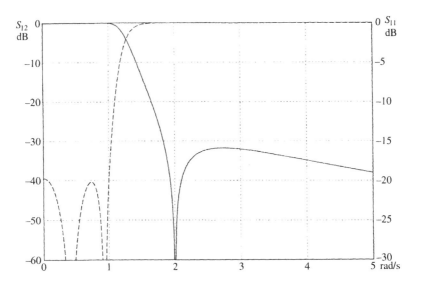

Figure 3.23 Simulated response of a fourth-degree cross-coupled filter

3.8.2 Synthesis of generalised Chebyshev prototypes with ladder-type networks

The previous cross-coupled ladder networks are only available for narrowband applications as they use inverters. Alternatively symmetrically located real frequency transmission zeros may be realised using the circuit shown in Figure 3.24, provided N is odd. The seventh-degree case shown has a minimum of one transmission zero at infinity.

As an example we will consider a degree 3 filter with 20 dB return loss, a single transmission zero at infinity and a pair of zeros at $\omega = \pm 2$. Hence

$$p_1 = \infty$$

$$p_2 = j_2 \tag{3.248}$$

$$p_3 = -j_2$$

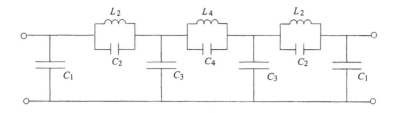

Figure 3.24 Symmetrical generalised Chebyshev filter of degree 7

In this case from (3.138) $F_N(p)$ is given by

$$\frac{j}{p^2+4}\{[p+(1+p^2)^{1/2}][1+4p^2+3(1+p^2)+4\sqrt{3}p(1+p^2)^{1/2}]$$

$$+[p-(1+p^2)^{1/2}][1+4p^2+3(1+p^2)-4\sqrt{3}p(1+p^2)^{1/2}]\}$$

$$=\frac{j}{p^2+4}[p(4+4\sqrt{3})+p^3(7+4\sqrt{3})] \tag{3.249}$$

The left half-plane roots of $S_{11}(p)$ are the left half-plane roots of

$$1+\varepsilon^2 F_N^2(p)=0 \tag{3.250}$$

These can be solved numerically giving

$$p_1 = -1.48368 \tag{3.251}$$

$$p_{2,3} = -0.38464 \pm j1.33334 \tag{3.252}$$

For $N=3$, Y_o is of degree 2 and Y_e of degree 1. The zeros of $1+Y_o$ are the zeros of $(p-p_2)(p-p_3)$, i.e.

$$(p+0.38464-j1.33334)(p+0.38464+j1.33334)$$

$$= p^2+0.76928p+1.92574$$

$$= E(p)+O(p) \tag{3.253}$$

Now forming

$$Y_o(p)=\frac{E(p)}{O(p)} \tag{3.254}$$

then

$$Y_o(p)=\frac{p^2+1.92574}{0.76928p} \tag{3.255}$$

The complete network for $N=3$ is of the form shown in Figure 3.25 and by

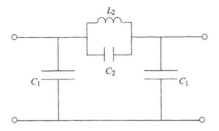

Figure 3.25   Generalised Chebyshev filter, ladder type, $N = 3$

analysis of the circuit

$$Y_o(p) = (C_1 + 2C_2)p + \frac{2}{L_2 p}$$

$$= \frac{p^2 + 2/L_2(C_1 + 2C_2)}{p/(C_1 + 2C_2)} \tag{3.256}$$

Thus from (3.255) and (3.256)

$$\frac{1.92574}{0.76928} = \frac{2}{L_2} \tag{3.257}$$

Therefore

$$L_2 = 0.7989 \tag{3.258}$$

Now from Figure 3.25

$$Y_e(p) = C_1 p \tag{3.259}$$

and from (3.251)

$$Y_e(p) = \frac{p}{1.48368} \tag{3.260}$$

Hence

$$C_1 = 0.6739 \tag{3.261}$$

and from (3.255) and (3.256)

$$C_1 + 2C_2 = \frac{1}{0.76928} \tag{3.262}$$

Therefore

$$C_2 = 0.3130 \tag{3.263}$$

As a check the series resonant circuit should be resonant at ω_0 and

$$L_2 C_2 = 1/\omega_0^2 = 4 \tag{3.264}$$

The element values of the final circuit are

$$\begin{aligned} C_1 &= 0.6739 \\ C_2 &= 0.3130 \\ L_2 &= 0.7989 \end{aligned} \tag{3.265}$$

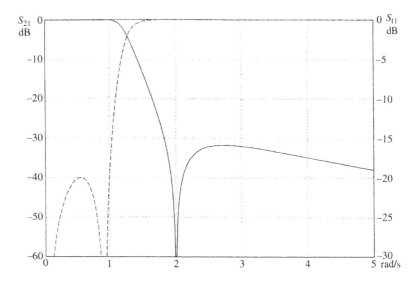

Figure 3.26 Simulated response of a lumped generalised Chebyshev filter

The simulated response of this circuit is shown in Figure 3.26.

The general synthesis for filters of this type of degree 5 or above is slightly more complicated. For example consider a prototype of degree 7 with a single transmission zero at infinity and three pairs of symmetrically located real frequency transmission zeros. One possible network realisation is shown in Figure 3.27. In this case

$$Y_0(p) = \frac{E_4(p)}{O_3(p)} \qquad (3.266)$$

where E_4 is a fourth-degree even polynomial and O_3 is a third-degree odd polynomial. Also although a pole exists at $p = \infty$ this is not completely removed. First we observe that at the resonant frequency ω_1 of the first resonator the input admittance is given by

$$Y_0(j\omega_1) = j\omega C_1 \qquad (3.267)$$

Thus we extract C_1 by a zero shifting procedure such that

$$C_1 = \frac{Y_0(j\omega_1)}{j\omega_1} \qquad (3.268)$$

Hence

$$Y_1(p) = Y_0(p) - C_1 p \qquad (3.269)$$

and inverting

$$Z_1(p) = 1/Y_1(p) \qquad (3.270)$$

90 *Theory and design of microwave filters*

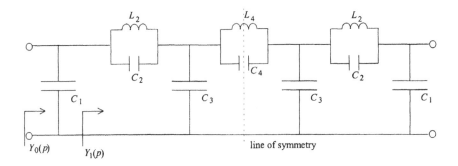

Figure 3.27 Generalised Chebyshev filter, ladder-type realisation

Now $Z_1(p)$ contains a factor $p^2 + w_1^2$ in the denominator, corresponding to the transmission zeros at $p = \pm jw_1$. $Z_1(p)$ should be synthesised into the series connected parallel tuned circuit and a remaining impedance. The procedure can then be repeated. An example of this procedure is given in Chapter 5.

3.8.3 Asymmetrically located transmission zeros

In many applications it is required that a filter should be more selective on one side of the passband than the other. For example in cellular communications (see Chapter 1) a transmit filter should have high rejection in the receive band. This may be elegantly achieved by using a generalised Chebyshev prototype with asymmetrically located transmission zeros.

As an example we will synthesise a third-degree network with two transmission zeros at infinity and one at $\omega = 2$, with $\varepsilon = 0.1$. Hence

$$p_1 = p_2 = \infty \tag{3.271}$$

and

$$p_3 = j2 \tag{3.272}$$

and

$$F_N(p) = \frac{j}{2} \frac{[p + (1+p^2)^{1/2}]^2[1 + j2p + j\sqrt{3}(1+p^2)^{1/2}] + [p - (1+p^2)^{1/2}]^2[1 + j2p - j\sqrt{3}(1+p^2)^{1/2}]}{p - j2}$$

$$= \frac{j}{p - j_2}\{1 + 2p^2 + j[(2 + 2\sqrt{3})p + (4 + 2\sqrt{3})p^3]\} \tag{3.273}$$

Hence for $p = j\omega$

$$F_N(\omega) = \frac{1 - (2 + 2\sqrt{3})\omega - 2\omega^2 + (4 + 2\sqrt{3})\omega^3}{\omega - 2} \tag{3.274}$$

$$|S_{11}(j\omega)|^2 = \frac{\varepsilon^2 F_N^2(\omega)}{1 + \varepsilon^2 F_N^2(\omega)} \tag{3.275}$$

Therefore

$$|S_{11}(j\omega)|^2 = \frac{\varepsilon^2(1 - 5.46410\omega - 2\omega^2 + 7.46410\omega^3)^2}{(\omega - 2)^2 + \varepsilon^2(1 - 5.46410\omega - 2\omega^2 + 7.46410\omega^3)^2} \quad (3.276)$$

$S_{11}(p)$ can be formed from the left half-plane roots of the denominator of (3.276), i.e.

$$(p - j2)^2 - \varepsilon^2[1 + 2p^2 + j(5.46410p + 76410p^3)]^2 = 0 \quad (3.277)$$

The roots are

$$p_{1,2} = \pm 0.8609917 - j1.41446 \quad (3.278)$$

$$p_{3,4} = \pm 1.1734578 + j0.4313449 \quad (3.279)$$

$$p_{5,6} = \pm 0.31246609 + j1.2510504 \quad (3.280)$$

In this particular case the transfer function may be synthesised by a symmetrical network. However, in general when the transmission zeros are placed asymmetrically this is not necessarily true. Thus although the above transfer function may be synthesised using even- and odd-mode networks we will use a more generally applicable method.

Formulating the denominator of $S_{11}(p)$ from the left half-plane roots we obtain

$$D(p) = (p + 0.860992 + j1.41446)(p + 1.1734578 - j0.431345)$$

$$\times (p + 0.312466 - j1.251050) \quad (3.281)$$

The numerator is formed from $\varepsilon F_N(p)$ multiplied by a constant such that $S_{11}(p)$ is equal to unity when $p = \infty$, to ensure zero transmission at infinity. Hence

$$S_{11}(p) = \frac{p^3 - j0.267949p^2 + 0.732051p - j0.133975}{p^3 + 2.346916p^2 - j0.267949p^2 + 3.486057p - j0.949665p + 2.118199 - j1.624683}$$

$$= \frac{N(p)}{D(p)} \quad (3.282)$$

Now formulating the input admittance from

$$Y(p) = \frac{1 + S_{11}(p)}{1 - S_{11}(p)}$$

$$= \frac{D(p) + N(p)}{D(p) - N(p)} \quad (3.283)$$

$$Y(p) = \frac{2p^3 + 2.346916p^2 - j0.535898p^2 + 4.218108p - j0.949605p + 2.118199 - j1.758658}{2.346916p^2 + 2.754006p - j0.949605p + 2.118199 - j1.490708}$$

$$(3.284)$$

$Y(p)$ has a pole at $p = \infty$. Thus we can extract a capacitor C_1 as follows:

$$C_1 = \left.\frac{Y(p)}{p}\right|_{p=\infty} = 0.852182 \tag{3.285}$$

Hence

$$Y_1(p) = Y(p) - C_1 p$$

$$= \frac{0.273338jp^2 + 2.413017p + j0.320749p + 2.118199 - j1.758658}{2.346916p^2 + 2.754006p - j0.949605p + 2.118199 - j1.490708} \tag{3.286}$$

As in the previous case a frequency-invariant reactance may now be extracted. In this case we must extract an element such that the remaining admittance possesses a zero at $p = j2$. $Y_1(p)$ is a reactance function; thus if we extract a susceptance of value equal to $Y_1(j2)$ then the remainder must be zero at $p = j2$. Thus

$$Y_1(p) = jB_1 + Y_2(p) \tag{3.287}$$

where

$$jB_1 = Y_1(j2) = -j0.36758 \tag{3.288}$$

and

$$Y_2(j2) = 0 \tag{3.289}$$

Hence

$$Y_1(p) = -j0.36758 + \frac{(p - j2)(jAp + B + jC)}{2.346916p^2 + 2.754006p - j0.949605 + 2.118199 - j1.490708} \tag{3.290}$$

Now equating coefficients of powers of p in (3.286) and (3.290)

coefficients of $jp^2 \Rightarrow 0.27338 = -0.36758 \times 2.346916 + A$ \hfill (3.291)

coefficients of $p \Rightarrow 2.413017 = -0.36758 \times 0.9499605 + B + 2A$ \hfill (3.292)

coefficients of $jp \Rightarrow 0.320749 = -0.36758 \times 2.754006 + C$ \hfill (3.293)

Hence

$$A = 1.136017 \qquad B = 0.49002 \qquad C = 1.333076 \tag{3.294}$$

The remaining admittance $Y_2(p)$ possesses a zero at $p = j2$, and hence its impedance possesses a pole at $p = j2$, i.e.

$$Z_2(p) = \frac{1}{Y_2(p)} = \frac{2.346916p^2 + 2.754006p - j0.949605 + 2.118199 - j1.490708}{(p - j2)(1.136017jp + 0.49002 + j1.333076)} \tag{3.295}$$

$Z_2(p)$ may be synthesised into a series bandstop resonator (to produce the pole)

Lumped lowpass prototype networks 93

and a remaining impedance by using a partial fraction expansion. Hence

$$Z_2(p) = \frac{K}{p-j2} + \frac{Ep + F + jG}{1.136017jp + 0.49002 + j1.333076} \tag{3.296}$$

where

$$K_1 = Z_1(p)(p-j2)|_{p=j2} \tag{3.297}$$

Hence

$$K_1 = 3.01358 \tag{3.298}$$

E, F and G can be evaluated by equating powers of p in (3.295) and (3.296).

$$\text{coefficients of } p^2 \Rightarrow 2.346916 = E \tag{3.299}$$

$$\text{coefficients of } p \Rightarrow 2.754006 = 0.49002 + F \tag{3.300}$$

$$\text{coefficients of } jp \Rightarrow -0.949605 = 1.136017K_1 - 2E + G \tag{3.301}$$

Hence

$$E = 2.346916 \quad F = 2.754006 \quad G = 0.320749 \tag{3.302}$$

Thus

$$Z_2(p) = \frac{3.01358}{p-j2} + \frac{2.346916p + 2.754006 + j0.320749}{1.136017jp + 0.49002 + j1.333076}$$

$$= \frac{3.01358}{p-j2} + Z_3(p) \tag{3.303}$$

The first term in (3.303) consists of a capacitor C_2 in parallel with a frequency-invariant susceptance B_2. where

$$C_2 = \frac{1}{3.01358} = 0.33183 \tag{3.304}$$

and

$$B_2 = \frac{-2}{3.01358} = -0.66366 \tag{3.305}$$

Now

$$Z_3(p) = \frac{2.346916p + 2.754006 + j0.320749}{1.136017jp + 0.49002 + j1.333076} \tag{3.306}$$

A frequency-invariant reactance must now be extracted from $Z_2(p)$ such that the remaining impedance has a pole at $p = \infty$. (Note that frequency-invariant reactances do not exist in reality but may be approximated over a narrow bandwidth by capacitors or inductors.) Thus

$$Z_4(p) = Z_3(p) - jX \tag{3.307}$$

94 Theory and design of microwave filters

where

$$X = \left.\frac{Z_3(p)}{j}\right|_{p=\infty} = -2.06592 \tag{3.308}$$

The remaining impedance $Z_4(p)$ is

$$Z_4(p) = Z_3(p) + j2.06592 = \frac{j1.33309}{1.136017jp + 0.49002 + j1.33308} \tag{3.309}$$

Inverting $Z_4(p)$ to form its admittance $Y_4(p)$ we obtain

$$Y_4(p) = 0.852168p - j0.36758 + 1 \tag{3.310}$$

This is a parallel combination of a capacitor C_3, a frequency-independent susceptance B_3 and a 1Ω load resistor. Here the load resistor is not exactly equal to unity because of numerical errors building up in the synthesis procedure. Synthesis typically loses one or two significant figures for each cycle of the process, although this depends on the method used. Thus it is important to use sufficient significant figures at the start of the process, especially for higher degree networks.

The complete cycle is shown in Figure 3.28. The synthesis process may be checked by analysing the final circuit. The simulated response is shown in Figure 3.29.

The final network shown in Figure 3.28 is not necessarily the most useful for bandpass applications. A bandpass transformation of this circuit would result in two shunt bandpass resonators shunted by susceptances. These are not a problem as they can be absorbed into the resonators resulting in a simple change in resonant frequency. However, the series branch would become a series bandstop resonator in series with a further reactance which would be difficult to realise in any microwave structure other than by using lumped elements.

It is usually more practical to convert the network into a cross-coupled array as follows [16]. First the series branch consisting of a bandstop resonator in series with a reactance may be converted into a parallel connection of a bandpass resonator and a reactance, as shown in Figure 3.30. Equating the admittances of the two circuits in Figure 3.30 we obtain, for the series branch

$$Y = \frac{1}{\dfrac{1}{Cp + jB} + jX} = \frac{Cp + jB}{1 - XB + jXCp} \tag{3.311}$$

and for the parallel branch

$$Y = jB' + \frac{1}{Lp + jX'} = \frac{1 - X'B' + jB'Lp}{Lp + jX'} \tag{3.312}$$

Thus

$$\frac{Cp + jB}{1 - XB + jXCp} = \frac{1 - X'B' + jB'Lp}{Lp + jX'} \tag{3.313}$$

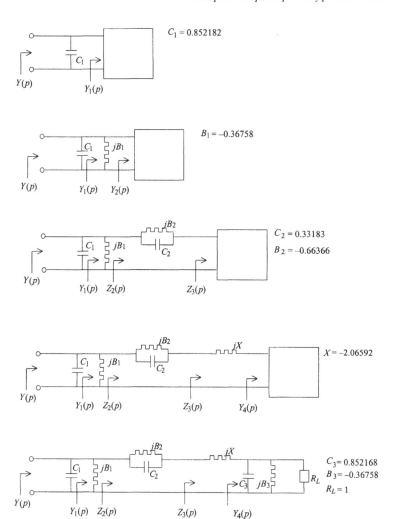

Figure 3.28 Complete synthesis cycle for a generalised Chebyshev filter with $N = 3$, two transmission zeros at infinity and one at $p = j2$

Equating coefficients of p we obtain

coefficients of $p^2 \Rightarrow LC = B'LXC$ (3.314)

Hence

$$B' = -1/X \tag{3.315}$$

coefficients of $jp \Rightarrow BL + X'C = XC(1 - X'B') + B'L(1 - XB)$ (3.316)

96 Theory and design of microwave filters

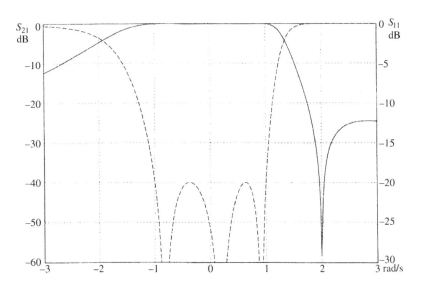

Figure 3.29 Simulated response of a generalised Chebyshev prototype with asymmetric frequency response

Hence

$$L = X^2 C \qquad (3.317)$$

coefficients of $p^0 \Rightarrow BX' = (1 - XB)(1 - X'B') \qquad (3.318)$

Hence

$$X' = X^2 B - X \qquad (3.319)$$

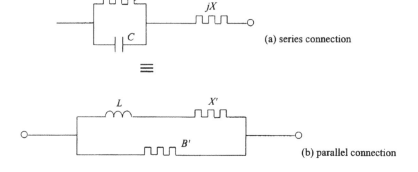

Figure 3.30 Equivalent of the series branch in a generalised Chebyshev prototype

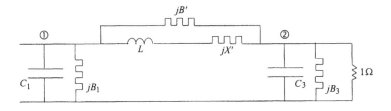

Figure 3.31   First circuit transformation

$C = C_2 = 0.33183$

$B = B_2 = -0.66366$ (3.320)

$X = -2.06592$

and from (3.315), (3.317) and (3.319)

$B' = 0.48404$

$L = 1.41625$ (3.321)

$X' = -0.76659$

The circuit is thus transformed into Figure 3.31.

Next we form an admittance inverter between the input and output of value $K = B'$, as shown in Figure 3.32. Note that this involves adding shunt susceptances at nodes (1) and (2).

Finally we observe that the series inductor and frequency-invariant reactance can be replaced by a cascade of an inverter, a shunt resonator and an inverter of opposite sign, as shown in Figure 3.33.

The proof of this equivalence is found by analysis of the transfer matrices of the two circuits. For the series resonator

$$[T] = \begin{bmatrix} 1 & Lp + jX \\ 0 & 1 \end{bmatrix}$$ (3.322)

For the inverter-coupled shunt resonator

$$[T] = \begin{bmatrix} 0 & j \\ j & 0 \end{bmatrix} \begin{bmatrix} 1 & 0 \\ Lp + jX & 1 \end{bmatrix} \begin{bmatrix} 0 & -j \\ -j & 0 \end{bmatrix}$$

$$= \begin{bmatrix} 1 & Lp + jX \\ 0 & 1 \end{bmatrix}$$ (3.323)

The final transformed network is shown in Figure 3.34.

Scrutiny of Figure 3.34 shows that the resultant network is a ladder network

98 *Theory and design of microwave filters*

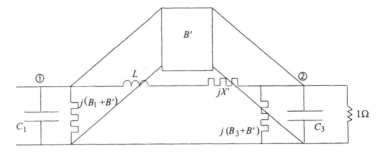

Figure 3.32   Second circuit transformation

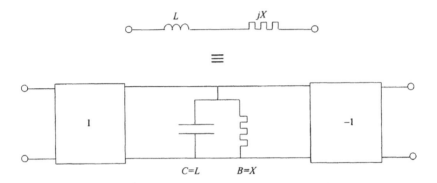

Figure 3.33   Equivalence of the series resonator

with a cross-coupling inverter from input to output. This inverter couples across three nodes. In general coupling around three nodes will produce a single transmission zero on the imaginary axis. We could change the value of K_{23} to +1 provided we change the sign of K_{13}. This will not change the response of the network. Thus we can say that if all the main couplings are positive then a negative cross-coupling across three nodes will produce a transmission zero on the high side of the passband. Conversely a positive cross-coupling would produce a transmission zero on the low side of the passband.

3.9 Summary

The synthesis techniques described in Chapter 2 have been built on so that lowpass prototype networks may be designed with prescribed amplitude, phase, or time domain characteristics. These include filters with maximally flat, Chebyshev and elliptic function amplitude characteristics. Specified phase responses included the maximally flat and equidistant approximations to linear phase. Filters with combined amplitude/phase and time domain

Lumped lowpass prototype networks 99

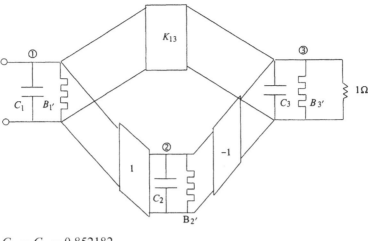

$C_1 = C_3 = 0.852182$

$B'_1 = B'_3 = B_1 + B' = 0.11646$

$C_2 = L - 1.41625$

$B'_2 = X = -0.76659$

$K_{12} = 1, K_{23} = -1, K_{13} = B' = 0.48404$

Figure 3.34 *Final cross-coupled realisation of an $N = 3$ generalised Chebyshev filter*

characteristics are also discussed. The generalised Chebyshev prototype is one of the most useful as it enables equiripple amplitude characteristics to be combined with arbitrary placement of transmission zeros in the complex plane. The synthesis of various realisations of this prototype is dealt with extensively. The material in this chapter leads naturally into the following chapters on specific hardware realisations of filters.

3.10 References

1 BUTTERWORTH, S.: 'On the theory of filter amplifiers', *Wireless Engineer*, 1930, **7**, pp. 536–41
2 NORTON, E.L.: 'Constant resistance networks with application to filter groups', *Bell System Technical Journal*, 1937, **16**, pp. 178–93
3 RHODES, J.D.: 'Theory of electrical filters' (Wiley, New York, 1976) pp. 40–50
4 CHEBYSHEV, P.L.: 'Complete works', An. USSR, 1947 (in Russian)
5 GREEN, E.: 'Synthesis of ladder networks to give Butterworth or Chebyshev response in the passband', IEE Monograph 88 (IEE, Stevenage, 1954)

6 WEINBERG, L.: 'Explicit formulas for Tschebyshev and Butterworth ladder networks', *Journal of Applied Physics*, 1957, **28**, pp. 1155–60
7 ABRAMOWITZ, M., and STEGUN, I.A.: 'Handbook of mathematical functions' (Dover, New York, 1970) pp. 567–626
8 RHODES, J.D.: 'Explicit formulas for element values in elliptic function prototype networks', *IEEE Transactions on Circuit Theory*, 1971, **CT-18**, pp. 264–76
9 SLEIGH, P.D.: 'Asymmetric filter design for satellite communications applications', IEE Colloquium on *Microwave Filters*, IEEE Colloquium Digest 1982/4, 1982
10 CAMERON, R.J.: 'Fast generation of Chebyshev filter prototypes with asymmetrically-prescribed transmission zeros', *ESA Journal*, 1982, **6** (1), pp. 83–95
11 THOMPSON, W.E.: 'Delay networks having maximally-flat frequency characteristics', *Proceedings of the IEE*, 1949, **96**, pp. 487–90
12 HENK, T.: 'The generation of arbitrary phase polynomials by recurrence formulae', *International Journal of Circuit Theory and Applications*, 1981, **9** (4), pp. 461–78
13 RHODES, J.D., and ZABALAWI, I.H.: 'Design of selective linear-phase filters with equiripple amplitude characteristics', *IEEE Transactions on Circuits and Systems*, 1978, **CAS-25** (12), pp. 989–1000
14 RHODES, J.D.: 'Prototype filters with a maximally flat impulse response', *Int. J. Circuit Theory & Appl.*, 1989, **17** (4), pp. 421–428
15 ALSEYAB, S.A.: 'A novel class of generalised Chebyshev lowpass prototype for suspended substrate filters', *IEEE Transactions on Microwave Theory and Techniques*, 1982, **MTT-31** (9), pp. 1341–47
16 CAMERON, R.J.: 'General prototype network synthesis methods for microwave filters', *ESA Journal*, 1982, **6**, pp. 193–206

Chapter 4
Circuit transformations on lumped prototype networks

4.1 Introduction

The lumped lowpass prototype filters discussed so far are restricted to a 1 Ω impedance level and a cut-off frequency of $\omega_c = 1$ rad/s. In reality we would like to design filters working into arbitrary impedance levels with arbitrary cut-off frequencies. We may require lowpass, highpass, bandpass or bandstop filters. Various circuit transformations to achieve this are described in the next section. Methods of realising impedance inverters and scaling internal circuit impedances to arbitrary levels are also described. In addition the effect of losses in real circuit elements and other practical issues are discussed.

4.2 Impedance transformations

The lowpass prototypes normally have a system impedance of 1 Ω, i.e. both the generator impedance Z_G and load impedance Z_L are 50 Ω (Figure 4.1). Most, but not all, microwave filters operate in a 50 Ω system. Historically 50 Ω was chosen as a compromise between the losses and power handling capacity of coaxial cable. To convert from a 1 Ω impedance level to an impedance level of Z_0 Ω we simply scale the impedances of all the circuit elements in the filter by 50 Ω (see Figure 4.2).

Thus for inductors

$$Z = Lp \Rightarrow Z_0 Lp = (Z_0 L)p \qquad (4.1)$$

That is,

$$L \Rightarrow Z_0 L \qquad (4.2)$$

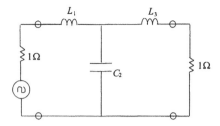

Figure 4.1 Degree 3 prototype ladder network

Thus the inductances are multiplied by Z_0.

For the capacitors

$$Z = \frac{1}{Cp} \Rightarrow \frac{Z_0}{Cp} = \frac{1}{\left(\dfrac{C}{Z_0}\right)p} \qquad (4.3)$$

That is,

$$C \Rightarrow C/Z_0 \qquad (4.4)$$

Thus the capacitances are divided by Z_0.

For the impedance inverters of characteristic impedance K

$$K \Rightarrow Z_0 K \qquad (4.5)$$

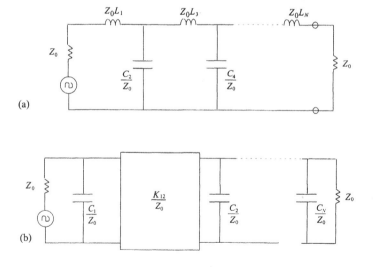

Figure 4.2 Impedance scaling of a lowpass prototype: (a) ladder; (b) admittance inverter coupled

4.2.1 Example

Design a degree 3 maximally flat prototype filter to operate in a 50 Ω system.
From Chapter 3

$$L_1 = L_3 = 1\,\text{H} \tag{4.6}$$

$$C_2 = 2\,\text{F} \tag{4.7}$$

Hence

$$L_1 \Rightarrow 50\,\text{H} \tag{4.8}$$

$$C_2 \Rightarrow 2/50\,\text{F} \tag{4.9}$$

4.3 Lowpass to arbitrary cut-off frequency lowpass transformation

The lowpass prototype networks normally have a band-edge or cut-off frequency of $\omega = 1$. We require a transformation to convert this cut-off to an arbitrary frequency ω_c, as shown in Figure 4.3.

Given a lowpass transmission characteristic of the form

$$|S_{12}(j\omega)|^2 = \frac{1}{1 + F_N^2(\omega)} \tag{4.10}$$

the transformation is

$$\omega \Rightarrow \omega/\omega_c \tag{4.11}$$

Hence

$$|S_{12}(j\omega)|^2 \Rightarrow \frac{1}{1 + F_N^2(\omega/\omega_c)} \tag{4.12}$$

Hence $F_n(\omega/\omega_c)$ has the same value at $\omega = \omega_c$ as $F_N(\omega)$ has at $\omega = 1$.

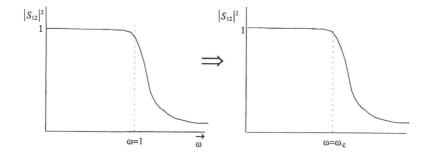

Figure 4.3   Lowpass to lowpass transformation

Applying this transformation to inductors we obtain

$$Z = Lp \tag{4.13}$$

$$Z(j\omega) = j\omega L \Rightarrow \frac{j\omega L}{\omega_c} \tag{4.14}$$

That is,

$$L \Rightarrow L/\omega_c \tag{4.15}$$

Similarly for capacitors

$$Z = \frac{1}{Cp} \tag{4.16}$$

$$Z(j\omega) = \frac{-j}{\omega C} \Rightarrow \frac{-j}{(\omega/\omega_c)C} \tag{4.17}$$

That is,

$$C \Rightarrow C/\omega_c \tag{4.18}$$

Inverters are frequency independent and are unaffected by this transformation.

4.3.1 Example

Transform the previous example into a filter with a band-edge frequency of 100 MHz.

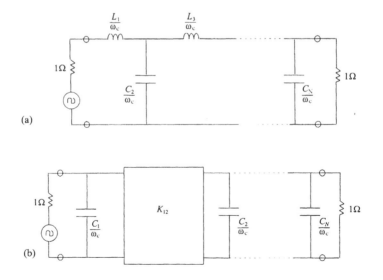

Figure 4.4 Frequency-scaled lowpass prototype: (a) ladder; (b) inverter coupled

Thus

$$\omega_c = 2\pi \times 10^8 = 6.283 \times 10^8 \tag{4.19}$$

$$L \Rightarrow \frac{50}{6.283 \times 10^8} = 79.57\,\text{nH} \tag{4.20}$$

$$C \Rightarrow \frac{2}{50 \times 6.283 \times 10^8} = 63.66\,\text{pF} \tag{4.21}$$

4.4 Lowpass to highpass transformation

We require a transformation to convert the lowpass prototype to a highpass filter with arbitrary band-edge frequency ω_c (Figure 4.5). Given

$$|S_{12}(j\omega)|^2 = \frac{1}{1 + F_N^2(\omega)} \tag{4.22}$$

the transformation is

$$\omega \Rightarrow \frac{-\omega_c}{\omega} \tag{4.23}$$

This maps d.c. to infinite frequency and vice versa, giving

$$|S_{12}(j\omega)|^2 \Rightarrow \frac{1}{1 + F_N^2(-\omega_c/\omega)} \tag{4.24}$$

Applying this transformation to inductors we obtain

$$Z(j\omega) = j\omega L \Rightarrow \frac{-j\omega_c L}{\omega}$$

$$= -j\bigg/\omega\left(\frac{1}{\omega_c L}\right)$$

$$= -j/\omega C' \tag{4.25}$$

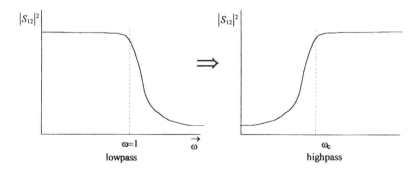

Figure 4.5 Lowpass to highpass transformation

where

$$C' = \frac{1}{\omega_c L} \tag{4.26}$$

Hence the inductors are transformed into capacitors (Figure 4.6). Applying this transformation to capacitors we obtain

$$Z(j\omega)) = \frac{-j}{\omega C} \Rightarrow \frac{j\omega}{\omega_c C}$$

$$= j\omega L' \tag{4.27}$$

where

$$L' = \frac{1}{\omega_c C} \tag{4.28}$$

The capacitors are transformed into inductors. Again the inverters are unaffected by the transformation. Note that the highpass transformation has the effect of shifting the transmission zeros of the network from $\omega = \infty$ to $\omega = 0$.

4.4.1 Example

Design a degree 3 maximally flat filter for 50 Ω system impedance and a highpass response with 100 MHz cut-off frequency.

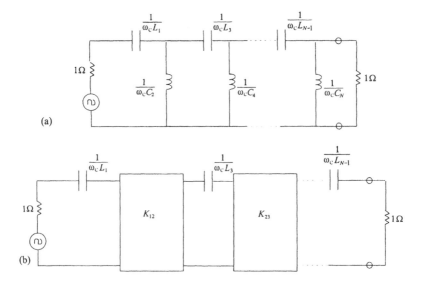

Figure 4.6 Arbitrary cut-off frequency highpass: (a) ladder; (b) inverter coupled

Starting with the impedance scaled lowpass prototype

$$L_1 = L_3 = 50 \tag{4.29}$$

$$C_2 = 2/50 \tag{4.30}$$

Applying equations (4.26) and (4.28)

$$C_1' = C_3' = \frac{1}{2\pi \times 10^8 \times 50} = 31.83\,\text{pF} \tag{4.31}$$

$$L_2' = \frac{1}{2\pi \times 10^8 \times 2/50} = 39.78\,\text{nH} \tag{4.32}$$

4.5 Lowpass to bandpass transformation

We require a transformation to convert the lowpass prototype into a bandpass filter with arbitrary centre frequency and bandwidth, as shown in Figure 4.7. The band-edges at $\omega = \pm 1$ in the lowpass prototype must map into the band-edges of the bandpass filter at ω_1 and ω_2. The transmission zeros at infinity in the lowpass must now occur at both $\omega = 0$ and $\omega = \infty$. The midband of the lowpass prototype at $\omega = 0$ must map into the centre of the passband in the bandpass filter.

This can be achieved by the following transformation:

$$\omega \to \alpha\left(\frac{\omega}{\omega_0} - \frac{\omega_0}{\omega}\right) \tag{4.33}$$

For $\omega = -1$ and $\omega = +1$ to map to ω_1 and ω_2 then

$$-1 = \alpha\left(\frac{\omega_1}{\omega_0} - \frac{\omega_0}{\omega_1}\right) \tag{4.34}$$

$$+1 = \alpha\left(\frac{\omega_2}{\omega_0} - \frac{\omega_0}{\omega_2}\right) \tag{4.35}$$

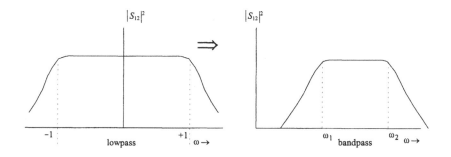

Figure 4.7 Lowpass to bandpass transformation

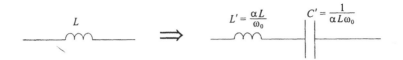

Figure 4.8 Bandpass transformation of an inductor

Solving (4.34) and (4.35) simultaneously yields

$$\omega_0 = (\omega_1 \omega_2)^{1/2} \tag{4.36}$$

$$\alpha = \frac{\omega_0}{\omega_2 - \omega_1} \tag{4.37}$$

ω_0 is thus the geometric midband frequency; α is known as the bandwidth scaling factor.

Applying this transformation to a series inductor we obtain

$$Z = j\omega L \Rightarrow j\alpha L \left(\frac{\omega}{\omega_0} - \frac{\omega_0}{\omega} \right)$$

$$= j\left(\frac{\alpha L}{\omega_0} \right)\omega - \frac{j}{\omega(1/\alpha L \omega_0)} \tag{4.38}$$

The resulting impedance is that of the series connected LC circuit shown in Figure 4.8.

Applying the transformation to a capacitor of admittance $j\omega C$ we obtain

$$Y = j\omega C \Rightarrow j\alpha C \left(\frac{\omega}{\omega_0} - \frac{\omega_0}{\omega} \right)$$

$$= j\left(\frac{\alpha C}{\omega_0} \right)\omega - \frac{j}{\omega(1/\alpha C \omega_0)} \tag{4.39}$$

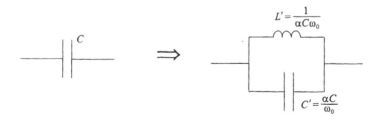

Figure 4.9 Bandpass transformation of a capacitor

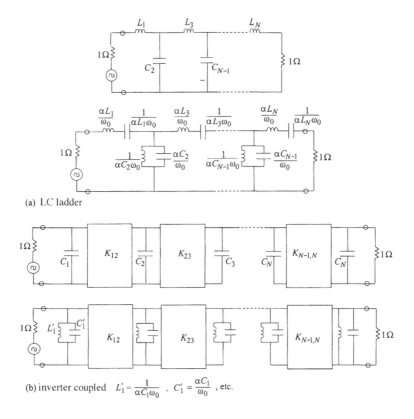

Figure 4.10 Bandpass transformation of a lowpass prototype: (a) ladder; (b) inverter coupled

The resulting admittance is that of the parallel connected LC circuit shown in Figure 4.9.

Again the inverters are invariant under the transformations. The complete transformation of a lowpass prototype to a bandpass filter is shown in Figure 4.10, where the use of impedance inverters becomes apparent. The bandpass transformation of the LC ladder results in a bandpass filter with both series and shunt connected resonators. This can be inconvenient when it comes to practical realisation. When inverters are used there are only shunt or series connected resonators, depending on whether the lowpass prototype has shunt capacitors or series inductors. This leaves the problem of how to realise the inverters.

Consider the pi network shown in Figure 4.11. This network consists of shunt negative susceptances of value $-jB$ connected by a series positive susceptance of value $+jB$. The transfer matrix of this network is given by

110 Theory and design of microwave filters

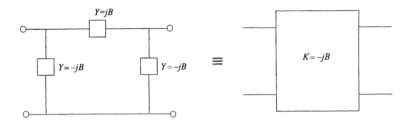

Figure 4.11 Realisation of an inverter by a pi network of reactances

$$[T] = \begin{bmatrix} 1 & 0 \\ -jB & 1 \end{bmatrix} \begin{bmatrix} 1 & -j/B \\ 0 & 1 \end{bmatrix} \begin{bmatrix} 1 & 0 \\ -jB & 1 \end{bmatrix}$$

$$= \begin{bmatrix} 1 & -j/B \\ -jB & 0 \end{bmatrix} \begin{bmatrix} 1 & 0 \\ -jB & 1 \end{bmatrix}$$

$$= \begin{bmatrix} 0 & -j/B \\ -jB & 0 \end{bmatrix}$$

$$= \begin{bmatrix} 0 & j/K \\ jK & 0 \end{bmatrix} \qquad (4.40)$$

where

$$K = -B \qquad (4.41)$$

Thus the pi network of reactance elements equates exactly to an inverter of characteristic admittance $K = -B$ [1].

In the real world ideal reactive elements do not exist but we can replace them by series capacitors for example (Figure 4.12). Now for a capacitor

$$Y = jB = j\omega C \qquad (4.42)$$

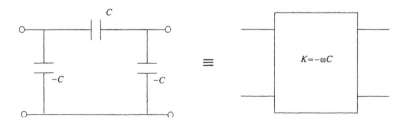

Figure 4.12 Narrowband approximation to an inverter

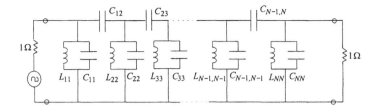

Figure 4.13 Capacitively coupled bandpass filter

Hence

$$K = -\omega C \tag{4.43}$$

The value of K is now frequency dependent; however, if the filter is sufficiently narrowband then K will not vary significantly across the passband. The negative sign of the capacitive inverter is of no significance, it only affects the phase response of the filter. Replacing the inverters in Figure 4.10(b) with capacitive pi sections we obtain the network shown in Figure 4.13. Here the shunt negative capacitances have been absorbed into the positive capacitances of the resonators.

The value of the rth shunt inductor is L_{rr} where

$$L_{rr} = \frac{1}{\alpha C_r \omega_0} \tag{4.44}$$

The value of the rth shunt capacitor is C_{rr} where

$$C_{rr} = \alpha C_r/\omega_0 - C_{r-1,r} - C_{r,r+1} \tag{4.45}$$

and

$$C_{r,r+1} = \frac{K_{r,r+1}}{\omega_0} \tag{4.46}$$

It is interesting to note the behaviour of the capacitively coupled filter at frequencies above the passband. As ω increases eventually the series capacitors all short together and the network behaves like a single shunt capacitor. The network thus has a single transmission zero at infinity and 2_{N-1} transmission zeros at d.c. Consequently the filter is slightly more selective on the low frequency side of the passband than on the high frequency side.

This asymmetry could be reversed by inductively coupling the resonators. Also the response can be made symmetrical by alternating inductive and capacitive coupling. This capacitively coupled type of filter is predominantly useful for narrowband applications, typically with bandwidths of less then 10 per cent of centre frequency. As the bandwidth is increased the response becomes progressively more asymmetric.

There may also be problems with this design when extremely narrow bandwidths are required. From (4.39) the admittance of the rth shunt resonator section prior to forming the capacitive inverters is given by

$$Y_r(j\omega) = j\left[\left(\frac{\alpha C_r}{\omega_0}\right)\omega - \frac{1}{(1/\alpha C_r \omega_0)\omega}\right] \qquad (4.47)$$

where

$$\alpha = \frac{\omega_0}{\Delta \omega} \qquad (4.48)$$

$$\Delta \omega = \omega_2 - \omega_1 \qquad (4.49)$$

Hence the inductance of the rth shunt inductor is given by

$$L_{rr} = \frac{1}{\alpha C_r \omega_0} \qquad (4.50)$$

This inductance is inversely proportional to α and for very narrow bandwidths may be too small to be physically realisable. To avoid this, the entire admittance of the filter (including source and load) may be scaled by $1/\alpha$. The element values of the inductors and capacitors then become independent of the filter bandwidth. An impedance transformer must be inserted between the filter and its terminations (Figure 4.14).

It is difficult to make the ideal transformers at high frequencies. However, a narrowband equivalent can be made using the circuit shown in Figure 4.15. Here

$$Y(j\omega) = j\omega C_a + \frac{1}{1 - j/\omega C_b}$$

$$= j\omega C_a + \frac{1 + j/\omega C_b}{1 + 1/\omega^2 C_b^2} \qquad (4.51)$$

The real part of $Y(j\omega)$ is given by

$$\operatorname{Re} Y(j\omega) = \frac{1}{1 + 1/\omega^2 C_b^2} \qquad (4.52)$$

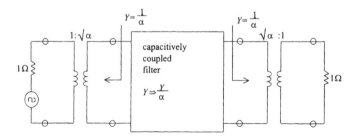

Figure 4.14 Impedance scaling of a capacitively coupled filter

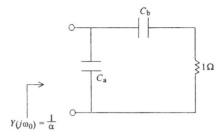

Figure 4.15 Narrowband impedance transformer

Hence equating to $1/\alpha$ at $\omega = \omega_0$

$$1 + \frac{1}{\omega_0^2 C_b^2} = \alpha \tag{4.53}$$

Hence

$$C_b = \frac{1}{\omega_0(\alpha - 1)^{1/2}} \tag{4.54}$$

The imaginary part of $Y(j\omega)$ is given by

$$\operatorname{Im} Y(j\omega) = \omega C_a + \frac{1/\omega C_b}{1 + 1/\omega^2 C_b^2} \tag{4.55}$$

This must be zero at $\omega = \omega_0$. Hence

$$\omega_0 C_a = \frac{-1/\omega_0 C_b}{1 + 1/\omega_0^2 C_b^2} \tag{4.56}$$

$$C_a = \frac{-(\alpha - 1)^{1/2}}{\omega_0 \alpha} \tag{4.57}$$

C_b becomes the first and last series capacitor coupling into and out of the network. The negative C_a is absorbed into the capacitance of the first and last resonators. The element values of the network are now given by

$$C_{01} = C_{N.N+1} = \frac{1}{\omega_0(\alpha - 1)^{1/2}} \tag{4.58}$$

$$C_{r.r+1} = \frac{K_{r.r+1}}{\alpha \omega_0} \qquad (r = 1, \ldots, N - 1) \tag{4.59}$$

$$C_{11} = \frac{C_1}{\omega_0} - \frac{(\alpha - 1)^{1/2}}{\omega_0 \alpha} - C_{12} \tag{4.60}$$

$$C_{NN} = \frac{C_N}{\omega_0} - \frac{(\alpha - 1)^{1/2}}{\omega_0 \alpha} - C_{N-1.N} \tag{4.61}$$

114   Theory and design of microwave filters

$$C_{rr} = \frac{C_r}{\omega_0} - C_{r-1,r} - C_{r,r+1} \qquad (r = 2, \ldots, N-1) \qquad (4.62)$$

$$L_{rr} = \frac{1}{C_r \omega_0} \qquad (r = 1, \ldots, N) \qquad (4.63)$$

4.5.1 Example

Design a capacitively coupled Chebyshev bandpass filter to meet the following specification.

Centre frequency (f_0)	1 GHz
Passband bandwidth (ΔF)	50 MHz
Passband return loss	≥ 20 dB
Stopband insertion loss	> 40 dB at $f_0 \pm 100$ MHz
System impedance	50 Ω

First we must evaluate the degree of the lowpass prototype. From (3.71)

$$N \geq \frac{L_A + L_R + 6}{20 \log_{10}[S + (S^2 - 1)^{1/2}]} \qquad (4.64)$$

where

$$L_A = 40 \text{ and } L_R = 20 \qquad (4.65)$$

S is the selectivity and is the ratio of stopband to passband bandwidth. Hence

$$S = \tfrac{200}{50} = 4 \qquad (4.66)$$

$$N \geq 3.682 \qquad (4.67)$$

That is, a degree 4 transfer function at least must be used.

The element values must now be calculated. The ripple level ε is

$$\varepsilon = (10^{L_R/10} - 1)^{-1/2} = 0.1005 \qquad (4.68)$$

Hence

$$\eta = \sinh\left[\frac{1}{N} \sinh^{-1}(1/\varepsilon)\right]$$

$$= 0.8201 \qquad (4.69)$$

and the element values are

$$C_r = \frac{2}{\eta} \sin\left[\frac{(2r-1)\pi}{2N}\right] \qquad (4.70)$$

$$K_{r,r+1} = \frac{[\eta^2 + \sin^2(r\pi/N)]^{1/2}}{\eta} \qquad (4.71)$$

$$C_1 = C_4 = 0.9332$$
$$C_2 = C_3 = 2.2531$$
$$K_{12} = K_{34} = 1.3204 \qquad (4.72)$$
$$K_{23} = 1.5770$$

Now
$$\omega_0 = 2\pi f_0 = 6.2832 \times 10^9 \qquad (4.73)$$
and
$$\alpha = \frac{f_0}{\Delta f} = 20 \qquad (4.74)$$

$$C_{01} = C_{45} = 36.512 \, \text{pF}$$
$$C_{12} = C_{34} = 10.507 \, \text{pF}$$
$$C_{23} = 12.549 \, \text{pF}$$
$$C_{11} = C_{44} = 103.33 \, \text{pF} \qquad (4.75)$$
$$C_{22} = C_{33} = 335.5 \, \text{pF}$$
$$L_{11} = L_{44} = 0.1705 \, \text{nH}$$
$$L_{22} = L_{33} = 0.07064 \, \text{nH}$$

Finally, scaling impedances by 50 Ω we obtain the circuit of Figure 4.16. The element values are

$$C_{01} = C_{45} = 0.7302 \, \text{pF}$$
$$C_{12} = C_{34} = 0.210 \, \text{pF}$$
$$C_{23} = 0.251 \, \text{pF}$$
$$C_{11} = C_{44} = 2.066 \, \text{pF} \qquad (4.76)$$
$$C_{22} = C_{33} = 6.71 \, \text{pF}$$
$$L_{11} = L_{44} = 8.525 \, \text{nH}$$
$$L_{22} = L_{33} = 3.53 \, \text{nH}$$

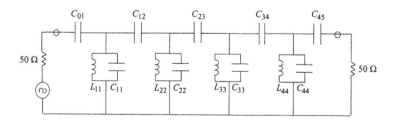

Figure 4.16   Fourth-degree Chebyshev capacitively coupled bandpass filter

116 Theory and design of microwave filters

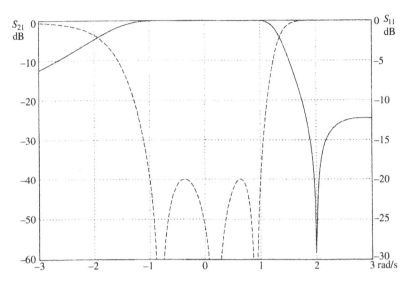

Figure 4.17 Simulated response of a capacitively coupled bandpass filter

The simulated response of this circuit is shown in Figure 4.17; note the asymmetry in the frequency response. This circuit is perfectly realisable using lumped element technology. It is worth noting, however, that L_{22} and L_{33} are slightly too small for a good high Q realisation using wire-wound inductors. It could also be better from a manufacturing point of view if all the inductors were made to have the same value.

4.5.2 Nodal admittance matrix scaling

We can force all the inductors (for example) to have the same value by the following procedure. Consider the ladder network with $N+2$ nodes shown in Figure 4.18. This has an admittance matrix Y where

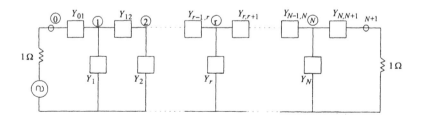

Figure 4.18 Ladder network

$$[Y] = \begin{bmatrix} Y_{00} & -Y_{01} & 0 & 0 & \cdots & & & & \\ -Y_{01} & Y_{11} & -Y_{12} & 0 & & & & & \\ 0 & -Y_{12} & Y_{22} & & & & & & \\ 0 & & & \ddots & & & & & \\ \vdots & & & & -Y_{r-1,r} & & & & \\ & & & -Y_{r-1,r} & Y_{rr} & -Y_{r,r+1} & & & \\ & & & & -Y_{r,r+1} & & & & \\ & & & & & \ddots & -Y_{N-1,N} & & \\ & & & & & -Y_{N-1,N} & Y_{NN} & -Y_{N,N+1} & \\ & & & & & & -Y_{N,N+1} & Y_{N+1,N+1} \end{bmatrix}$$

(4.77)

where

$$Y_{rr} = Y_r + Y_{r-1,r} + Y_{r,r+1} \tag{4.78}$$

The internal nodal admittances in the circuit can be scaled without affecting the terminal characteristics of the network. This is achieved by multiplying the rth row and column by a constant α_r. This operation cannot be performed on nodes 0 and $N+1$ unless the source and loads are appropriately scaled. Hence

$$Y_{rr} \rightarrow \alpha_r^2 Y_{rr} \tag{4.79}$$

$$Y_{r-1,r} \rightarrow \alpha_r Y_{r-1,r} \tag{4.80}$$

$$Y_{r,r+1} \rightarrow \alpha_r Y_{r,r+1} \tag{4.81}$$

In this manner by progressive scaling of nodes the impedance level to ground in each node can be adjusted to any required level. In the case of the previous example all the shunt inductors (or capacitors) could be made to have the same value. As an example consider the third-order Butterworth lowpass prototype filter shown in Figure 4.19.

The nodal admittance matrix of this network is given by

$$[Y] = \begin{bmatrix} p & -j & 0 \\ -j & 2p & -j \\ 0 & -j & p \end{bmatrix} \tag{4.82}$$

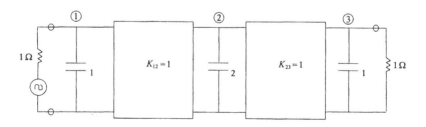

Figure 4.19 Third-order Butterworth filter

118 Theory and design of microwave filters

Scaling row 2 and column 2 by $1/\sqrt{2}$ we obtain

$$[Y]' = \begin{bmatrix} p & -j/\sqrt{2} & 0 \\ -j/\sqrt{2} & p & -j/\sqrt{2} \\ 0 & -j/\sqrt{2} & p \end{bmatrix} \tag{4.83}$$

All the capacitors are now equal with a value of unity and the inverters have a value of $1/\sqrt{2}$. This procedure could be performed at any stage in the design.

4.6 Lowpass to bandstop transformation

We require a transformation to convert the lowpass prototype into a bandstop filter with arbitrary centre frequency and bandwidth as shown in Figure 4.20. In this case the transmission zeros at infinity in the lowpass prototype must be mapped to ω_0, the centre of the stopband of the bandstop filter. The transformation is

$$\omega \to \frac{-1}{\alpha\left(\dfrac{\omega}{\omega_0} - \dfrac{\omega_0}{\omega}\right)} \tag{4.84}$$

where

$$\omega_0 = (\omega_1 \omega_2)^{1/2} \tag{4.85}$$

$$\alpha = \frac{\omega_0}{\omega_2 - \omega_1} = \frac{\omega_0}{\Delta\omega} \tag{4.86}$$

Hence for capacitors of admittance $Y(j\omega) = j\omega C$ then

$$Y(j\omega) \Rightarrow \frac{-jC}{\alpha\left(\dfrac{\omega}{\omega_0} - \dfrac{\omega_0}{\omega}\right)} \tag{4.87}$$

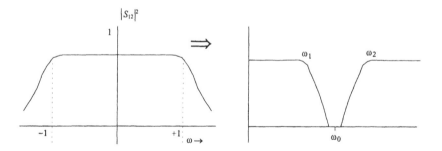

Figure 4.20 Lowpass to bandstop transformation

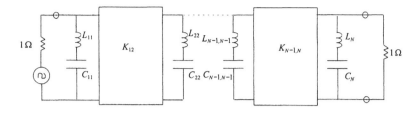

Figure 4.21   Bandstop filter

Thus the capacitor is converted into a series combination of an inductor L' and capacitor C' where

$$L' = \frac{\alpha}{\omega_0 C} \tag{4.88}$$

$$C' = \frac{C}{\alpha \omega_0} \tag{4.89}$$

Again inverters are unaffected by the transformation. Thus a shunt capacitor, inverter coupled prototype is converted into the bandstop filter shown in Figure 4.21. It is left to the reader to show that the transformation converts an inductor into a parallel tuned circuit.

The realisation of inverters is somewhat different in the case of a bandstop filter. Normally we require a broad passband and so a narrowband approximation to an inverter is of little use. Instead inverters are usually constructed from unit elements of transmission line which are one quarter wavelength long at ω_0. Since for a length of line the transfer matrix is given by

$$[T] = \begin{bmatrix} \cos(\theta) & jZ_0 \sin(\theta) \\ \dfrac{j\sin(\theta)}{Z_0} & \cos(\theta) \end{bmatrix} \tag{4.90}$$

if $\theta = \pi/2$ then

$$[T] = \begin{bmatrix} 0 & jZ_0 \\ j/Z_0 & 0 \end{bmatrix} = \begin{bmatrix} 0 & j/K \\ jK & 0 \end{bmatrix} \tag{4.91}$$

where

$$K = 1/Z_0 \tag{4.92}$$

This is a relatively broadband approximation to an inverter and of course a matched transmission line will pass energy at all frequencies.

In the case of Chebyshev bandstop filters it is best to scale the admittance matrix so that all the inverters have the same value. They can then be constructed from a single length of uniform transmission line with minimum discontinuities.

In narrowband bandstop filter design we take a different approach from that

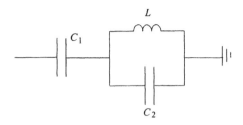

Figure 4.22 Capacitively coupled resonator

used for bandpass filters. In the bandpass case we retain realisable element values for narrow bandwidths by scaling the network so that the couplings become relatively weak. However, in the bandstop case the impedance inverters must remain at unity impedance so that the filter has a broad passband. In order to control the impedance levels we use a capacitively coupled resonator as shown in Figure 4.22.

The approach is to equate the resonant frequency and the differential of the reactance of the resonator to that of the original resonator [2] which has an impedance given by

$$Z(j\omega) = \frac{\alpha}{C_r}\left(\frac{\omega}{\omega_0} - \frac{\omega_0}{\omega}\right) \tag{4.93}$$

with

$$\frac{dZ(j\omega)}{d\omega} = \frac{\alpha}{C_r}\left(\frac{1}{\omega_0} + \frac{\omega_0}{\omega^2}\right) \tag{4.94}$$

and

$$\left.\frac{dZ(j\omega)}{d\omega}\right|_{\omega=\omega_0} = \frac{2\alpha}{C_r\omega_0} \tag{4.95}$$

Now the impedance of the capacitively coupled resonator is

$$Z(p) = \frac{1}{C_1 p} + \frac{1}{C_2 p + 1/Lp}$$

$$= \frac{1 + L(C_1 + C_2)p^2}{C_1 p(1 + LC_2 p^2)} \tag{4.96}$$

Hence

$$Z(j\omega) = \frac{-j[1 - \omega^2 L(C_1 + C_2)]}{\omega C_1(1 - \omega^2 LC_2)} \tag{4.97}$$

and

$$\left.\frac{dZ(j\omega)}{d\omega}\right|_{\omega=\omega_0} = \frac{2L(C_1 + C_2)}{C_1(1 - \omega_0^2 LC_2)} \tag{4.98}$$

with

$$\omega_0 = \frac{1}{[L(C_1 + C_2)]^{1/2}} \quad (4.99)$$

Substituting (4.99) into (4.98) and equating with (4.95) yields

$$\frac{\alpha}{C_r \omega_0} = \frac{L(C_1 + C_2)^2}{C_1^2} \quad (4.100)$$

The value of L can be chosen for physical realisability and C_1 and C_2 are then found by solving (4.99) and (4.100) simultaneously.

From (4.99)

$$C_1 + C_2 = \frac{1}{\omega_0^2 L} \quad (4.101)$$

and from (4.100) and (4.101)

$$(C_1 + C_2)^2 = \frac{\alpha C_1^2}{LC_r \omega_0} = \frac{1}{\omega_0^4 L^2} \quad (4.102)$$

Hence

$$C_{1r} = \left(\frac{C_r}{\omega_0^3 L_r \alpha}\right)^{1/2} \quad (4.103)$$

and

$$C_{2r} = \frac{1}{\omega_0^2 L_r} - C_{1r} \quad (4.104)$$

where C_{1r} and C_{2r} are the values of the capacitors in the rth resonator.

It is useful to note that this method can also be used with prototypes using frequency-invariant reactances, such as the elliptic function filter.

4.6.1 Design example

Design a bandstop filter with the following specification

Centre frequency	900 MHz
Passband bandwidth	d.c.–880 MHz, 920 MHz–2 GHz
Passband return loss	20 dB
Stopband	890–910 MHz
Stopband insertion loss	> 30 dB
System impedance	50 Ω

The passband bandwidth is 40 MHz and the stopband bandwidth is 20 MHz. Thus $S = 2$, $L_A = 30$ dB, $L_R = 20$ dB and

$$N \geq \frac{L_A + L_R + 6}{20 \log_{10}[S + (S^2 - 1)^{1/2}]} \geq 4.895 \quad (4.105)$$

Hence $N = 5$. Since $L_R = 20\,\text{dB}$, then $\eta = 0.8201$. Now

$$C_r = \frac{2}{\eta}\sin\left[\frac{(2r-1)\pi}{2N}\right] \tag{4.106}$$

$$K_{r,r+1} = \frac{[\eta^2 + \sin^2(r\pi/N)]^{1/2}}{\eta} \tag{4.107}$$

giving

$$\begin{aligned} C_1 &= C_5 = 0.7536 \\ C_2 &= C_4 = 1.9730 \\ C_3 &= 2.4387 \\ K_{12} &= K_{45} = 1.2303 \\ K_{23} &= K_{34} = 1.5313 \end{aligned} \tag{4.108}$$

We now scale the rows and columns of the nodal admittance matrix as follows:

$$[Y] = \begin{bmatrix} C_1 p & -jK_{12} & 0 & 0 & 0 \\ -jK_{12} & C_2 p & -jK_{23} & 0 & 0 \\ 0 & -jK_{23} & C_3 p & -jK_{34} & 0 \\ 0 & 0 & -jK_{34} & C_4 p & -jK_{45} \\ 0 & 0 & 0 & -jK_{45} & C_5 p \end{bmatrix} \begin{matrix} \\ \leftarrow \alpha \\ \leftarrow B \\ \leftarrow \alpha \\ \end{matrix}$$

$$\quad \downarrow \alpha \quad \downarrow B \quad \downarrow \alpha$$

$$= \begin{bmatrix} C_1 p & -j\alpha K_{12} & 0 & 0 & 0 \\ -j\alpha K_{12} & \alpha^2 C_2 p & -j\alpha B K_{23} & 0 & 0 \\ 0 & -j\alpha B K_{23} & B^2 C_3 p & -j\alpha B K_{34} & 0 \\ 0 & 0 & -j\alpha B K_{34} & \alpha^2 C_4 p & -j\alpha K_{45} \\ 0 & 0 & 0 & -j\alpha K_{45} & C_5 p \end{bmatrix} \tag{4.109}$$

Hence for unity admittance inverters

$$\alpha K_{12} = 1 \quad (= \alpha K_{45}) \tag{4.110}$$

$$B\alpha K_{23} = 1 \quad (= \alpha B K_{34}) \tag{4.111}$$

Hence

$$\alpha = 0.8128 \tag{4.112}$$

and

$$B = 0.8034 \tag{4.113}$$

and the new element values are

$$C_1 = C_5 = 0.7536$$
$$C_2 = C_4 = 1.3034$$
$$C_3 = 1.5741$$
$$K_{r,r+1} = 1$$
(4.114)

Now
$$\omega_0 = 5.6548 \times 10^9 \qquad (4.115)$$
and
$$\alpha = 900/40 = 22.5 \qquad (4.116)$$

Choosing all the inductors to have a physically realisable value of 10 nH, then in a 1 Ω system $L = 2 \times 10^{-10}$. Then using (4.103) and (4.104) and scaling to 50 Ω we obtain

$$C_{11} = 0.6086\,\text{pF} = C_{15}$$
$$C_{12} = 0.8005\,\text{pF} = C_{14} \qquad (4.117)$$
$$C_{13} = 0.8797\,\text{pF}$$

$$C_{21} = 2.5187\,\text{pF} = C_{25}$$
$$C_{22} = 2.3267\,\text{pF} = C_{24} \qquad (4.118)$$
$$C_{23} = 2.2475\,\text{pF}$$

The final circuit is shown in Figure 4.23. All the unit elements are quarter wavelength long at the centre frequency and have a characteristic impedance of 50 Ω.

The simulated frequency response of the circuit is shown in Figure 4.24. Scrutiny of the simulated response of the bandstop filter shows a slight asymmetry in the frequency response with the filter being more selective on the high frequency side of the stopband. This is explained by examining the expression for the resonator impedance.

$$Z(j\omega) = \frac{-j[1 - \omega^2 L(C_1 + C_2)]}{\omega C_1 (1 - \omega^2 L C_2)} \qquad (4.119)$$

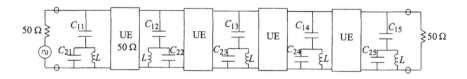

Figure 4.23 Narrowband bandstop filter (UE, unit element)

124 *Theory and design of microwave filters*

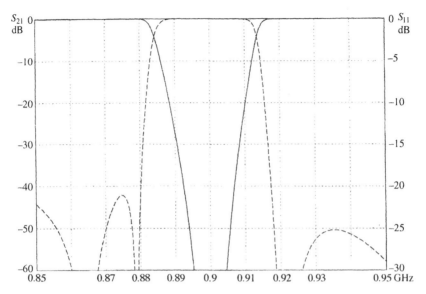

Figure 4.24 Simulated response of the bandstop filter

The impedance has a zero at ω_0 where

$$\omega_0 = \frac{1}{[L(C_1 + C_2)]^{1/2}} \tag{4.120}$$

It has a pole at $\omega = 0$ and a second pole at ω_p where

$$\omega_p = \frac{1}{(LC_2)^{1/2}} \tag{4.121}$$

This pole occurs at a frequency which is higher than, but close to, ω_0 rather than at $\omega = \infty$ as would be more desirable. Thus the rate of change of reactance of the resonator is greater on the high frequency side of the passband than the lower side, explaining the asymmetry. This effect can be compensated for by slightly altering the phase lengths between the resonators. There is a theoretical procedure for doing this [3], but in reality it is quite effective to shorten the phase length by a few degrees until the simulated response is symmetrical.

The wideband response of the filter is shown in Figure 4.25. Here we see that the return loss response deteriorates above 2 GHz. This is a consequence of the resonators loading the through line. From (4.119) we have

$$Z(j\omega) = \frac{-j[1 - \omega^2 L(C_1 + C_2)]}{\omega C_1(1 - \omega^2 LC_2)} \tag{4.122}$$

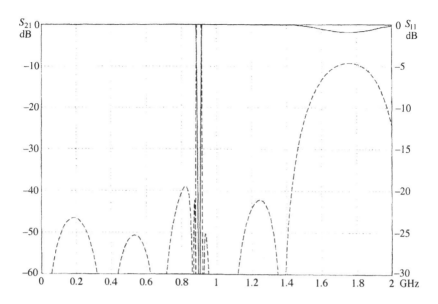

Figure 4.25   Simulated broadband response of a bandstop filter

As ω becomes very large

$$Z(j\omega) \to \frac{-j(C_1 + C_2)}{\omega C_1 C_2} \quad (4.123)$$

That is, the inductive part of the resonator becomes open circuited and the resonator behaves as a pure capacitor. Thus at high frequencies the filter behaves as a periodically capacitively loaded transmission line; hence the deterioration in response.

4.7 Effects of losses on bandpass filters

Up to this point design procedures have assumed lossless lowpass prototypes, thus yielding lossless bandpass and bandstop filters. Real filters, however, use components with finite resistance which will produce a degradation in performance. The effects of this resistance can be related directly to the inherent quality or Q factor of individual components used in the filter design.

The Q factor for a circuit is defined as [4]

$$Q = \frac{2\pi \times \text{maximum energy stored in a cycle}}{\text{energy dissipated per cycle}} \quad (4.124)$$

For example, an inductor with finite resistance is shown in Figure 4.26. The

Figure 4.26   Inductor with finite resistance

maximum energy stored in the inductor is given by

$$E = \tfrac{1}{2} L I_0^2 \qquad (4.125)$$

where I_0 is the peak current. The dissipated power is given by

$$P_D = \frac{I_0^2 R}{2} \qquad (4.126)$$

and the dissipated energy per cycle of period τ is given by

$$E_D = P_D \tau = P_D \frac{2\pi}{\omega} = \frac{I_0^2 R}{2} \frac{2\pi}{\omega} \qquad (4.127)$$

Hence from (4.124), (4.125) and (4.127)

$$Q = \frac{2\pi L I_0^2}{2} \bigg/ \frac{I_0^2 R}{2} \frac{2\pi}{\omega} \qquad (4.128)$$

Therefore

$$Q = \frac{\omega L}{R} \qquad (4.129)$$

For a capacitor with shunt leakage conductance G we have

$$Q_C = \frac{\omega C}{G} \qquad (4.130)$$

Hence for the capacitor

$$G = \frac{\omega C}{Q_C} \qquad (4.131)$$

and for the inductor

$$R = \frac{\omega L}{Q_L} \qquad (4.132)$$

Now consider the effect of finite losses on the third-order bandpass filter arising from the LC ladder prototype. Applying the bandpass transformation

$$\omega \to \alpha \left(\frac{\omega}{\omega_0} - \frac{\omega_0}{\omega} \right) \qquad (4.133)$$

we obtain the bandpass circuit shown in Figure 4.27.

Now let us assume that the dominant loss mechanism in the series resonant circuits is from the series resistance associated with the inductors. Similarly we

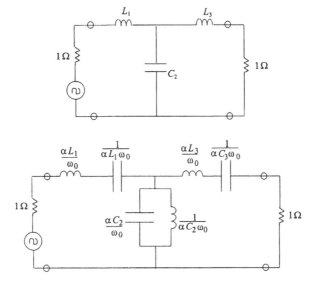

Figure 4.27   LC ladder prototype and its bandpass equivalent

assume the dominant loss mechanism in the shunt resonators is associated with the shunt capacitors. Hence for the rth series circuit

$$R_r = \frac{\omega \alpha L_r}{\omega_0 Q_r} \tag{4.134}$$

Now assuming uniform dissipation, i.e. all the resonators have the same unloaded Q,

$$Q_r = Q \qquad r = 1, \ldots, N \tag{4.135}$$

and evaluating at the midband frequency ω_0,

$$R_r = \frac{\alpha L_r}{Q} \tag{4.136}$$

and similarly for the shunt elements

$$R_r = \frac{Q}{\alpha C_r} \tag{4.137}$$

Now at the midband frequency of the filter the reactive parts of the series resonators become short circuited and the shunt resonators are open circuited. The bandpass filter has the equivalent circuit shown in Figure 4.28.

128 *Theory and design of microwave filters*

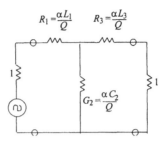

Figure 4.28 Equivalent circuit of a lossy bandpass filter evaluated at midband frequency

The midband insertion loss of the filter can be evaluated from the transfer matrix of the filter $[T]$.

$$[T] = \begin{bmatrix} 1 & R_1 \\ 0 & 1 \end{bmatrix} \begin{bmatrix} 1 & 0 \\ G_2 & 1 \end{bmatrix} \begin{bmatrix} 1 & R_3 \\ 0 & 0 \end{bmatrix}$$

$$= \begin{bmatrix} 1 + R_1 G_2 & R_1 + R_3 + R_1 G_2 R_3 \\ G_2 & 1 + G_2 R_3 \end{bmatrix} \tag{4.138}$$

Now let

$$\alpha_r = \frac{\alpha g_r}{Q} \tag{4.139}$$

where g_r is the value of the rth element in the lowpass prototype. Hence

$$[T] = \begin{bmatrix} 1 + \alpha_1 \alpha_2 & \alpha_1 + \alpha_3 + \alpha_1 \alpha_2 \alpha_3 \\ \alpha_2 & 1 + \alpha_2 \alpha_3 \end{bmatrix} \tag{4.140}$$

and

$$|S_{12}|^2 \text{ dB} = 10 \log_{10} \left[\left(\frac{A + B + C + D}{2} \right)^2 \right]$$

$$= 10 \log_{10} \left[\left(\frac{2 + \alpha_1 + \alpha_2 + \alpha_3 + \alpha_1 \alpha_2 + \alpha_2 \alpha_3 + \alpha_1 \alpha_2 \alpha_3}{2} \right)^2 \right] \tag{4.141}$$

Now for a relatively low loss filter the resonator Q must be greater than the bandwidth scaling factor, i.e.

$$Q \gg \alpha \tag{4.142}$$

Hence

$$\alpha_r \ll 1 \tag{4.143}$$

Then we can ignore second-order and higher terms in (4.141) and

$$|S_{12}|^2 \, \text{dB} = 10 \log_{10} \left[\left(1 + \frac{\alpha_1 + \alpha_2 + \alpha_3}{2} \right)^2 \right]$$

$$\approx 10 \log_{10}(1 + \alpha_1 + \alpha_2 + \alpha_3)$$

$$= 4.343 \log_e(1 + \alpha_1 + \alpha_2 + \alpha_3) \tag{4.144}$$

Now

$$\log_e(1 + X) \approx X \; [X \ll 1] \tag{4.145}$$

Hence the midband insertion loss L is given approximately by

$$L = 4.343(\alpha_1 + \alpha_2 + \alpha_3)$$

$$= 4.343 \frac{\alpha}{Q}(g_1 + g_2 + g_3)$$

$$= \frac{4.343 f_0}{\Delta f Q_u}(g_1 + g_2 + g_3) \tag{4.146}$$

In general the analysis can be extended to an nth-degree filter and it may be shown that

$$L = \frac{4.343 f_0}{\Delta f Q_u} \sum_{r=1}^{N} g_r \tag{4.147}$$

A slightly different solution is given in Reference 3.

As an example consider a third-order maximally flat filter with 1 GHz centre frequency and 10 MHz passband bandwidth and a resonator Q of 1000. Hence

$$L = \frac{4.343 \times 1000}{10 \times 1000}(1 + 2 + 1)$$

$$\approx 1.72 \, \text{dB} \tag{4.148}$$

This is the midband insertion loss of the filter. The group delay will increase near the band-edge causing a further increase in insertion loss. Typically for Chebyshev filters of degree 8 the insertion loss at band-edge will be approximately twice the midband value.

Scrutiny of (4.147) shows that the main effects of finite dissipation on bandpass filters may be summarised as follows:

- The midband insertion loss is inversely proportional to the unloaded Q of the resonators.
- The midband insertion loss is inversely proportional to the passband bandwidth.
- The insertion loss is approximately proportional to the degree of the filter.

The above analysis is meant as a design guide. A more accurate measure of loss is obtained by using a circuit analysis package to analyse the filter with finite resonator Q factors. Plots of the degree 4 Chebyshev bandpass filter design

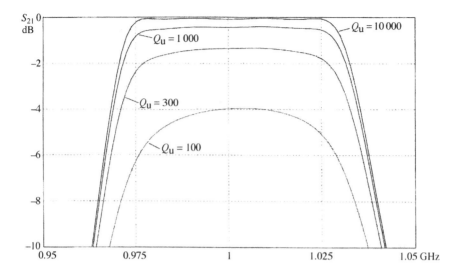

Figure 4.29 Simulation of the bandpass filter with various resonator Q factors

example response with various Q factors are shown in Figure 4.29. We see that as the Q factor is reduced the midband loss increases, as does the band-edge loss. The loss variation across the passband increases. If a maximum passband insertion loss specification must be met then the filter bandwidth may have to be increased. This means that the degree of the filter may have to be increased; hence the losses increase and the process rapidly becomes self-defeating. Usually it is necessary to increase the Q factor of the resonators if the insertion loss is too high.

Lossy circuit elements also cause a deterioration in the performance of bandstop filters. Consider the inverter coupled LC bandstop filter shown in Figure 4.30. The resonant circuits should produce transmission zeros at the mid stopband frequency of the filter. The effect of finite losses in the inductors is to shift the transmission zeros onto the real axis, i.e. the resonant circuits do

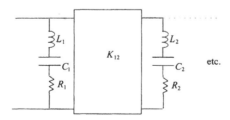

Figure 4.30 Lossy bandstop filter

not produce perfect short circuits at resonance. This has the effect of reducing the ultimate stopband insertion loss. It also causes a roll-off of insertion loss towards the edges of the passband.

4.8 Practical procedures

In the practical development of filters it is useful to have systematic procedures to obtain the correct couplings between resonators etc. The most important procedures are measurement of input couplings, measurement of the coupling between two resonators and measurement of the unloaded Q of a resonator.

4.8.1 Measurement of input coupling

First we will develop a method for measuring the input coupling into a bandpass resonator using reflected group delay. Consider the bandpass resonator shown in Figure 4.31. The element values in the resonator are related to the element values in the lowpass prototype. Assuming that the resonator is the first resonator in the filter and that the prototype consists of shunt capacitors separated by inverters, then

$$C = \frac{\alpha C_1}{\omega_0} \tag{4.149}$$

$$L = \frac{1}{\alpha C_1 \omega_0} \tag{4.150}$$

and

$$\alpha = \frac{\omega_0}{\Delta \omega} \tag{4.151}$$

C_1 is the first capacitor in the lowpass prototype. If this is considered in isolation then its admittance is

$$Y(j\omega) = j\omega C_1 \tag{4.152}$$

with

$$S_{11}(j\omega) = \frac{1 - j\omega C_1}{1 + j\omega C_1} = \frac{1 - \omega^2 C_1^2 - j2\omega C_1}{1 + \omega^2 C_1^2} \tag{4.153}$$

The phase of S_{11} is

$$\psi(\omega) = \tan^{-1}\left(\frac{2\omega C_1}{\omega^2 C_1^2 - 1}\right) \tag{4.154}$$

Figure 4.31 Bandpass resonator

Now in the bandpass resonator

$$\omega' = \alpha\left(\frac{\omega}{\omega_0} - \frac{\omega_0}{\omega}\right) \tag{4.155}$$

and

$$\psi'(\omega) = \tan^{-1}\left(\frac{2\omega' C_1}{\omega'^2 C_1^2 - 1}\right) \tag{4.156}$$

The reflected group delay of the bandpass resonator is thus

$$T_g(\omega) = \frac{-d\psi'(\omega)}{d\omega}$$

$$= -\left[\frac{d}{d\omega'}\tan^{-1}\left(\frac{2\omega' C_1}{\omega'^2 C_1^2 - 1}\right)\right]\frac{d\omega'}{d\omega}$$

$$= \frac{2C_1}{1 + \omega'^2 C_1^2}\,\alpha\left(\frac{1}{\omega_0} + \frac{\omega_0}{\omega^2}\right) \tag{4.157}$$

The reflected delay is a maximum at the resonant frequency when ω' is zero and $\omega = \omega_0$. Thus

$$T_{g\,\text{max}} = \frac{4\alpha C_1}{\omega_0} \tag{4.158}$$

or

$$T_{g\,\text{max}} = \frac{4C_1}{\Delta\omega} \tag{4.159}$$

The reflected delay can thus be computed from (4.159) and the actual delay of a single resonator can be measured using a network analyser. The input coupling to the first resonator can then be adjusted until the theoretical and measured couplings agree.

As an example, if $C_1 = 1$ and the filter has 10 MHz bandwidth then

$$T_{g\,\text{max}} = 63.66\,\text{ns} \tag{4.160}$$

From a practical point of view the delay measurement should ideally be made on a single resonator. The other resonators should be detuned or removed from the filter.

The analysis assumes a lossless resonator; in fact the delay is independent of resonator losses provided the unloaded Q is high. A good rule of thumb is

$$Q_u > 10\alpha \tag{4.161}$$

Thus for a 1 GHz filter with 10 MHz bandwidth the unloaded Q should be greater than 1000 for the measurement to be valid. This would normally be the case for a low loss filter. The reflected group delay response for the example is shown in Figure 4.32.

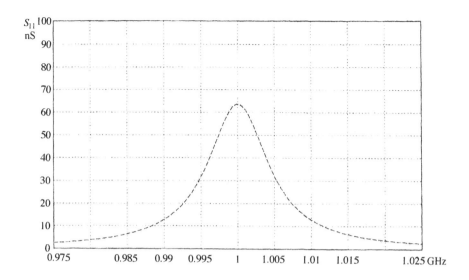

Figure 4.32 Reflected group delay for a bandpass resonator

4.8.2 Measurement of inter-resonator coupling

A procedure for measuring the couplings between two resonators can also be developed. Consider a section of lowpass prototype consisting of two shunt capacitors separated by an inverter shown in Figure 4.33. The input impedance of the circuit is

$$Z(j\omega) = \frac{j\omega C_2}{K_{12}^2 - \omega^2 C_1 C_2} \tag{4.162}$$

The poles of $Z(j\omega)$ occur when

$$\omega_{a,b} = \pm \frac{K_{12}}{(C_1 C_2)^{1/2}} \tag{4.163}$$

Now applying the bandpass transformation

$$\omega \to \alpha \left(\frac{\omega}{\omega_0} - \frac{\omega_0}{\omega} \right) \tag{4.164}$$

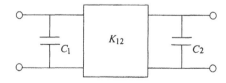

Figure 4.33 A section of a lowpass prototype

134 *Theory and design of microwave filters*

then

$$\alpha\left(\frac{\omega_a}{\omega_0} - \frac{\omega_0}{\omega_a}\right) = \frac{-K_{12}}{(C_1 C_2)^{1/2}} \tag{4.165}$$

and

$$\alpha\left(\frac{\omega_b}{\omega_0} - \frac{\omega_0}{\omega_b}\right) = \frac{K_{12}}{(C_1 C_2)^{1/2}} \tag{4.166}$$

Hence

$$\omega_a^2 + \frac{K_{12}\omega_0}{\alpha(C_1 C_2)^{1/2}} - \omega_0^2 = 0 \tag{4.167}$$

$$\omega_b^2 - \frac{K_{12}\omega_0}{\alpha(C_1 C_2)^{1/2}} - \omega_0^2 = 0 \tag{4.168}$$

Solving these two quadratic equations and subtracting the solutions we obtain

$$\omega_b - \omega_a = \frac{K_{12}\omega_0}{\alpha(C_1 C_2)^{1/2}} \tag{4.169}$$

or

$$\Delta C = \frac{K_{12}\Delta\omega}{(C_1 C_2)^{1/2}} \tag{4.170}$$

ΔC is known as the coupling bandwidth and from (4.170) it is directly related to the element values in the lowpass prototype and the bandwidth $\Delta\omega$ of the bandpass filter. The coupling bandwidth can be computed and then measured for a pair of resonators using a network analyser. Again the other resonators should be removed or detuned.

4.8.3 Measurement of resonator Q factor

The unloaded Q factor of a resonator can also be measured experimentally. Consider the lumped bandpass resonator with finite unloaded Q coupled via an inverter, as shown in Figure 4.34. The unloaded Q of the resonator is

$$Q_u = \frac{\omega_0 C}{G} \tag{4.171}$$

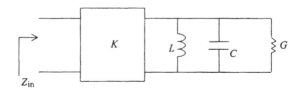

Figure 4.34 *Bandpass resonator with finite Q_u*

and the input impedance of the resonator is

$$Z(j\omega) = \frac{G + j(\omega C - 1/\omega L)}{K^2}$$

$$= \frac{\dfrac{\omega_0 C}{Q_u} + j\left(\omega C - \dfrac{\omega_0^2 C}{\omega}\right)}{K^2} \tag{4.172}$$

We now adjust the input coupling K so that the resonator is perfectly matched (critically coupled) at resonance, i.e. $Z_{in}(j\omega)$ is equal to unity. Thus

$$K^2 = \frac{\omega_0 C}{Q_u} \tag{4.173}$$

Hence

$$Z_{in}(j\omega) = 1 + jQ_u\left[\frac{\omega}{\omega_0} - \frac{\omega_0}{\omega}\right] \tag{4.174}$$

Having matched the resonator at ω_0 the 3 dB frequencies ω_a and ω_b are measured. The imaginary part of Z is equal to 2 at these frequencies. Thus

$$Q_u\left[\frac{\omega_a}{\omega_0} - \frac{\omega_0}{\omega_a}\right] = -2 \tag{4.175}$$

$$Q_u\left[\frac{\omega_b}{\omega_0} - \frac{\omega_0}{\omega_b}\right] = 2 \tag{4.176}$$

Again quadratic equations may be generated and the difference between their solutions is

$$\omega_b - \omega_a = \frac{2\omega_0}{Q_u} \tag{4.177}$$

or

$$Q_u = \frac{2\omega_0}{\Delta\omega} \tag{4.178}$$

Thus the unloaded Q_u of the critically coupled resonator is equal to the centre frequency divided by half the 3 dB bandwidth. From a practical point of view the level of return loss at resonance should be at least 35 dB for an accurate measurement.

4.9 Summary

Starting from lowpass prototype networks a series of transformations are used to convert to arbitrary cut-off frequency lowpass, highpass, bandpass and bandstop filters with arbitrary impedance terminations. Procedures are developed for narrowband bandpass and bandstop filters where the inverters are approximated by pi sections of capacitors and quarter wave transmission lines respectively. These procedures are illustrated by design examples which also introduce the concept of nodal admittance matrix scaling. The effect of losses in filters is described with particular emphasis on bandpass filters so that the designer can compute the midband insertion loss of a particular design. Finally various practical procedures for measuring resonator couplings and Q factors are described.

4.10 References

1 HELSAJN, J.: 'Synthesis of lumped element, distributed and planar filters' (McGraw-Hill, New York, 1990) pp. 221–42
2 MATTHAEI, G., YOUNG. L., and JONES, E.M.T.: 'Microwave filters, impedance matching networks and coupling structures' (Artech House, Norwood, MA, 1980) pp. 427–34
3 HUNTER, I.C., and RHODES, J.D.: 'Electronically tunable microwave bandstop filters', *IEEE Transactions on Microwave Theory and Techniques*, 1982, **30**, (9) (Special Issue on Microwave Filters) pp. 1361–67
4 CHENG, D.K.: 'Field and wave electromagnetics' (Addison-Wesley, Reading, MA, 1989) pp. 586–87

Chapter 5
TEM transmission line filters

5.1 Commensurate distributed circuits

The previous chapters have concentrated on the theory and design of lumped element filters. By definition lumped elements are zero-dimensional, i.e. they have no physical dimensions which are significant with respect to the wavelength at the operating frequency. One of the great advantages of restricting oneself to lumped elements is that circuits may be completely described in terms of one complex frequency variable.

As we increase frequency into the microwave spectrum it is easy to see that lumped element theories will not suffice, e.g. the wavelength at 10 GHz is only 3 cm and circuit elements may easily have dimensions in excess of a quarter wavelength. Furthermore, as we have already seen, narrowband filters with low insertion loss require high Q resonators. This implies physically large resonators, again meaning that dimensions become significant fractions of a wavelength. It is thus necessary to have design theories which are pertinent to these 'distributed' circuits.

In general, networks consisting of arbitrary connections of distributed circuit elements do not have a unified design theory. Although analysis of such circuits may be accomplished by solving Maxwell's equations using, for example, finite element analysis, this is not the same as having a design theory. As an example a circuit consisting of an interconnection of transmission lines of different lengths would require a theoretical approach using more than one complex variable. Work in this area has been extremely limited. To simplify the design theories we usually restrict ourselves to the case where distributed circuits consist of interconnections of transmission lines of equal length, i.e. commensurate distributed networks. This enables us to work with a single complex frequency variable, thus simplifying the design process.

The simplest commensurate distributed networks consist of interconnections of lossless transmission lines of equal length, each supporting a pure TEM mode of propagation. This mode is particularly useful as it is supported in coaxial cables and has zero cut-off frequency. A basic section of lossless line is called a 'unit element' (UE) and has the following transfer matrix:

$$[T] = \begin{bmatrix} \cos(\beta\ell) & jZ_0 \sin(\beta\ell) \\ jY_0 \sin(\beta\ell) & \cos(\beta\ell) \end{bmatrix} \quad (5.1)$$

(sinusoidal excitation is assumed). Here ℓ is the length of the line and Z_0 is its characteristic impedance. β is the propagation constant of the line where

$$\beta = \frac{2\pi}{\lambda} \quad (5.2)$$

Alternatively since

$$\beta\ell = \frac{2\pi\ell}{\lambda} \quad (5.3)$$

and

$$v = f\lambda \quad (5.4)$$

(where v is the velocity of propagation)

$$\beta\ell = \frac{\omega\ell}{v} \quad (5.5)$$

or

$$\beta\ell = a\omega = \theta \quad (5.6)$$

Thus

$$[T] = \begin{bmatrix} \cos(a\omega) & jZ_0 \sin(a\omega) \\ jY_0 \sin(a\omega) & \cos(a\omega) \end{bmatrix} \quad (5.7)$$

and for complex frequencies

$$[T] = \begin{bmatrix} \cosh(ap) & Z_0 \sinh(ap) \\ Y_0 \sinh(ap) & \cosh(ap) \end{bmatrix}$$

$$= \frac{1}{(1-t^2)^{1/2}} \begin{bmatrix} 1 & Z_0 t \\ Y_0 t & 1 \end{bmatrix} \quad (5.8)$$

where

$$t = \tanh(ap) \quad (5.9)$$

A circuit consisting of interconnections of commensurate lines can be described by rational polynomial functions of t, although multiples of $(1-t^2)^{1/2}$ may also occur.

Given that commensurate distributed networks can be described in terms of the complex frequency variable t, it is possible to borrow from lumped theory to design certain classes of filter. This can be achieved by applying the Richards' transformation [1] as follows:

$$p \rightarrow \alpha \tanh(ap) \tag{5.10}$$

or

$$\omega \rightarrow \alpha \tan(a\omega) \tag{5.11}$$

Applying (5.10) to a capacitor we obtain

$$Y(p) = Cp \Rightarrow \alpha C \tanh(ap) \tag{5.12}$$

Therefore

$$Y(p) \Rightarrow Y_0 \tanh(ap) \tag{5.13}$$

and

$$Y(j\omega) \Rightarrow jY_0 \tan(a\omega) \tag{5.14}$$

where

$$Y_0 = \alpha C \tag{5.15}$$

The transformation converts a capacitor into an open circuited stub.
Similarly, applying the transformation to an inductor L, we obtain

$$Z(p) = Lp \Rightarrow \alpha L \tanh(ap) \tag{5.16}$$

Therefore

$$Z(p) \Rightarrow Z_0 \tanh(ap) \tag{5.17}$$

and

$$Z(j\omega) \Rightarrow jZ_0 \tan(a\omega) \tag{5.18}$$

where

$$Z_0 = \alpha L \tag{5.19}$$

The transformation converts an inductor into a short circuited stub.
These transformations are shown in Figure 5.1. Note that there is no lumped element equivalent to a UE of transmission line.

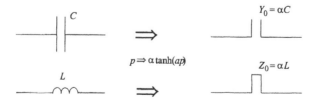

Figure 5.1 Richards transformation of lumped elements

140 *Theory and design of microwave filters*

The Richards transformation may be applied to a filter transfer function. Thus if

$$|S_{12}(j\omega)|^2 = \frac{1}{1 + F_N^2(\omega)} \tag{5.20}$$

then

$$|S_{12}(j\omega)|^2 \Rightarrow \frac{1}{1 + F_N^2[\alpha \tan(a\omega)]} \tag{5.21}$$

For a Chebyshev lowpass prototype we have

$$|S_{12}(j\omega)|^2 \Rightarrow \frac{1}{1 + \varepsilon^2 T_N^2[\alpha \tan(a\omega)]} \tag{5.22}$$

The transmission zeros which occur at infinite frequency in the original prototype are mapped into odd multiples of the quarter wave frequency. The passband centre at $\omega = 0$ in the original prototype is mapped to even multiples of the half wave frequency. The passband edge at $\omega = 1$ is mapped to ω_1 as follows:

$$1 \Rightarrow \alpha \tan(a\omega_1) \tag{5.23}$$

Thus

$$\alpha = \frac{1}{\tan(a\omega_1)} \tag{5.24}$$

The resultant response is a quasi-lowpass or bandstop response with stopbands repeating *ad infinitum* at odd multiples of the quarter wave frequency

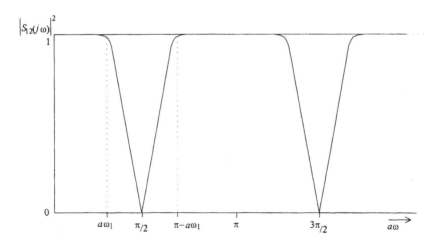

Figure 5.2 *Distributed quasi-lowpass response*

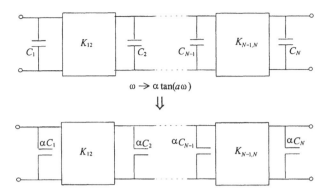

Figure 5.3   Distributed quasi-lowpass transformation

(Figure 5.2). The lowpass prototype is transformed into the distributed circuit shown in Figure 5.3. Again the inverters are invariant under the frequency transformation.

The distributed quasi-lowpass filter consists of shunt open circuited stubs separated by inverters. The inverters may be realised as sections of quarter wave line, i.e. UEs one quarter wave long at ω_0. The circuit is perfectly realisable for moderate bandwidths but is not suitable for very narrow bandwidths. In that case as ω_1 approaches ω_0, $\tan(a\omega_1)$ approaches infinity and α becomes very small. The impedance of the shunt stubs will then become unrealisably high. In any case, as we shall see, there are better distributed lowpass filters available to the designer.

We can also apply the highpass Richards transformation as follows:

$$p \to \frac{1}{\alpha \tanh(ap)} \tag{5.25}$$

or

$$\omega \to \frac{-1}{\alpha \tan(a\omega)} \tag{5.26}$$

Applying this to capacitors yields

$$Y(p) = Cp \Rightarrow \frac{C}{\alpha \tanh(ap)} \tag{5.27}$$

$$Y(j\omega) = j\omega C \Rightarrow \frac{-jC}{\alpha \tan(a\omega)} \tag{5.28}$$

Relation (5.28) converts capacitors C into short circuited stubs of impedance Z_0 where

$$Z_0 = \frac{\alpha}{C} \tag{5.29}$$

142 Theory and design of microwave filters

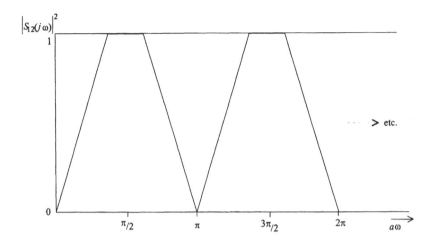

Figure 5.4 Response of a distributed quasi-highpass filter

Similarly inductors are converted into open circuited stubs of admittance

$$Y_0 = \frac{\alpha}{L} \tag{5.30}$$

Applying the Richards highpass transformation to a Chebyshev lowpass prototype we obtain

$$|S_{12}(j\omega)|^2 = \frac{1}{1 + \varepsilon^2 T_N^2\{1/[\alpha \tan(a\omega)]\}} \tag{5.31}$$

The transmission zeros at infinity in the original prototype are mapped to d.c. and even multiples of the quarter wave frequency. $\omega = 0$ maps into odd multiples of the quarter wave frequency. The resultant response is a quasi-highpass or bandpass response as shown in Figure 5.4.

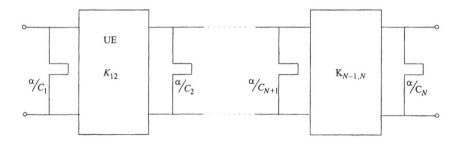

Figure 5.5 Distributed quasi-highpass filter

TEM transmission line filters 143

The band-edge frequency in the lowpass prototype at $\omega = 1$ maps to ω_1 with

$$\alpha = \frac{1}{\tan(a\omega_1)} \tag{5.32}$$

The quasi-highpass filter is realisable as shunt short circuited stubs separated by impedance inverters. These may again be approximated by UEs one quarter wave long at ω_0 (Figure 5.5). Again, this circuit is not suitable for very narrow-band applications. However, if the inverters were realised by reactive elements instead of UEs then the network could be scaled using the procedures developed in Chapter 4, and a narrowband design could be realised.

A design example for a relatively broadband bandpass filter is now presented. The specification is as follows:

Prototype	Degree 4 Chebyshev
Passband return loss	≥ 20 dB
Centre frequency	4 GHz
System impedance	50 Ω

From Chapter 3 we obtain the element values of the lowpass prototype filter:

$$\begin{aligned} C_1 = C_4 = 0.9332 \\ C_2 = C_3 = 2.2531 \end{aligned} \tag{5.33}$$

Now

$$a\omega_1 = \frac{\omega_1}{\omega_0} \frac{\pi}{2} \tag{5.34}$$

where $\omega_0 = 8\pi \times 10^9$ and $\omega_1 = 4\pi \times 10^9$. Hence $a\omega_1 = \pi/4$ and $\alpha = 1$. Now

$$Z_r = \frac{50\alpha}{C_r} \Omega \tag{5.35}$$

and

$$Z_1 = Z_4 = 53.27 \, \Omega \tag{5.36}$$
$$Z_2 = Z_3 = 22.2 \, \Omega \tag{5.37}$$

Realising the inverters as UEs we obtain

$$Z_{r,r+1} = 50/K_{r,r+1} \tag{5.38}$$

Therefore

$$Z_{12} = Z_{34} = 37.86 \, \Omega \tag{5.39}$$
$$Z_{23} = 31.71 \, \Omega \tag{5.40}$$

All the transmission lines in the circuit are commensurate and are one quarter wave long at 4 GHz, i.e. 1.875 cm long.

The simulated response of the circuit is shown in Figure 5.6. Here we see that

144 Theory and design of microwave filters

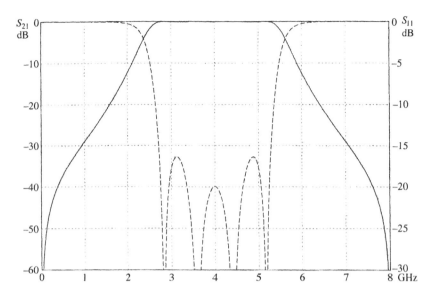

Figure 5.6 Simulated frequency response of the distributed bandpass filter

the bandwidth is narrower than desired and the return loss is only 16.5 dB. This is because the series UEs are a poor approximation to inverters over broad bandwidths. An accurate design procedure is available using the method described for interdigital filters.

5.2 Stepped impedance unit element prototypes

We have seen in the previous sections that microwave filters can be constructed using interconnections of stubs and inverters. Over narrow bandwidths the inverters can be replaced by UEs of transmission line. However, these UEs are themselves frequency dependent and it is possible to design useful filters consisting entirely of a cascade of UEs.

The transfer matrix of a UE is given by

$$[T] = \frac{1}{(1-t^2)^{1/2}} \begin{bmatrix} 1 & Zt \\ Yt & 1 \end{bmatrix} \quad \text{(where } t = \tanh(ap)\text{)} \quad (5.41)$$

Given a cascade of a pair of UEs of characteristic impedances Z_1 and Z_2, we obtain

$$[T] = \frac{1}{1-t^2} \begin{bmatrix} 1 + Z_1 Y_2 t^2 & (Z_1 + Z_2)t \\ (Y_1 + Y_2)t & 1 + Y_1 Z_2 t^2 \end{bmatrix} \quad (5.42)$$

Hence

$$S_{12}(t) = \frac{1-t^2}{1+At+Bt^2} \quad (5.43)$$

(A and B are constants of no significance). Hence

$$|S_{12}[j\tan(\omega)]|^2$$

$$= \frac{[1+\tan^2(\omega)]^2}{1+(A^2-2B)\tan^2(\omega)+B^2\tan^4(\omega)}$$

$$= \frac{1}{[1-\sin^2(\omega)]^2 + (A^2-2B)[1-\sin^2(\omega)]\sin^2(\omega) + B^2\sin^4(\omega)}$$

$$= \frac{1}{1+\alpha\sin^2(\omega)+\beta\sin^4(\omega)} \quad (5.44)$$

Thus $|S_{12}|^2$ is a polynomial in $\sin^2(\omega)$.

In general the cascade of N UEs shown in Figure 5.7 has the properties that [2, 3]

$$S_{12}(t) = \frac{(1-t^2)^{N/2}}{D_N(t)} \quad (5.45)$$

where $D_N(t)$ is a Hurwitz polynomial in t and $|S_{12}[j\tan(\omega)]|^2 \leq 1$. Furthermore,

$$|S_{12}[j\tan(\omega)]|^2 = \frac{1}{1+F_N[\sin^2(\omega)]} \quad (5.46)$$

For a maximally flat response $|S_{12}|^2$ must be of the form

$$|S_{12}[j\tan(\omega)]|^2 = \frac{1}{1+[\sin(\omega)/\alpha]^{2N}} \quad (5.47)$$

where $\alpha = \sin\omega_0$ and ω_0 is the 3 dB frequency. The minimum value of $|S_{12}|^2$ occurs when $\omega = \pi/2$, but this is not a transmission zero. Furthermore, the first $2N-1$ derivatives of (5.47) are zero at $\omega = 0$ but only the first derivative is zero at $\omega = \pi/2$.

For a Chebyshev response we have

$$|S_{12}[j\tan(\omega)]|^2 = \frac{1}{1+\varepsilon^2 T_N^2[\sin(\omega)/\alpha]} \quad (5.48)$$

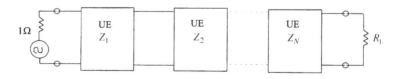

Figure 5.7 *Cascade of N unit elements*

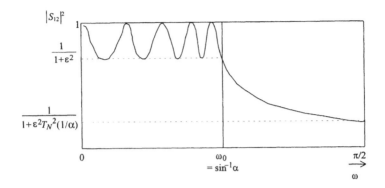

Figure 5.8 Frequency response of a Chebyshev unit element filter

The frequency response is shown in Figure 5.8.

Explicit formulae for these 'stepped impedance' filters have been developed and the results for a Chebyshev response are given below [4].

$$Z_r = 1/g_r \quad (r \text{ odd}) \tag{5.49}$$

$$Z_r = g_r \quad (r \text{ even}) \tag{5.50}$$

For $r = 1, \ldots, N$

$$g_r = A_r \left(\frac{2\sin[(2r-1)\pi/2N]}{\alpha} \right.$$

$$\left. - \frac{\alpha}{4} \left\{ \frac{\eta^2 + \sin^2(r\pi/N)}{\sin[(2r+1)\pi/2N]} + \frac{\eta^2 + \sin^2[(r-1)\pi/N]}{\sin[(2r-3)\pi/2N]} \right\} \right) \tag{5.51}$$

where

$$A_r = \frac{\{\eta^2 + \sin^2[(r-2)\pi/N]\}\{\eta^2 + \sin^2[(r-4)\pi/N]\} \ldots}{\{\eta^2 + \sin^2[(r-1)\pi/N]\}\{\eta^2 + \sin^2[(r-3)\pi/N]\} \ldots} \tag{5.52}$$

and with the term $\eta^2 + \sin^2(0)$ replaced by η, i.e.

$$A_2 = \frac{\eta}{\eta^2 + \sin^2(\pi/N)} \tag{5.53}$$

$$R_L = 1 \quad (N \text{ odd}) \tag{5.54}$$

$$R_L = \frac{(1+\varepsilon^2)^{1/2} - \varepsilon}{(1+\varepsilon^2)^{1/2} + \varepsilon} \quad (N \text{ even}) \tag{5.55}$$

and

$$\eta = \sinh\left[\frac{1}{N}\sinh^{-1}\left(\frac{1}{\varepsilon}\right)\right] \tag{5.56}$$

As an example we will design a degree 5, 20 dB return loss Chebyshev filter to operate in a 50 Ω system. Although the selectivity of the filter response close to the passband edge is controlled by the degree of the network, the ultimate stopband rejection is controlled by α. At $\omega = \pi/2$ the value of $|S_{12}(j\omega)|^2$ is given by

$$|S_{12}(j\pi/2)|^2 = \frac{1}{1+\varepsilon^2 T_N^2(1/\alpha)}$$

$$= \frac{1}{1+\varepsilon^2 \cosh^2[N\cosh^{-1}(1/\alpha)]} \quad (5.57)$$

Choosing the electrical length of the UEs to be 30° at band-edge we obtain

$$\alpha = \sin(\omega_0) = \sin(30°) = 0.5 \quad (5.58)$$

From (5.57) and (5.58) we find that the ultimate stopband insertion loss is 31.2 dB.

We calculate the element values as follows. First

$$\eta = \sinh\left[\frac{1}{N}\sinh^{-1}\left(\frac{1}{\varepsilon}\right)\right] \quad (5.59)$$

For $N = 5$ and $\varepsilon = 0.1005$ we obtain

$$\eta = 0.635 \quad (5.60)$$

From (5.52)

$$A_1 = \frac{1}{\eta} = A_5 = 1.5748 \quad (5.61)$$

$$A_2 = \frac{\eta}{\eta^2 + \sin^2(\pi/5)} = A_4 = 0.8481 \quad (5.62)$$

$$A_3 = \frac{\eta^2 + \sin^2(\pi/5)}{\eta[\eta^2 + \sin^2(2\pi/5)]} = 0.90158 \quad (5.63)$$

From (5.51)

$$g_1 = g_5 = 0.4947 \quad (5.64)$$

$$g_2 = g_4 = 2.3490 \quad (5.65)$$

$$g_3 = 0.3084 \quad (5.66)$$

and using (5.49) and (5.50), in a 50 Ω system we obtain

$$Z_1 = Z_5 = 24.74\,\Omega \quad (5.67)$$

$$Z_2 = Z_4 = 117.45\,\Omega \quad (5.68)$$

$$Z_3 = 15.43\,\Omega \quad (5.69)$$

The final circuit is shown in Figure 5.9. The simulated frequency response of the filter is shown in Figure 5.10. Note that as in the case of LC ladder lowpass

148 Theory and design of microwave filters

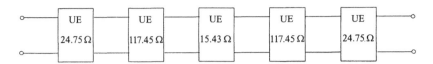

Figure 5.9   N = 5 Chebyshev stepped impedance lowpass filter

prototype networks the dual prototype network could have been used. In this case the first element would have been a high impedance rather than a low impedance line.

It should be noted that the distributed nature of the filter gives rise to repeating passbands. In this case the electrical length of the UEs was 30° at a band-edge frequency of 1 GHz. The second passband band-edge occurs at 5 GHz. If a broader stopband were required then a shorter electrical length design could have been used. This means a larger value of α and a more extreme variation in element values. As the electrical length at band-edge becomes shorter the design degenerates into a lumped element design as follows.

The transfer matrix of a UE is given (for sinusoidal excitation) by

$$[T] = \begin{bmatrix} \cos(\theta) & jZ_0 \sin(\theta) \\ j\sin(\theta)/Z_0 & \cos(\theta) \end{bmatrix} \tag{5.70}$$

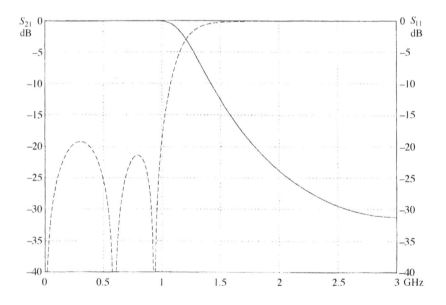

Figure 5.10   Simulated frequency response of the N = 5 Chebyshev stepped impedance lowpass filter

and for short electrical lengths

$$\theta \ll \frac{\pi}{2} \tag{5.71}$$

Hence

$$[T] \approx \begin{bmatrix} 1 & jZ_0\theta \\ j\theta/Z_0 & 1 \end{bmatrix} \tag{5.72}$$

and for a high impedance line

$$[T] \approx \begin{bmatrix} 1 & jZ_0\theta \\ 0 & 1 \end{bmatrix} \tag{5.73}$$

where

$$\theta = \frac{\omega \ell}{\nu} \tag{5.74}$$

Hence

$$[T] \approx \begin{bmatrix} 1 & jZ_0\ell\omega/\nu \\ 0 & 1 \end{bmatrix} \tag{5.75}$$

Thus a short section of high impedance line approximates to a series inductor. Similarly a short section of low impedance line approximates to a shunt capacitor. Thus for short electrical lengths the design could be accomplished by approximating to the LC ladder prototypes presented in Chapter 3 [5]. The method used in this chapter, however, is more accurate and enables longer electrical lengths to be used.

5.2.1 Physical realisation of stepped impedance lowpass filters

The stepped impedance lowpass filter is often used to 'clean up' harmonics in amplifier circuits, or as an IF filter in a mixer. Alternatively the filter may be used to improve the stopband performance of a microwave filter. This is

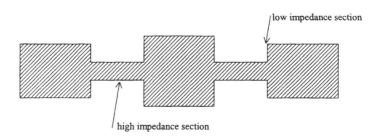

Figure 5.11 Typical microstrip circuit pattern for an $N = 5$ stepped impedance lowpass filter

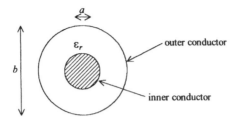

Figure 5.12   Coaxial transmission line

particularly important in the case of dielectric resonator filters which have poor spurious performance. In amplifier and mixer applications the insertion loss of the filter is not of prime importance and an MIC (usually microstrip) realisation is often used. A typical layout for a microstrip stepped impedance filter is shown in Figure 5.11. Design equations for calculating the widths and lengths of the microstrip lines are readily available in specialist texts on microstrip [6].

Microstrip circuits have relatively low unloaded Q factors and if a lower loss lowpass filter is required a coaxial realisation is more suitable. In this case the stepped impedance line may be realised as a coaxial line with a stepped inner conductor. Design equations for the coaxial line shown in Figure 5.12 are very simple.

$$Z_0 = \frac{60}{\sqrt{\varepsilon_r}} \log_e \left(\frac{b}{a}\right) \tag{5.76}$$

or

$$\frac{b}{a} = \exp(\sqrt{\varepsilon_r} Z_0/60) \tag{5.77}$$

Choice of dimension b is determined mainly by the maximum loss allowable in the lowpass filter. The Q factor of a coaxial line, normalised to ground plane spacing in centimetres and frequency in gigahertz, is shown as a function of characteristic impedance in Figure 5.13.

To choose b we first synthesise the element values of the filter. Second, we analyse the equivalent circuit to determine the minimum Q factor which enables the maximum insertion loss specification to be met. The minimum ground plane spacing can then be determined from Figure 5.13. It is important, however, not to choose too large a value for b in case this gives rise to unwanted waveguide mode resonances in the coaxial line.

As an example, say we require a minimum Q_u of 700 at 1 GHz. From (5.77) assuming $\varepsilon_r = 1$, for a 117.5 Ω impedance UE

$$\frac{b}{a} = \exp\left(\frac{117.5}{60}\right) = 7.087 \tag{5.78}$$

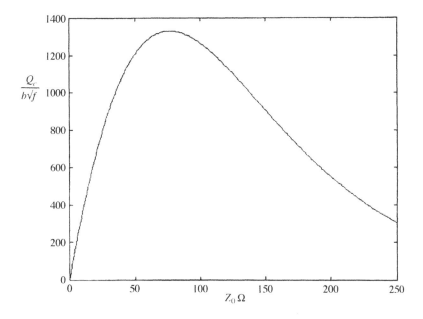

Figure 5.13   Q factor of coaxial line

Now from Figure 5.13, for a 117 Ω line

$$\frac{Q}{b\sqrt{f}} = 1160 \tag{5.79}$$

Hence for $Q = 700$ and $f = 1\,\text{GHz}$

$$b = 0.6\,\text{cm} \tag{5.80}$$

5.3 Broadband TEM filters with generalised Chebyshev characteristics

In certain applications wideband filters with extreme selectivity are required. For example in radar warning receivers broadband multiplexers require octave plus bandwidths with at least 60 dB stopband insertion loss within 10 per cent of the band-edge frequency. Cellular radio base stations often use low loss dielectric resonator filters, and the spurious modes of these devices often occur at frequencies only 25 per cent above the passband. Thus 'clean-up' lowpass filters must have high selectivity combined with low loss and small size. Such severe specifications are not easily achievable using all-pole transfer functions and more selective generalised Chebyshev characteristics are often required.

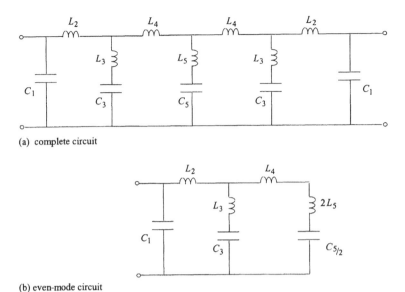

Figure 5.14 Generalised Chebyshev lowpass prototype, degree 9

As an example we will design a TEM lowpass filter from a ninth-degree generalised Chebyshev prototype. The particular prototype will have 20 dB return loss and three transmission zeros at infinity. The remaining transmission zeros are all at the same frequency $\omega_0 = 1.32599$. The choice of ω_0 was determined after analysis of the transfer function given in Chapter 3 in order to achieve a minimum stopband insertion loss of 60 dB. The lumped element prototype network is shown in Figure 5.14.

The location of the poles of the transfer function are

$$p_{1,9} = -0.03033 \pm j1.02275 \tag{5.81}$$

$$p_{2,8} = -0.10604 \pm j0.96344 \tag{5.82}$$

$$p_{3,7} = -0.22511 \pm j0.80937 \tag{5.83}$$

$$p_5 = -0.455417 \tag{5.84}$$

Now we form a polynomial $P(p)$ from alternating poles, i.e.

$$\begin{aligned} P(p) &= (p - p_1)(p - p_9)(p - p_3)(p - p_7)(p - p_5) \\ &= 0.45541 + 1.3169p + 1.7930p^2 + 2.72394p^3 + 1.30779p^4 \\ &\quad + 1.35330p^5 \end{aligned} \tag{5.85}$$

Forming $Y_e(p)$ from the even and odd parts of (5.85),

$$Y_e(p) = \frac{1.35330p^5 + 2.72394p^3 + 1.31690p}{1.30779p^4 + 1.7930p^2 + 0.45541} \tag{5.86}$$

We extract a capacitor by removing the pole at infinity

$$C_1 = \left.\frac{Y_e(p)}{p}\right|_{p=\infty} = 1.03487 \tag{5.87}$$

The remaining admittance is given by

$$Y_1(p) = Y_e(p) - C_1 p$$

$$= \frac{0.84555p + 0.86842p^3}{0.45542 + 1.79300p^2 + 1.30779p^4} \tag{5.88}$$

The series inductance L_2 must then be extracted so that the remaining impedance has a zero at ω_0. Thus

$$Z_1(p) = \frac{1}{Y_1(p)} \tag{5.89}$$

and

$$L_2 = \frac{Z_1(j\omega_0)}{j\omega_0} = 1.12352 \tag{5.90}$$

Also

$$Z_2(p) = Z_1(p) - L_2 p$$

$$= \frac{0.45542 + 0.84299p^2 + 0.33210p^4}{0.84555p + 0.86842p^3} \tag{5.91}$$

and

$$Y_2(p) = \frac{1}{Z_2(p)}$$

$$= \frac{0.84555p + 0.86842p^3}{0.45542 + 0.84299p^2 + 0.33210p^4} \tag{5.92}$$

$Y_2(p)$ has a pole at $p = j\omega_0$. Thus

$$Y_2(p) = \frac{Ap}{p^2 + \omega_0^2} + Y_3(p)$$

$$= \frac{Ap}{p^2 + 1.75825} + \frac{Bp}{0.3310p^2 + 0.25902} \tag{5.93}$$

Hence

$$0.3321 A + B = 0.86842 \tag{5.94}$$

154 Theory and design of microwave filters

and
$$0.25902A + 1.75825B = 0.84555 \tag{5.95}$$

Solving (5.94) and (5.95) simultaneously yields
$$A = 2.0969 \qquad B = 0.17204 \tag{5.96}$$

The first term in (5.93) represents the admittance of the resonant circuit containing L_3 and C_3, i.e.
$$\frac{2.0969p}{p^2 1.75825} = \frac{p/L_3}{p^2 + 1/L_3 C_3} \tag{5.97}$$

Thus
$$L_3 = 0.47688 \tag{5.98}$$

and
$$C_3 = 1.19263 \tag{5.99}$$

After extracting this resonant circuit the remaining admittance consists of the series inductor L_4 and the resonant circuit containing $2L_5$ and $C_5/2$ shown in Figure 5.14(b). These may be obtained by repeating the procedure and the final element values for the circuit are

$$\begin{aligned} C_1 &= 1.03487 \\ L_2 &= 1.12352 \\ L_3 &= 0.47688 \\ C_3 &= 1.19263 \\ L_4 &= 1.07413 \\ L_5 &= 0.42818 \\ C_5 &= 1.32834 \end{aligned} \tag{5.100}$$

The prototype network can be converted to a TEM microwave network by applying the Richards transformation [7]:
$$p \to \alpha \tanh(ap) \tag{5.101}$$

Applying this to the series resonators of impedance $Z(p)$ where
$$Z(p) = Lp + \frac{1}{Cp} \tag{5.102}$$

then
$$Z(p) \Rightarrow \alpha L \tanh(ap) + \frac{1}{\alpha C \tanh(ap)}$$
$$= \frac{\alpha^2 LC \tanh^2(ap) + 1}{\alpha C \tanh(ap)} \tag{5.103}$$

Now
$$\tanh(2x) = \frac{2\tanh(x)}{1+\tanh^2(x)} \qquad (5.104)$$
and if we let $\alpha = \omega_0$ then
$$\alpha^2 LC = 1 \qquad (5.105)$$
Thus
$$Z(p) = \frac{2}{\alpha C \tanh(2ap)} \qquad (5.106)$$
and
$$Y(p) = \frac{\alpha C}{2} \tanh(2ap) \qquad (5.107)$$

Equation (5.107) represents the admittance of an open circuited stub of characteristic admittance $\alpha C/2$. The length of the stub is one quarter wavelength at ω_0 (Figure 5.15).

Now applying the Richards transformation to the series inductors we obtain
$$Z(p) = Lp \Rightarrow \alpha L \tanh(ap) \qquad (5.108)$$
The series inductors become series short circuited stubs of impedance αL. Similarly the shunt capacitors become shunt open circuited stubs of admittance αC.

The transformation converts the lumped lowpass prototype circuit into a distributed circuit of band-edge frequency ω_c. Thus
$$1 \Rightarrow \alpha \tan(a\omega_c) \qquad (5.109)$$
and
$$a = \frac{1}{\omega_c} \tan^{-1}(1/\alpha) \qquad (5.110)$$

Equation (5.110) determines the length of the stubs in the distributed circuit. Note that the shunt stubs associated with the resonators are twice the length of the other stubs. The complete distributed circuit is shown in Figure 5.16.

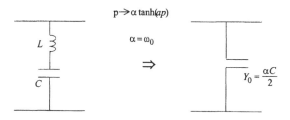

Figure 5.15 Richards transformation of a resonator

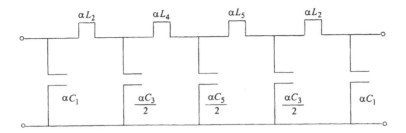

Figure 5.16   Distributed lowpass generalised Chebyshev filter, $N = 9$

At this point one would think that the design of the filter was finished. However, the realisation of series short circuited stubs is impractical. Instead we approximate the series stubs by lengths of high impedance transmission line. From (5.75) we know that an electrically short transmission line is equivalent to an inductor of value

$$L = \frac{Z_0 \ell}{\nu} \qquad (5.111)$$

or more accurately

$$\omega L = Z_0 \sin\left(\frac{\omega \ell}{\nu}\right) \qquad (5.112)$$

where ℓ is the length of the line. We can equate the impedances of the series short circuited stubs to be equal to the impedance of the high impedance line at the band-edge frequency ω_c. Thus

$$Z_0 \sin\left(\frac{\omega_c \ell_r}{\nu}\right) = \alpha L_r \tan(a\omega_c) \qquad (r = 2, 4, \ldots) \qquad (5.113)$$

Thus choosing a suitably high value for Z_0 we can calculate the length of the line from

$$\ell_r = \frac{\nu}{\omega_c} \sin^{-1}\left[\frac{\alpha L_r \tan(a\omega_c)}{Z_0}\right] \qquad (5.114)$$

As a design example we will design a distributed filter from the ninth-degree prototype already discussed. The band-edge frequency is to be 4 GHz and the system impedance level is 50 Ω.

First compute

$$\alpha = \omega_0 = 1.32599 \qquad (5.115)$$

and

$$a = \frac{1}{\omega_c} \tan^{-1}\left(\frac{1}{\alpha}\right) = 2.57096 \times 10^{-11} \qquad (5.116)$$

The admittance of the first (and last) shunt open circuited stub is

$$Y = \alpha C_1 = 1.37223 \tag{5.117}$$

and its impedance in a 50 Ω system is 36.43 Ω.
The length of the stub is given from

$$a = \frac{\ell}{\nu} \qquad \ell = a\nu \tag{5.118}$$

If the relative permittivity of the medium of propagation is unity then $\nu = 3 \times 10^8$ m/s and

$$\ell = 7.713 \text{ mm} \tag{5.119}$$

The lengths of the resonator stubs are 2ℓ, i.e. 15.425 mm. Their admittances are

$$Y_r = \frac{\alpha C_r}{2} \qquad (r = 3, 5, \ldots) \tag{5.120}$$

Thus

$$Y_3 = 0.7907 \tag{5.121}$$

and

$$Y_5 = 0.8806 \tag{5.122}$$

Their impedances in a 50 Ω system are 63.23 Ω and 56.78 Ω.

A reasonable choice for the high impedance series lines is $Z_0 = 120$ Ω. Thus in a 1 Ω system $Z_0 = 2.4$ Ω. The lengths of the series lines are given by

$$\ell_r = \frac{3 \times 10^8}{\omega_c} \sin^{-1}\left[\frac{\alpha L_r \tan(a\omega_c)}{Z_0}\right] \tag{5.123}$$

and from (5.116)

$$\tan(a\omega_c) = \frac{1}{\alpha} \tag{5.124}$$

Hence

$$\ell_r = \frac{3 \times 10^8}{\omega_c} \sin^{-1}\left(\frac{L_r}{Z_0}\right) \tag{5.125}$$

Thus

$$\ell_2 = 5.815 \text{ mm} \tag{5.126}$$

and

$$\ell_4 = 5.539 \text{ mm} \tag{5.127}$$

The longest line, ℓ_2, is quarter wave long at 12.89 GHz, which is electrically short at the band-edge frequency. The complete circuit is shown in Figure 5.17.

158 Theory and design of microwave filters

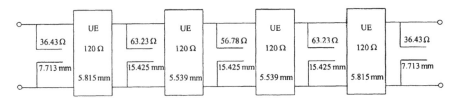

Figure 5.17 Generalised Chebyshev distributed lowpass filter

The simulated response of the circuit is shown in Figure 5.18. Here we see an almost exact equiripple passband response. This could easily be optimised to be perfectly equiripple. A broadband plot of the filter is shown in Figure 5.19.

The stopband of the filter shows spurious responses between the first and second passbands. These are caused because the series short circuited stubs have been approximated by high impedance UEs. Unlike the stubs these do not produce transmission zeros at their quarter wave frequencies. The stopband performance could be improved by distributing the transmission zero frequencies throughout the stopband of the filter.

One of the best methods for physically realising the filter is to use suspended substrate stripline. This is a microwave integrated circuit structure consisting of a thin printed circuit suspended between parallel ground planes (Figure 5.20). In

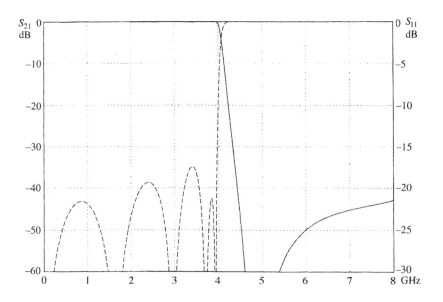

Figure 5.18 Simulated response of a generalised Chebyshev distributed lowpass filter

TEM transmission line filters 159

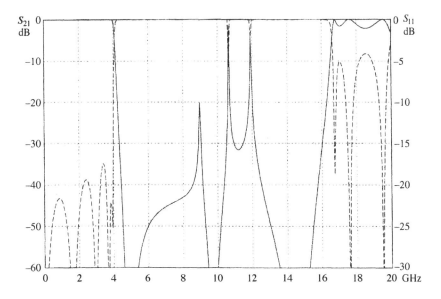

Figure 5.19 Simulated broadband response of a generalised Chebyshev distributed lowpass filter

the suspended stripline configuration the majority of the fields are in the air cavity and the substrate has little effect on the Q factor or the effective permittivity of the elements. It is relatively straightforward to calculate the dimensions of the individual circuit elements within the filter.

Consider a single transmission line of width w and thickness t suspended between ground planes with spacing b, as shown in Figure 5.21. Note that the thin dielectric substrate has been ignored because normally these are made from Teflon with relatively low dielectric constant ($\varepsilon_r < 3$) and are very thin, typically less than 0.25 mm.

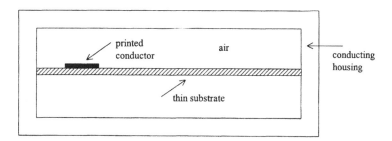

Figure 5.20 Suspended substrate stripline

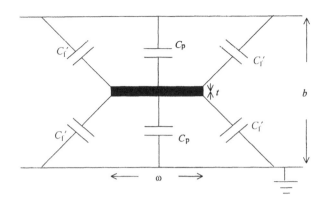

Figure 5.21 Single suspended transmission line

The impedance of a TEM transmission line is related to its static capacitance to ground per unit length by

$$Z_0\sqrt{\varepsilon_r} = \frac{377}{C/\varepsilon} \tag{5.128}$$

where ε_r is the dielectric constant of the medium and C/ε is the normalised static capacitance per unit length of the transmission line. Now from Figure 5.21 we obtain

$$\frac{C}{\varepsilon} = 2C_p + \frac{4C_f'}{\varepsilon} \tag{5.129}$$

and

$$C_p = \frac{w}{(b-t)/2} \tag{5.130}$$

For $b \ll t$

$$\frac{C}{\varepsilon} = \frac{4w}{b} + \frac{4C_f'}{\varepsilon} \tag{5.131}$$

C_f' is the fringing capacitance to ground which is plotted in Figure 5.22 [8].
For a printed circuit we can assume t is zero and hence from Figure 5.22

$$\frac{C_f'}{\varepsilon} = 0.46 \tag{5.132}$$

Thus from (5.131)

$$\frac{C}{\varepsilon} = \frac{4w}{b} + 1.84 \tag{5.133}$$

TEM transmission line filters 161

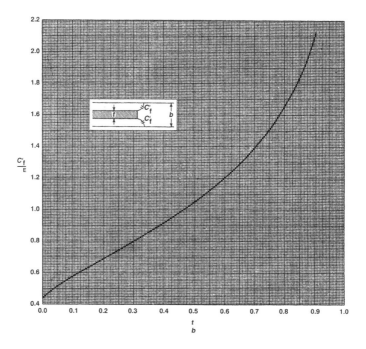

Figure 5.22 *Fringing capacitance from a rectangular bar*
(*Source*: Getsinger, W.J.: 'Coupled rectangular bars between parallel plates', *IEEE Transactions on Microwave Theory and Techniques*, 1962, **10** (1), pp. 54–72; © 1962 IEEE)

and from (5.128) and (5.133), assuming ε_r is unity, we obtain

$$w = \frac{b}{4}\left(\frac{377}{Z_0} - 1.84\right) \tag{5.134}$$

Thus for a 120 Ω line with $b = 2$ mm

$$w = \frac{2}{4}\left(\frac{377}{120} - 1.84\right) = 0.65 \text{ mm} \tag{5.135}$$

for a 36.43 Ω stub $w = 4.254$ mm
for a 63.23 Ω stub $w = 2.061$ mm   (5.136)
for a 56.78 Ω stub $w = 2.400$ mm

In order to compute the exact dimensions of the filter account must be taken of the reference plane locations associated with the interconnected transmission lines. Data on these are available in Reference 9.

The layout of the filter is shown in Figure 5.23.

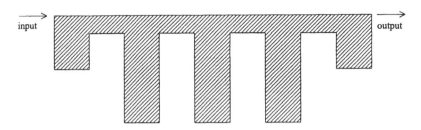

Figure 5.23 Circuit layout of a suspended substrate lowpass filter

5.3.1 Generalised Chebyshev highpass filters

Highpass generalised Chebyshev filters can also be constructed. In this case the Richards highpass transformation should be used and the transformed circuit has series open circuited stubs. These may be approximated by inhomogeneous coupled lines realised in suspended substrate. A typical seventh-degree lowpass prototype with a single transmission zero at infinity is shown in Figure 5.24.

After applying the Richards transformation

$$\omega \to \frac{-1}{\alpha \tan(a\omega)} \tag{5.137}$$

and for the series inductors

$$\omega L_r \Rightarrow \frac{-L_r}{\alpha \tan(a\omega)} \qquad (r = 1, 3, \ldots) \tag{5.138}$$

That is, the inductors L become open circuited stubs with admittances

$$Y_r = \frac{\alpha}{L_r} \tag{5.139}$$

The resonators in the prototype have an impedance

$$Z(j\omega) = j\omega L_r - \frac{j}{\omega C_r} \qquad (r = 2, 4, \ldots) \tag{5.140}$$

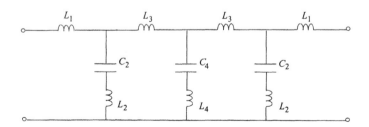

Figure 5.24 Seventh-degree generalised Chebyshev lowpass prototype network

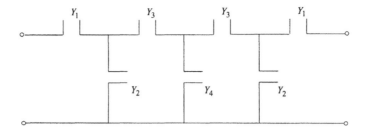

Figure 5.25 Distributed highpass generalised Chebyshev filter

and applying the Richards highpass transformation

$$Z(j\omega) \Rightarrow \frac{-jL_r}{\alpha \tan(a\omega)} + \frac{j\alpha \tan(a\omega)}{C_r}$$

$$= \frac{j[\alpha^2 \tan^2(a\omega) - L_r C_r]}{\alpha C_r \tan(a\omega)} \qquad (5.141)$$

If

$$\alpha^2 = L_r C_r = \frac{1}{\omega_0^2} \qquad (5.142)$$

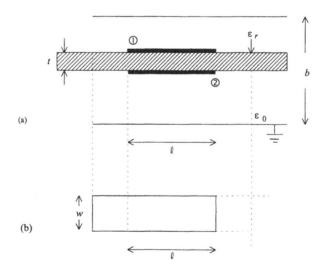

Figure 5.26 Inhomogeneous parallel coupled transmission lines: (a) cross-section; (b) top view

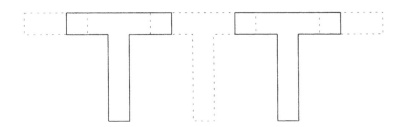

Figure 5.27 Circuit layout for a seventh-degree generalised Chebyshev highpass filter

then

$$Z(j\omega) = \frac{j\alpha[\tan^2(a\omega) - 1]}{C_r \tan(a\omega)}$$

$$= \frac{-j2\alpha}{C_r \tan(2a\omega)} \tag{5.143}$$

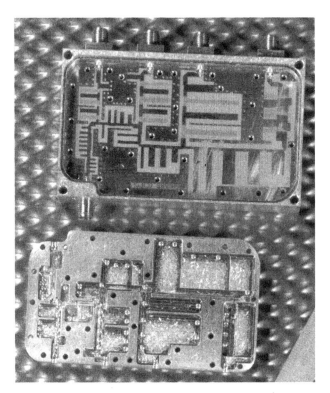

Figure 5.28 A typical suspended substrate device (photograph courtesy of Filtronic plc)

which is the impedance of an open circuited stub of characteristic admittance

$$Y_r = \frac{C_r}{2\alpha} \qquad (r = 2, 4, \ldots) \tag{5.144}$$

The distributed highpass filter is shown in Figure 5.25.

The series open circuited stubs in Figure 5.25 cannot be realised directly. However, they can be approximated by coupled parallel lines. The strength of the coupling is such that they are best realised by an overlap coupling through a thin dielectric substrate, as shown in Figure 5.26 [10–12].

A typical circuit layout for a seventh-degree suspended substrate highpass filter is shown in Figure 5.27, and a photograph of a typical suspended substrate device is shown in Figure 5.28.

5.4 Parallel coupled transmission lines

In addition to interconnections of stubs and transmission lines it is also possible to construct useful microwave filters with coupled transmission lines. Consider the network shown in Figure 5.29, consisting of an array of N parallel coupled commensurate lines and an associated ground plane.

It is assumed that the permittivity of the medium in which the lines are supported is homogeneous. By assuming that only TEM waves are supported in the structure we can assign input and output voltages and currents to each line. The standard incremental approach for analysing a single transmission line may then be applied to the complete network. It may be shown that the N wire

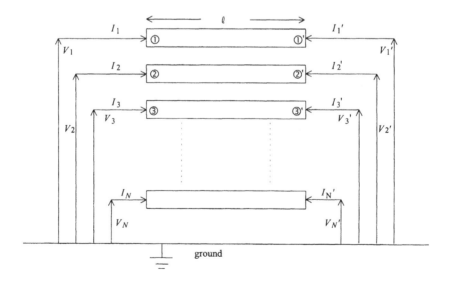

Figure 5.29 Parallel coupled N-wire line

line can be described by the admittance matrix equation [13]

$$\begin{bmatrix} I_1 \\ I_2 \\ I_3 \\ \vdots \\ I_N \\ I'_1 \\ I'_2 \\ I'_3 \\ \vdots \\ I'_N \end{bmatrix} = \frac{1}{t} \begin{bmatrix} [\eta] & -(1-t^2)^{1/2}[\eta] \\ -(1-t^2)^{1/2}[\eta] & [\eta] \end{bmatrix} \begin{bmatrix} V_1 \\ V_2 \\ V_3 \\ \vdots \\ V_N \\ V'_1 \\ V'_2 \\ V'_3 \\ \vdots \\ V'_N \end{bmatrix} \quad (5.145)$$

where

$$t = \tanh(ap) \quad (5.146)$$

and

$$a = \frac{\ell}{v} \quad (5.147)$$

$[\eta]$ is an $N \times N$ matrix called the characteristic admittance matrix of the line. This matrix is related to the static capacitance matrix of the N-wire line in a similar way to the relationship between capacitance per unit length and characteristic admittance of a single line. Thus

$$[\eta] = \begin{bmatrix} Y_{11} & -Y_{12} & -Y_{13} & \cdots \\ -Y_{12} & Y_{22} & -Y_{23} & \cdots \\ -Y_{13} & -Y_{23} & Y_{33} & \\ & & & \ddots \\ & & & & Y_{N-1,N-1} & -Y_{N-1,N} \\ & & & & -Y_{N-1,N} & Y_{NN} \end{bmatrix} \quad (5.148)$$

and

$$\frac{[C]}{\varepsilon} = 7.534[\eta] \quad (5.149)$$

(in a 1 Ω system). $[C]$ is the static capacitance matrix of the line given by

$$[C] = \begin{bmatrix} C_{11} & -C_{12} & -C_{13} & \cdots \\ -C_{12} & C_{22} & -C_{23} & \cdots \\ -C_{13} & -C_{23} & C_{33} & \\ \vdots & & & \ddots \\ & & & & C_{N-1,N-1} & -C_{N-1,N} \\ & & & & -C_{N-1,N} & C_{NN} \end{bmatrix} \quad (5.150)$$

where

$$C_{11} = C_1 + C_{12} + C_{23} + \ldots \tag{5.151}$$

$$C_{rr} = C_r + C_{1r} + C_{2r} + \ldots \tag{5.152}$$

The capacitances $C_1, C_2, \ldots, C_r$ are the capacitances per unit length to ground for each of the N lines. C_{12}, C_{23} etc. are the coupling capacitances per unit length between pairs of lines (Figure 5.30).

Equation (5.145) gives a complete description of the coupled-line structure in terms of its static capacitance matrix. In general any line may be coupled to each of its neighbours and the boundary conditions on the ports are arbitrary. Thus the most general matrix is rather complex. However, in reality we are interested in constructing filter networks with a single input port and a single output port. Also the number of couplings may be considerably restricted. In the next sections we will examine some specific cases yielding useful filtering devices.

5.5 The interdigital filter

The interdigital filter [14] is a device which may be constructed entirely from an N-wire coupled line. In this case the couplings are restricted to be between adjacent lines, i.e. non-adjacent line couplings are assumed to be zero. Thus

$$C_{r,r+1} \neq 0 \tag{5.153}$$

$$C_{r,r+2}, C_{r,r+3}, \ldots = 0 \tag{5.154}$$

Furthermore, the input to the network is on line 1 and the output on line N. Thus for an even number of lines the input is at port 1 and the output at port N'. For an odd degree network the output would be at port N.

Finally a short circuit is applied to alternate ends of each line with the other ends left floating. The N-wire line is shown in Figure 5.31. The characteristic

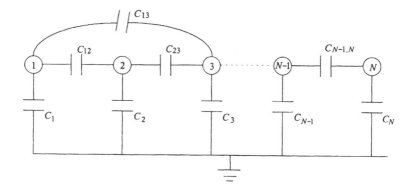

Figure 5.30   Static capacitances of an N-wire line

168 Theory and design of microwave filters

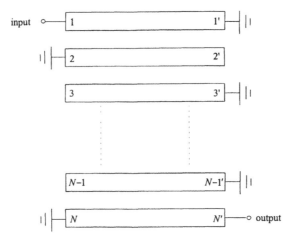

Figure 5.31   The interdigital filter (N even)

admittance matrix of the interdigital filter is thus given by

$$[\eta] = \begin{bmatrix} Y_{11} & -Y_{12} & 0 & \cdots & & & \\ -Y_{12} & Y_{22} & -Y_{23} & 0 & & & \\ 0 & -Y_{23} & Y_{33} & -Y_{34} & & & \\ 0 & 0 & -Y_{34} & Y_{44} & & & \\ \vdots & & & & \ddots & & 0 \\ & & & & & Y_{N-1,N-1} & -Y_{N-1,N} \\ & & & & 0 & -Y_{N-1,N} & Y_{N,N} \end{bmatrix} \quad (5.155)$$

Consider the three-wire interdigital line shown in Figure 5.32. Here

$$V'_1 = V_2 = V'_3 = 0 \quad (5.156)$$

Figure 5.32   Three-wire interdigital line

Hence from (5.145), (5.155) and (5.156)

$$
\begin{bmatrix} I_1 \\ I_2 \\ I_3 \\ I_1' \\ I_2' \\ I_3' \end{bmatrix} = \frac{1}{t} \begin{bmatrix} Y_{11} & -Y_{12} & 0 & -(1-t^2)^{1/2}Y_{11} & (1-t^2)^{1/2}Y_{12} & 0 \\ -Y_{12} & Y_{22} & -Y_{23} & (1-t^2)^{1/2}Y_{12} & -(1-t^2)^{1/2}Y_{22} & (1-t^2)^{1/2}Y_{23} \\ 0 & -Y_{23} & Y_{33} & 0 & (1-t^2)^{1/2}Y_{23} & -(1-t^2)^{1/2}Y_{33} \\ -(1-t^2)^{1/2}Y_{11} & (1-t^2)^{1/2}Y_{12} & 0 & Y_{11} & -Y_{12} & 0 \\ (1-t^2)^{1/2}Y_{12} & -(1-t^2)^{1/2}Y_{22} & (1-t^2)^{1/2}Y_{23} & -Y_{12} & Y_{22} & -Y_{23} \\ 0 & (1-t^2)^{1/2}Y_{23} & -(1-t^2)^{1/2}Y_{23} & 0 & -Y_{23} & Y_{33} \end{bmatrix}
$$
$$
\times \begin{bmatrix} V_1 \\ 0 \\ V_3 \\ 0 \\ V_2' \\ 0 \end{bmatrix} \tag{5.157}
$$

and rearranging (5.157) we obtain

$$
\begin{bmatrix} I_1 \\ I_2' \\ I_3 \end{bmatrix} = \frac{1}{t} \begin{bmatrix} Y_{11} & (1-t^2)^{1/2}Y_{12} & 0 \\ (1-t^2)^{1/2}Y_{12} & Y_{22} & (1-t^2)^{1/2}Y_{23} \\ 0 & (1-t^2)^{1/2}Y_{23} & Y_{33} \end{bmatrix} \begin{bmatrix} V_1 \\ V_2' \\ V_3 \end{bmatrix} \tag{5.158}
$$

Now the transfer matrix of a single-wire (two-port) transmission line of admittance Y is

$$
[T] = \frac{1}{(1-t^2)^{1/2}} \begin{bmatrix} 1 & t/Y \\ Yt & 1 \end{bmatrix} \tag{5.159}
$$

Conversion to an admittance matrix is given by

$$
[Y] = \begin{bmatrix} D/B & -\Delta/B \\ -1/B & A/B \end{bmatrix} \tag{5.160}
$$

where Δ is the determinant of the transfer matrix. Hence the Y matrix of a single line is given by

$$
[Y] = \frac{1}{t} \begin{bmatrix} Y & -(1-t^2)^{1/2}Y \\ -(1-t^2)^{1/2}Y & Y \end{bmatrix} \tag{5.161}
$$

Now from (5.158) the admittance matrix between nodes 1 and 2' is given by

$$
[Y] = \frac{1}{t} \begin{bmatrix} Y_1 + Y_{12} & (1-t^2)^{1/2}Y_{12} \\ (1-t^2)^{1/2}Y_{12} & Y_2 + Y_{12} \end{bmatrix} \tag{5.162}
$$

The equivalent circuit of (5.162) is shown in Figure 5.33.

Thus from (5.158) the equivalent circuit of the three-wire interdigital line is as shown in Figure 5.34. The ideal $1:-1$ transformers can be transformed to the output of the network and have no effect on the amplitude response of the filter. Hence the equivalent circuit of the general Nth-degree interdigital filter is as shown in Figure 5.35.

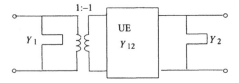

Figure 5.33 Equivalent circuit of (5.162)

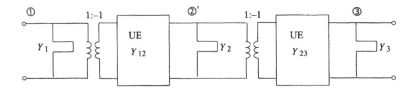

Figure 5.34 Equivalent circuit of a three-wire interdigital line

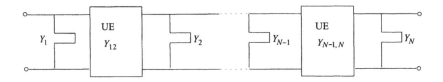

Figure 5.35 Equivalent circuit of an N-wire interdigital line

Having derived the equivalent circuit we can now develop a systematic design procedure for a bandpass filter. First we can make the UEs quarter wave long at ω_0 and they become inverters. The equivalent circuit then consists of resonant stubs separated by inverters. Richards transformation of the lowpass prototype would then establish the design procedure. However, this would only be accurate for very narrow bandwidths. A more accurate procedure for broader bandwidths can be obtained if we consider the UEs in more detail.

From (5.159) for sinusoidal excitation the transfer matrix of a UE is given by

$$[T] = \begin{bmatrix} \cos(\theta) & \dfrac{j\sin(\theta)}{Y} \\ jY\sin(\theta) & \cos(\theta) \end{bmatrix} \tag{5.163}$$

and this can be decomposed into

$$[T] = \begin{bmatrix} 1 & 0 \\ \dfrac{-jY}{\tan(\theta)} & 1 \end{bmatrix} \begin{bmatrix} 0 & \dfrac{j\sin(\theta)}{Y} \\ \dfrac{jY}{\sin(\theta)} & 0 \end{bmatrix} \begin{bmatrix} 1 & 0 \\ \dfrac{-jY}{\tan(\theta)} & 1 \end{bmatrix} \tag{5.164}$$

TEM transmission line filters 171

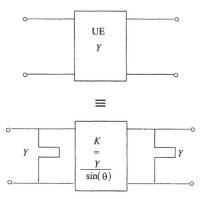

Figure 5.36 Equivalent circuit of a unit element

Equation (5.164) is the matrix of two shunt short circuited stubs of admittance Y separated by a frequency-dependent admittance inverter, as shown in Figure 5.36. Thus substituting for the equivalent circuit of the UEs into Figure 5.35 we obtain a new equivalent circuit for the interdigital filter (Figure 5.37).

The inverter coupling the rth to the $(r+1)$th stub is then given by

$$K_{r,r+1} = \frac{Y_{r,r+1}}{\sin(\theta)} \tag{5.165}$$

The frequency dependence of this inverter is relatively small as $\sin(60°) = 0.8666$ and the variation in admittance is small even over an octave band.

The rth resonator is now a short circuited stub of admittance

$$Y_{rr} = Y_r + Y_{r-1,r} + Y_{r,r+1} \tag{5.166}$$

Now we apply the Richards highpass transformation to the lowpass prototype shown in Figure 5.38. Then

$$\omega \to \frac{-1}{\alpha \tan(a\omega)} \tag{5.167}$$

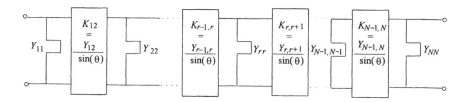

Figure 5.37 New equivalent circuit for an interdigital filter

172 Theory and design of microwave filters

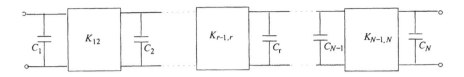

Figure 5.38 Lowpass prototype for an interdigital filter

and

$$\omega C_r \rightarrow \frac{-C_r}{\alpha \tan(a\omega)} \quad (5.168)$$

Hence

$$Y_{rr} = \frac{C_r}{\alpha} \quad (5.169)$$

where

$$\alpha = \frac{1}{\tan(a\omega_1)} = \frac{1}{\tan(\theta_1)} \quad (5.170)$$

Now from (5.165) assuming $\theta = 90°$ at ω_0 we obtain

$$Y_{r,r+1} = K_{r,r+1} \quad (5.171)$$

and from (5.166)

$$Y_r = Y_{rr} - Y_{r-1,r} - Y_{r,r+1} \quad (5.172)$$

Also, from (5.168), (5.169) and (5.172)

$$Y_r = \frac{C_r}{\alpha} - K_{r-1,r} - K_{r,r+1} \quad (5.173)$$

Hence the design equations for the Nth-degree interdigital filter are given by

$$Y_{r,r+1} = K_{r,r+1} \quad (r = 1, \ldots, N-1) \quad (5.174)$$

$$Y_1 = \frac{C_1}{\alpha} - K_{12} \quad (r = 1) \quad (5.175)$$

$$Y_r = \frac{C_r}{\alpha} - K_{r-1,r} - K_{r,r+1} \quad (r = 2, \ldots, N-1) \quad (5.176)$$

$$Y_N = \frac{C_N}{\alpha} - K_{N-1,N} \quad (r = N) \quad (5.177)$$

$$\alpha = \frac{1}{\tan(\theta_1)} \quad (5.178)$$

$$\theta_1 = \frac{90°\omega_1}{\omega_0} \quad (5.179)$$

TEM transmission line filters 173

where ω_1 is the lower band-edge frequency. C_r and $K_{r,r+1}$ are the element values for all-pole lowpass prototype networks given in Chapter 3 and shown in Figure 5.38.

5.5.1 Design example

As an example we will design a filter with a centre frequency of 2 GHz and a passband bandwidth of 1 GHz. We will use a degree 4 Chebychev prototype with 20 dB passband return loss. Thus

$$\theta_1 = \frac{90° \times 1.5}{2} = 67.5° \tag{5.180}$$

and

$$\alpha = \frac{1}{\tan(\theta_1)} = 0.4142 \tag{5.181}$$

The lowpass prototype elements values are

$$\begin{aligned} C_1 &= C_4 = 0.9314 \\ C_2 &= C_3 = 2.2487 \end{aligned} \tag{5.182}$$

$$\begin{aligned} K_{12} &= K_{34} = 1.3193 \\ K_{23} &= 1.5751 \end{aligned} \tag{5.183}$$

and from (5.174)

$$Y_{12} = Y_{34} = 1.3193 \tag{5.184}$$

$$Y_{23} = 1.5751 \tag{5.185}$$

From (5.175)

$$Y_1 = \tfrac{0.9314}{0.4142} - 1.3193 = 0.9293 \tag{5.186}$$

and since $C_1 = C_4$

$$Y_4 = Y_1 \tag{5.187}$$

From (5.176)

$$Y_2 = \tfrac{2.2487}{0.4142} - 1.3193 - 1.5751 = 2.5346 \tag{5.188}$$

$$Y_3 = Y_2 \tag{5.189}$$

Hence in a 50 Ω system

$$\begin{aligned} Z_1 &= Z_4 = 53.804\,\Omega \\ Z_2 &= Z_3 = 19.727\,\Omega \\ Z_{12} &= Z_{34} = 37.898\,\Omega \\ Z_{23} &= 31.744\,\Omega \end{aligned} \tag{5.190}$$

174 Theory and design of microwave filters

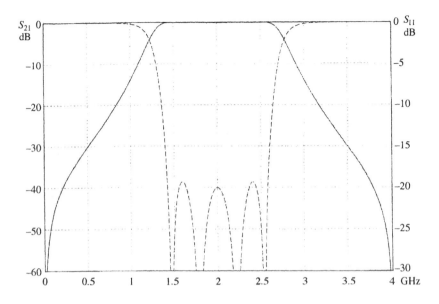

Figure 5.39 Simulated response of an interdigital filter

Analysis of the equivalent circuit gives the frequency response in Figure 5.39, showing an almost exact equiripple response. Simulation of the broadband response is shown in Figure 5.40. Here we see the passband repeating at three times the centre frequency.

5.5.2 Narrowband interdigital filters

From (5.176) we see that, as the bandwidth of the interdigital filter becomes small, α becomes small and the admittances become very large and unrealisable. We can solve this problem in a similar way to that described for narrowband bandpass lumped element filters in Chapter 4.

First we introduce a UE with admittance unity at the input and output of the filter. This does not change the amplitude response of the device. In addition we scale rows and columns of the admittance matrix so that

$$Y_{rr} \rightarrow n_r^2 Y_{rr} \qquad (5.191)$$

Hence the admittance matrix becomes

$$[Y] = \frac{1}{t} \begin{bmatrix} 1 & -n_1(1-t^2)^{1/2} & 0 & \\ -n_1(1-t^2)^{1/2} & n_1^2(1+Y_{11}) & -n_1 n_2(1-t^2)^{1/2} Y_{12} & \\ 0 & -n_1 n_2(1-t^2)^{1/2} Y_{12} & n_2^2 Y_{22} & \\ \vdots & \vdots & \vdots & \end{bmatrix} \qquad (5.192)$$

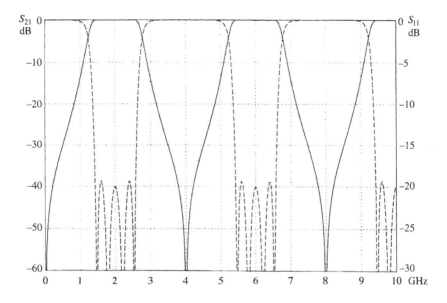

Figure 5.40   Simulated broadband response of the interdigital filter

Note that this is equivalent to introducing an extra interdigital coupled-line section at the input and output of the network.

Now from (5.169)

$$Y_{rr} = \frac{C_r}{\alpha} \tag{5.193}$$

and from (5.171)

$$Y_{r,r+1} = K_{r,r+1} \tag{5.194}$$

Hence

$$[Y] = \frac{1}{t} \begin{bmatrix} 1 & n_1(1-t^2)^{1/2} & 0 & \cdots \\ n_1(1-t^2)^{1/2} & n_1^2(1+C_1/\alpha) & -n_1 n_2(1-t^2)^{1/2} K_{12} & \\ 0 & -n_1 n_2(1-t^2)^{1/2} K_{12} & n_2^2 \dfrac{C_2}{\alpha} & \\ \vdots & \vdots & & \ddots \end{bmatrix} \tag{5.195}$$

Thus the new element values are

$$Y_{00} = Y_{NN} = 1 \tag{5.196}$$

$$Y_{01} = Y_{N-1,N} = n_1 \tag{5.197}$$

$$Y_{11} = Y_{NN} = n_1^2(1 + C_1/\alpha) \tag{5.198}$$

$$Y_{r,r} = \frac{n_r^2 C_r}{\alpha} \qquad (r = 2, \ldots, N-1) \tag{5.199}$$

$$Y_{r,r+1} = n_r n_{r+1} K_{r,r+1} \qquad (r = 1, \ldots, N-1) \tag{5.200}$$

We can now find the design equation for the admittances to ground Y_r and the coupling admittances $Y_{r,r+1}$. First let us choose Y_{rr} to be arbitrarily equal to unity, for physical convenience. Then from (5.198)

$$n_1 = n_N = \frac{1}{(1 + C_1/\alpha)^{1/2}} \tag{5.201}$$

and from (5.199)

$$n_r = \left(\frac{\alpha}{C_r}\right)^{1/2} \qquad (r = 2, \ldots, N-1) \tag{5.202}$$

Hence from (5.200)

$$Y_{r,r+1} = n_r n_{r+1} K_{r,r+1} \tag{5.203}$$

Hence

$$Y_0 = 1 - n_1 \tag{5.204}$$

$$Y_{01} = n_1 \tag{5.205}$$

$$Y_1 = Y_N = 1 - n_1 - n_1 n_2 K_{12} \tag{5.206}$$

$$Y_r = 1 - n_{r-1} n_r K_{r-1,r} - n_r n_{r+1} K_{r,r+1} \tag{5.207}$$

5.5.3 Design example

In this case the design will be identical to the previous example except that the bandwidth will be 40 MHz, i.e. 2 per cent bandwidth. Thus

$$\theta_1 = \frac{90° \times 1.98}{2} = 89.1° \tag{5.208}$$

and

$$\alpha = \frac{1}{\tan(\theta_1)} = 0.0157 \tag{5.209}$$

The element values for the lowpass prototype are given in (5.182) and (5.183).

From (5.201) we obtain

$$n_1 = \frac{1}{(1 + 0.9314/0.0157)^{1/2}} = 0.1287 = n_N \tag{5.210}$$

and from (5.202)

$$n_2 = n_3 = \left(\frac{0.0157}{2.2487}\right)^{1/2} = 0.0835 \tag{5.211}$$

From (5.203)

$$Y_{12} = Y_{34} = (0.1287)(0.0835)(1.1393) = 1.4177 \times 10^{-2} \tag{5.212}$$

$$Y_{23} = (0.0835)^2(1.5751) = 1.098 \times 10^{-2} \tag{5.213}$$

From (5.204)

$$Y_0 = 0.8713 \tag{5.214}$$

and from (5.205)

$$Y_{01} = 0.1287 \tag{5.215}$$

From (5.206) and (5.207)

$$Y_1 = Y_4 = 0.8571 \tag{5.216}$$

$$Y_2 = Y_3 = 0.9748 \tag{5.217}$$

The element values in a $50\,\Omega$ system are

$$\begin{aligned} Z_0 &= Z_5 = 57.38\,\Omega \\ Z_1 &= Z_4 = 58.34\,\Omega \\ Z_2 &= Z_3 = 51.29\,\Omega \\ Z_{01} &= Z_{45} = 388.5\,\Omega \\ Z_{12} &= Z_{34} = 3526\,\Omega \\ Z_{23} &= 4554\,\Omega \end{aligned} \tag{5.218}$$

The simulated response of the filter is shown in Figure 5.41. This shows a highly symmetrical frequency response which is characteristic of an interdigital filter. The introduction of the UEs at each end of the filter results in an extra interdigital section as shown in Figure 5.42.

5.5.4 Physical design of the interdigital filter

The physical dimension of the filter can be found from the static capacitances per unit length between each interdigital line, its nearest neighbours and ground, where

$$\frac{C}{\varepsilon} = \frac{377}{Z_0(\varepsilon_r)^{1/2}} \tag{5.219}$$

178 *Theory and design of microwave filters*

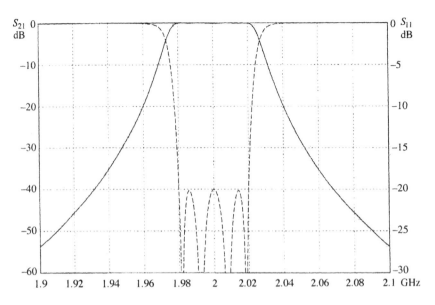

Figure 5.41 Simulated frequency response of a narrowband interdigital filter

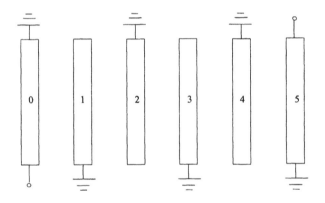

Figure 5.42 Narrowband interdigital filter ($N = 4$)

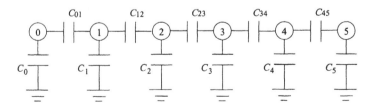

Figure 5.43 Static capacitances of interdigital filter

TEM transmission line filters 179

The various capacitances for our design example are shown in Figure 5.43.
From (5.218) and (5.219) we obtain

$$\frac{C_0}{\varepsilon} = \frac{C_5}{\varepsilon} = 6.5702$$

$$\frac{C_1}{\varepsilon} = \frac{C_4}{\varepsilon} = 6.4621$$

$$\frac{C_2}{\varepsilon} = \frac{C_3}{\varepsilon} = 7.3503$$

$$\frac{C_{01}}{\varepsilon} = \frac{C_{45}}{\varepsilon} = 6.5702 \qquad (5.220)$$

$$\frac{C_{12}}{\varepsilon} = \frac{C_{34}}{\varepsilon} = 0.1069$$

$$\frac{C_{23}}{\varepsilon} = 0.0828$$

We will assume that the filter is to be constructed from an array of rectangular bars between parallel ground planes as shown in Figure 5.44.

In Figure 5.44 we have two equal width bars of width w and thickness t spaced apart between parallel ground planes of spacing b. The parallel plate capacitance between one face of one bar and ground is C_p. The fringing capacitance from the isolated corner of one bar to ground is C_f'. This is shown in Figure 5.22 as a function of t/b.

The fringing capacitances from the coupled corners of the bars are denoted C_{fe}' and C_{fo}' depending on whether an even- or odd-mode excitation is applied. Thus C_{fe}' is the fringing capacitance with an open circuit along the line of symmetry and C_{fo}' is the fringing capacitance with a short circuit along the line of symmetry. The equivalent circuit of the pair of coupled lines can be represented as a pi network of capacitance to ground C_e and coupling capacitance ΔC as shown in Figure 5.45.

By applying a short circuit along the line of symmetry in Figures 5.43 and 5.44

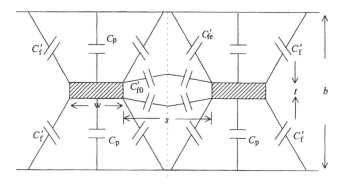

Figure 5.44 Coupled rectangular bars between parallel ground planes

180 Theory and design of microwave filters

we obtain
$$C_o = 2C_p + 2C'_f + 2C'_{fo} = C_e + 2\Delta C \tag{5.221}$$

and applying an open circuit
$$C_e = 2C_p + 2C'_f + 2C'_{fe} \tag{5.222}$$

Thus
$$\Delta C = \frac{C_o - C_e}{2}$$
$$= C'_{fo} - C'_{fe} \tag{5.223}$$

ΔC and C'_{fe} are shown in Figure 5.46 as a function of s/b.

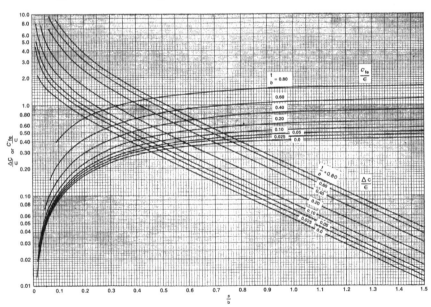

Figure 5.46 Even-mode fringing capacitance and coupling capacitance of coupled rectangular bars
(Source: Getsinger, W.J.: 'Coupled rectangular bars between parallel plates', IEEE Transactions on Microwave Theory and Techniques, **10** (1), pp. 65–72; (© 1962 IEEE).

TEM transmission line filters 181

Thus to calculate the dimensions of the interdigital filter first we calculate the normalised spacing between the bars. From (5.220) assuming $t/b = 0.4$

$$\frac{C_{01}}{\varepsilon} = \frac{C_{45}}{\varepsilon} = 6.5702 \qquad (5.224)$$

Therefore from Figure 5.45

$$\frac{S_{01}}{b} = \frac{S_{45}}{b} = 0.07 \qquad (5.225)$$

Similarly

$$\frac{S_{12}}{b} = \frac{S_{34}}{b} = 1.12 \qquad (5.226)$$

and

$$\frac{S_{23}}{b} = 1.17 \qquad (5.227)$$

Now from Figure 5.46 we can calculate the even-mode fringing capacitances, giving

$$C'_{\text{fe } 01} = C'_{\text{fe } 45} = 0.11 \qquad (5.228)$$

$$C'_{\text{fe } 12} = C'_{\text{fe } 34} = 0.88 \qquad (5.229)$$

$$C'_{\text{fe } 23} = 0.89 \qquad (5.230)$$

Now we can calculate the normalised widths for the bars from (5.222), i.e.

$$C_e = 2C_p + 2C'_f + 2C'_{\text{fe}} = \frac{4w}{b-t} + 2C'_f + 2C'_{\text{fe}} \qquad (5.231)$$

Thus

$$w_r = \frac{b-t}{4}\left(\frac{C_r}{\varepsilon} - 2C'_{\text{fe } r-1,r} - 2C'_{\text{fe } r,r+1}\right) \qquad (5.232)$$

Note that the isolated fringing capacitance is replaced by the even-mode fringing capacitance for the previous coupled-bar pair except for the first (and last) bars.

$$w_0 = \frac{b-t}{4}\left(\frac{C_0}{\varepsilon} - 2C'_f - 2C'_{f\,01}\right) \qquad (5.233)$$

$$C'_f = 0.91 \qquad (5.234)$$

$$C'_{\text{fe}} = 0.11 \qquad (5.235)$$

and

$$\frac{C_0}{\varepsilon} = 6.5702 \qquad (5.236)$$

Thus

$$w_0 = \frac{b-t}{4}(6.5702 - 0.22 - 1.82)$$

$$= 1.132(b-t) \quad (5.237)$$

Similarly

$$w_1 = 1.121(b-t) \quad (5.238)$$

$$w_2 = 0.953(b-t) \quad (5.239)$$

b must be chosen for physical realisability and to achieve a certain resonator Q factor.

From (4.147) the midband insertion loss of the filter is given by

$$L_A = 950/Q_u \quad (5.240)$$

Thus in our design, for 0.2 dB midband insertion loss we require a Q_u of 4750. The Q factor of a rectangular bar as a function of impedance is relatively constant and may be approximated by

$$\frac{Q}{b(f)^{1/2}} = 2000 - 7.5Z_0 \quad \left(0.1 < \frac{t}{b} < 0.5\right) \quad (5.241)$$

where b is in cm and f in GHz. Thus for our design with a typical Z_0 of 55 Ω we choose $b = 2$ cm giving $t = 0.8$ cm and the actual dimensions of the filter are

$$\begin{aligned}
w_0 &= w_5 = 1.36\,\text{cm} \\
w_1 &= w_4 = 1.34\,\text{cm} \\
w_2 &= w_3 = 1.14\,\text{cm} \\
w_{01} &= w_{45} = 0.14\,\text{cm} \\
w_{12} &= w_{34} = 1.76\,\text{cm} \\
w_{23} &= 1.75\,\text{cm}
\end{aligned} \quad (5.242)$$

The resonators are one quarter wavelength at 2 GHz, i.e. 3.75 cm long.

5.6 The combline filter

The interdigital filter has the advantages of a broad stopband and a highly symmetrical frequency response. From a physical viewpoint it has certain disadvantages. First, it is quite large: the resonators are quarter wavelength long and for narrow bandwidths they are physically well separated. Furthermore, tuning screws for final electrical alignment are on alternate opposite faces of the filter.

The combline filter [15] shown in Figure 5.47 overcomes these disadvantages at the expense of a slightly asymmetrical frequency response. It consists of any array of coupled TEM lines with couplings constrained to be between nearest

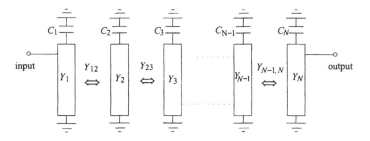

Figure 5.47 The combline filter

neighbours. The lines are all short circuited at the same end. The opposite ends of the lines are loaded with capacitors which are connected to ground.

The principle of operation of the combline filter is as follows. First, if the lumped capacitors were removed then the shunt lines would resonate at their quarter wave frequency. However, the couplings would also resonate at this frequency, producing an all-stop network. As the capacitors are increased the shunt lines behave as inductive elements and resonate with the capacitors at a frequency below the quarter wave frequency. The couplings would then be finite but relatively weak compared with an interdigital filter with the same resonator spacing. Thus the combline filter is compact, as the resonators may be significantly shorter than one quarter wavelength and are closer together than in an interdigital filter with the same bandwidth and ground plane spacing.

The equivalent circuit of the combline filter will now be derived. First consider the array of coupled lines all shorted at the same end, shown in Figure 5.48. The admittance matrix equation is

$$\begin{bmatrix} I_1 \\ I_2 \\ I_3 \\ \vdots \\ I_N \\ I'_1 \\ I'_2 \\ I'_3 \\ \vdots \\ I'_N \end{bmatrix} = \frac{1}{t} \begin{bmatrix} [\eta] & \dfrac{-1}{(1-t^2)^{1/2}}[\eta] \\ \dfrac{-1}{(1-t^2)^{1/2}}[\eta] & [\eta] \end{bmatrix} \begin{bmatrix} V_1 \\ V_2 \\ V_3 \\ \vdots \\ V_N \\ V'_1 \\ V'_2 \\ V'_3 \\ \vdots \\ V'_N \end{bmatrix} \quad (5.243)$$

However, as nodes $1', 2', \ldots, N'$ are all short circuited then

$$V'_1, V'_2 \ldots V'_N = 0 \tag{5.244}$$

184 Theory and design of microwave filters

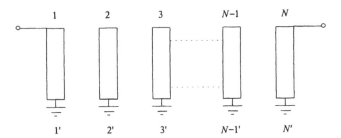

Figure 5.48 An array of coupled short circuited stubs

and the matrix equation reduces to

$$\begin{bmatrix} I_1 \\ I_2 \\ I_3 \\ \vdots \\ I_N \end{bmatrix} = \frac{1}{t}[\eta] \begin{bmatrix} V_1 \\ V_2 \\ V_3 \\ \vdots \\ V_N \end{bmatrix} \quad (5.245)$$

or

$$\begin{bmatrix} I_1 \\ I_2 \\ I_3 \\ \vdots \\ I_{N-1} \\ I_N \end{bmatrix} = \frac{1}{t} \begin{bmatrix} Y_{11} & -Y_{12} & 0 & 0 & \cdots & & \\ -Y_{12} & Y_{22} & -Y_{23} & 0 & & & \\ 0 & -Y_{23} & Y_{33} & -Y_{34} & & & \\ & & & & \ddots & & \\ & & & & & Y_{N-1,N-1} & -Y_{N-1,N} \\ & & & & & -Y_{N-1,N} & Y_{NN} \end{bmatrix} \begin{bmatrix} V_1 \\ V_2 \\ V_3 \\ \vdots \\ V_{N-1} \\ V_N \end{bmatrix}$$
(5.246)

Scrutiny of (5.246) shows that each nodal current is only related to its own nodal voltage and the voltage at the adjacent nodes. The equivalent circuit of the coupled-line array is thus an array of shunt short circuited stubs coupled via series short circuited stubs as shown in Figure 5.49.

The equivalent circuit of the combline filter is obtained simply by adding shunt capacitor C_r from the rth node to ground as shown in Figure 5.50. The equivalent circuit between the rth and $(r+1)$th nodes is given in Figure 5.51. Inspection of the equivalent circuit shows that an inverter can be formed from a pi section of the short circuited stubs, as in Figure 5.52. Here we see that an inverter can be formed in a similar way to that for lumped element bandpass filters. The pi network of stubs between the dotted lines has a transfer matrix

TEM transmission line filters 185

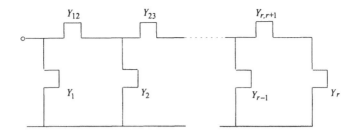

Figure 5.49 Equivalent circuit of an array of coupled short circuited lines

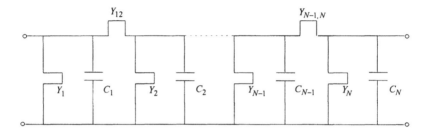

Figure 5.50 Equivalent circuit of the combline filter

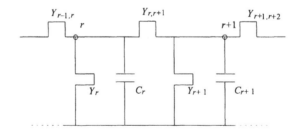

Figure 5.51 Equivalent circuit of the combline filter between the rth and $(r+1)$th nodes

given by

$$[T] = \begin{bmatrix} 1 & 0 \\ \dfrac{jY_{r,r+1}}{\tan(\theta)} & 1 \end{bmatrix} \begin{bmatrix} 1 & \dfrac{j\tan(\theta)}{Y_{r,r+1}} \\ 0 & 1 \end{bmatrix} \begin{bmatrix} 1 & 0 \\ \dfrac{jY_{r,r+1}}{\tan(\theta)} & 1 \end{bmatrix} = \begin{bmatrix} 0 & \dfrac{j\tan(\theta)}{Y_{r,r+1}} \\ \dfrac{jY_{r,r+1}}{\tan(\theta)} & 0 \end{bmatrix}$$

(5.247)

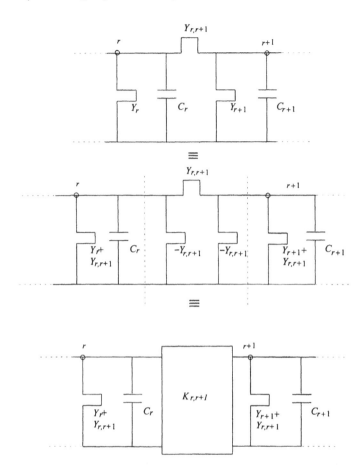

Figure 5.52 Formulation of inverters in the combline filter

Hence

$$K_{r,r+1} = \frac{Y_{r,r+1}}{\tan(\theta)} \tag{5.248}$$

The admittance of the rth resonator is given by

$$Y'_r = j\omega C - \frac{j(Y_r + Y_{r-1,r} + Y_{r,r+1})}{\tan(\theta)} \tag{5.249}$$

The equivalent circuit of the filter is shown in Figure 5.53 where Y_{rr} is given by

$$Y_{rr} = Y_r + Y_{r-1,r} + Y_{r,r+1} \tag{5.250}$$

It is convenient to scale the admittance of the entire network (including

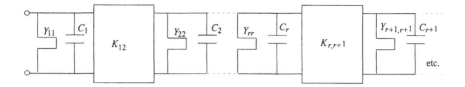

Figure 5.53   Equivalent circuit of the combline filter

source and load) by a factor $\tan(\theta)/\tan(\theta_0)$ where θ_0 is the electrical length of the resonators at the centre frequency ω_0 of the filter. This removes the frequency dependence from the inverters; hence

$$K_{r,r+1} = \frac{Y_{r,r+1}}{\tan(\theta_0)} \tag{5.251}$$

and

$$Y_r' = \frac{j}{\tan(\theta_0)} [\omega C_r \tan(\theta) - Y_{rr}] \tag{5.252}$$

We can now derive a frequency transformation from the lowpass prototype network to the combline resonators. For a shunt capacitor, inverter-coupled prototype we have

$$\omega C_{Lr} \rightarrow \left[\frac{\omega C_r \tan(\theta)}{\tan(\theta_0)} - \frac{Y_{rr}}{\tan(\theta_0)} \right] \tag{5.253}$$

where C_{Lr} is the rth capacitor in the lowpass prototype; thus

$$\omega \rightarrow \alpha[\beta \omega \tan(\theta) - 1] \tag{5.254}$$

where

$$\alpha = \frac{Y_{rr}}{C_{Lr} \tan(\theta_0)} \tag{5.255}$$

and

$$\beta = \frac{C_r}{Y_{rr}} \tag{5.256}$$

Since $\omega = 0$ in the lowpass prototype maps to ω_0 in the combline filter,

$$\beta = \frac{1}{\omega_0 \tan(\theta_0)} \tag{5.257}$$

The band-edges at ± 1 in the lowpass prototype map into the band-edges at ω_1 and ω_2 in the combline filter, i.e.

$$-1 = \alpha[\beta \omega_1 \tan(\theta_1) - 1] \tag{5.258}$$

$$+1 = \alpha[\beta \omega_2 \tan(\theta_2) - 1] \tag{5.259}$$

188 Theory and design of microwave filters

Now

$$\omega_1 = \omega_0 - \frac{\Delta\omega}{2} \tag{5.260}$$

and

$$\omega_2 = \omega_0 + \frac{\Delta\omega}{2} \tag{5.261}$$

where $\Delta\omega$ is the passband bandwidth. Hence

$$-1 = \alpha\left[\beta\left(\omega_0 - \frac{\Delta\omega}{2}\right)\tan\left(\theta_0 - \frac{a\Delta\omega}{2}\right) - 1\right] \tag{5.262}$$

$$+1 = \alpha\left[\beta\left(\omega_0 + \frac{\Delta\omega}{2}\right)\tan\left(\theta_0 + \frac{a\Delta\omega}{2}\right) - 1\right] \tag{5.263}$$

and for narrow bandwidths

$$a\Delta\omega \ll \theta_0 \tag{5.264}$$

Hence

$$\tan\left(\theta_0 + \frac{a\Delta\omega}{2}\right) \approx \frac{\tan(\theta_0) + a\Delta\omega/2}{1 - (a\Delta\omega/2)\tan(\theta_0)} \tag{5.265}$$

From (5.262), (5.263) and (5.265) we obtain

$$-1 = \alpha\left[\beta\left(\omega_0 - \frac{\Delta\omega}{2}\right)\left\{\tan(\theta_0) - \frac{a\Delta\omega}{2}[1+\tan^2(\theta_0)]\right\} - 1\right] \tag{5.266}$$

$$+1 = \alpha\left[\beta\left(\omega_0 + \frac{\Delta\omega}{2}\right)\left\{\tan(\theta_0) + \frac{a\Delta\omega}{2}[1+\tan^2(\theta_0)]\right\} - 1\right] \tag{5.267}$$

Solving (5.266) and (5.267) simultaneously and ignoring terms in $(a\Delta\omega)^2$ we obtain

$$\alpha = \frac{2}{\Delta\omega\beta\{\tan(\theta_0) + \theta_0[1+\tan^2(\theta_0)]\}} \tag{5.268}$$

or

$$\alpha = \frac{2\omega_0\tan(\theta_0)}{\Delta\omega\{\tan(\theta_0) + \theta_0[1+\tan^2(\theta_0)]\}} \tag{5.269}$$

From (5.269) we see that for narrow percentage bandwidths α will be large and from (5.255) we would obtain unrealisably high values for the shunt admittances. We can solve this by scaling the internal nodal admittances of the filter but before we do that we introduce redundant impedance transforming elements at the input and output of the filter. Initially we connect a frequency-independent phase shifter of unity impedance and phase length θ_0 between the source (and load) and the filter (Figure 5.54).

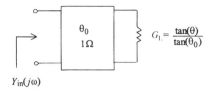

Figure 5.54 Introduction of a phase shifter at input and output of the filter

The transfer matrix of the phase shifter is

$$[T] = \begin{bmatrix} \cos(\theta_0) & j\sin(\theta_0) \\ j\sin(\theta_0) & \cos(\theta_0) \end{bmatrix} \quad (5.270)$$

Remembering that the source and load admittances have been scaled in order to remove the frequency variation of the load, then the effective source admittance after introducing the phase shifter is

$$\begin{aligned} Y_{in}(j\omega) &= \frac{\cos(\theta_0)\tan(\theta)/\tan(\theta_0) + j\sin(\theta_0)}{\cos(\theta_0) + j\sin(\theta_0)\tan(\theta)/\tan(\theta_0)} \\ &= \frac{\cos^2(\theta_0)\tan(\theta) + j\sin^2(\theta_0)}{\sin(\theta_0)\cos(\theta_0)[1 + j\tan(\theta)]} \\ &= \frac{[\cos^2(\theta_0)\tan(\theta) + j\sin^2(\theta_0)][1 - j\tan(\theta)]}{[\sin(2\theta_0)/2][1 + \tan^2(\theta)]} \end{aligned} \quad (5.271)$$

The real part of the effective load admittance is

$$\begin{aligned} \operatorname{Re} Y(j\omega) &= \frac{\tan(\theta)[\cos^2(\theta_0) + \sin^2(\theta_0)]\cos^2(\theta)}{\sin(2\theta_0)/2} \\ &= \frac{\sin(2\theta)}{\sin(2\theta_0)} \end{aligned} \quad (5.272)$$

The imaginary part of $Y(j\omega)$ is

$$\operatorname{Im} Y(j\omega) = j\left[\frac{\cos(2\theta) - \cos(2\theta_0)}{\sin(2\theta_0)}\right] \quad (5.273)$$

The real part varies slowly with frequency; for example if θ_0 is $45°$ then it varies by $0.866:1$ over an octave. The imaginary part is resonant at θ_0 and again varies slowly with frequency. Thus it can be said that introduction of the ideal phase shifter effectively removes the frequency variation of the coupling inverters from the filter.

The phase shifter may be represented by the introduction of an extra coupled-line section at the input and output of the filter as shown in Figure 5.55. The

190 Theory and design of microwave filters

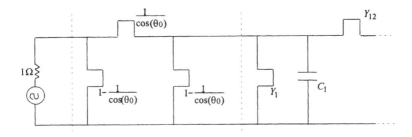

Figure 5.55 Input-output network

transfer matrix of this network is given by

$$[T] = \begin{bmatrix} 1 & 0 \\ \frac{1}{t}\left[1 - \frac{1}{\cos(\theta_0)}\right] & 1 \end{bmatrix} \begin{bmatrix} 1 & t\cos(\theta_0) \\ 0 & 1 \end{bmatrix} \begin{bmatrix} 1 & 1 \\ \frac{1}{t}\left[1 - \frac{1}{\cos(\theta_0)}\right] & 1 \end{bmatrix}$$

$$= \begin{bmatrix} \cos(\theta_0) & j\tan(\theta)\cos(\theta_0) \\ \dfrac{-j}{\tan(\theta)\{\cos(\theta_0) - 1/\cos(\theta_0)\}} & \cos(\theta_0) \end{bmatrix} \quad (5.274)$$

and after scaling the admittance by $\tan(\theta)/\tan(\theta_0)$

$$[T] = \begin{bmatrix} \cos(\theta_0) & j\sin(\theta_0) \\ j\sin(\theta_0) & \cos(\theta_0) \end{bmatrix} \quad (5.275)$$

which is the matrix of the ideal unity impedance phase shifter.
  The admittance matrix of the combline filter is now given by

$$[Y] = \begin{bmatrix} \dfrac{1}{t} & -\dfrac{1}{t\cos(\theta_0)} & 0 & \cdots \\ -\dfrac{1}{t\cos(\theta_0)} & \dfrac{1}{t} + \dfrac{Y_{11}}{t} + C_1 p & \dfrac{-Y_{12}}{t} & \\ 0 & \dfrac{-Y_{12}}{t} & \dfrac{Y_{22}}{t} + C_2 p & \\ & & & \ddots \\ & & & \text{etc.} \end{bmatrix} \quad (5.276)$$

and after scaling internal rows and columns we obtain the equivalent circuit shown in Figure 5.56.
  The element values can then be obtained as follows. First we choose all the capacitors C_r to be equal of value C. We also choose θ_0, the electrical length at

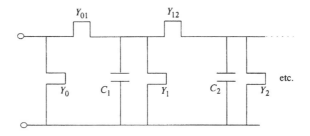

Figure 5.56 Equivalent circuit of the combline filter after the introduction of the input transformers

ω_0, e.g. $\theta_0 = 45°$. Then calculate α from

$$\alpha = \frac{2\omega_0 \tan(\theta_0)}{\Delta\omega\{\tan(\theta_0) + \theta_0[1 + \tan^2(\theta_0)]\}} \tag{5.277}$$

and

$$\alpha = \frac{n_r^2 Y_{rr}}{C_{Lr} \tan(\theta_0)} \tag{5.278}$$

$$\beta = \frac{1}{\omega_0 \tan(\theta_0)} = \frac{C}{Y_{rr}} \tag{5.279}$$

Hence

$$Y_{rr} = C\omega_0 \tan(\theta_0) \tag{5.280}$$

$$n_r = \left[\frac{\alpha C_{Lr} \tan(\theta_0)}{Y_{rr}}\right]^{1/2} \quad (r = 1, \ldots, N) \tag{5.281}$$

$$Y_{r,r+1} = \frac{K_{r,r+1} \tan(\theta_0)}{n_r n_{r+1}} \quad (r = 1, \ldots, N-1) \tag{5.282}$$

$$Y_r = Y_{rr} - Y_{r-1,r} - Y_{r,r+1} \quad (r = 2, \ldots, N-1) \tag{5.283}$$

$$Y_1 = Y_N = Y_{11} - Y_{12} + \frac{1}{n_1^2} - \frac{1}{n_1 \cos(\theta_0)} \quad (r = 1 \text{ and } N) \tag{5.284}$$

$$Y_0 = Y_{N+1} = 1 - \frac{1}{n_1 \cos(\theta_0)} \tag{5.285}$$

$$Y_{01} = Y_{N,N+1} = \frac{1}{n_1 \cos(\theta_0)} \tag{5.286}$$

5.6.1 Design example

We shall design a filter to the same specification as for the narrowband interdigital filter, i.e. a degree 4 Chebyshev filter centred at 2 GHz with 40 MHz bandwidth. The lowpass prototype element values are

$$C_{L1} = C_{L4} = 0.9314$$
$$C_{L2} = C_{L3} = 2.2487$$
$$K_{12} = K_{34} = 1.3193$$
$$K_{23} = 1.5751$$
(5.287)

Hence

$$\omega_0 = 1.2566 \times 10^{10} \tag{5.288}$$

$$\Delta\omega = 2.5132 \times 10^8 \tag{5.289}$$

Choosing $\theta_0 = 50°$, i.e. 0.8726 radians, we obtain from (5.277)

$$\alpha = 36.075 \tag{5.290}$$

From (5.279)

$$\beta = \frac{1}{\omega_0 \tan(\theta_0)} = 6.678 \times 10^{-11} \tag{5.291}$$

Also from (5.279)

$$Y_{rr} = \frac{C}{\beta} \tag{5.292}$$

Choosing $Y_{rr} = 1$, then

$$C = \beta = 6.678 \times 10^{-11} \tag{5.293}$$

and from (5.281)

$$n_1 = n_4 = 6.3275 \tag{5.294}$$
$$n_2 = n_3 = 9.8318 \tag{5.295}$$

From (5.282)

$$Y_{12} = Y_{34} = \frac{K_{12} \tan(\theta_0)}{n_1 n_2} = 0.0253 \tag{5.296}$$

and

$$Y_{23} = \frac{K_{23} \tan(\theta_0)}{n_2^2} = 0.0194 \tag{5.297}$$

From (5.284)
$$Y_1 = Y_4 = 1 - Y_{12} + \frac{1}{n_1^2} - \frac{1}{n_1 \cos(\theta_0)} = 0.7538 \tag{5.298}$$

and from (5.283)
$$Y_2 = Y_3 = 1 - Y_{12} - Y_{23} = 0.9553 \tag{5.299}$$

From (5.285) and (5.286)
$$Y_0 = Y_5 = 1 - \frac{1}{n_1 \cos(\theta_0)} = 0.7542 \tag{5.300}$$

$$Y_{01} = Y_{45} = \frac{1}{n_1 \cos(\theta_0)} = 0.2458 \tag{5.301}$$

After scaling to 50 Ω the impedances of the elements are as follows:

$$\begin{aligned}
Z_0 &= Z_5 = 66.295\,\Omega \\
Z_1 &= Z_4 = 66.33\,\Omega \\
Z_2 &= Z_3 = 52.34\,\Omega \\
Z_{01} &= Z_{45} = 203.42\,\Omega \\
Z_{12} &= Z_{34} = 1976.3\,\Omega \\
Z_{23} &= 2577.3\,\Omega
\end{aligned} \tag{5.302}$$

and
$$C = 1.3356\,\text{pF} \tag{5.303}$$

The equivalent circuit of the filter is shown in Figure 5.57.

The length of the resonators is 50° at 2 GHz, i.e. 20.83 mm. The filter is considerably more compact than an interdigital filter with the same ground plane spacing. The resonators are shorter and also closer together. For example Z_{12} is 1976 Ω in the combline filter and 3526 Ω in the interdigital filter. The value of S_{12}/b is 0.88 in the combline filter compared with 1.12 for the interdigital filter. The simulated frequency response of the combline filter is shown in Figure 5.58.

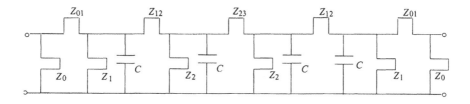

Figure 5.57   Equivalent circuit of a degree 4 combline filter

194 Theory and design of microwave filters

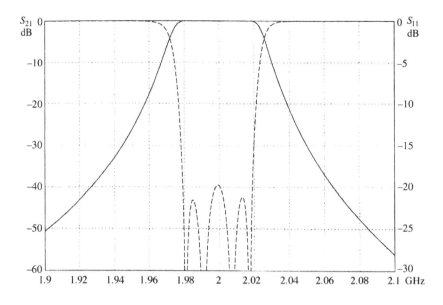

Figure 5.58 Simulated response of the combline filter

5.6.2 Tunable combline filters

The centre frequency of the filter may be altered by synchronously tuning the capacitors in each resonator [16]. It is worth noting that the frequency dependence of the inverters is relatively small over octave bandwidths; thus the return loss of the filter will remain fairly constant over this bandwidth. Furthermore, the bandwidth of the filter is given by

$$\Delta \omega = \frac{2\omega_0 \tan(\theta_0)}{\alpha \{\tan(\theta_0) + \theta_0[1 + \tan^2(\theta_0)]\}} \tag{5.304}$$

This is a maximum when $\theta_0 = 52.885°$. $\Delta\omega$ as a function of θ_0 is as follows:

θ_0	30°	40°	50°	60°	70°
$\Delta\omega$	0.237	0.288	0.314	0.306	0.254

The bandwidth of the filter remains approximately constant over a broad tuning range.

5.7 The parallel coupled-line filter

The parallel coupled-line filter consists of a cascade of pairs of parallel coupled open circuited lines [17]. The lines are quarter wave long at the centre frequency

TEM transmission line filters 195

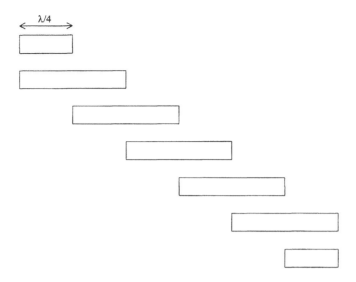

Figure 5.59 Parallel coupled-line filter of degree 5

of the filter. There are $N + 1$ coupled-line sections (including input and output transformers) in an Nth-degree filter (Figure 5.59).

The parallel coupled-line filter is often used in microstrip subassemblies as it is easy to fabricate due to the absence of short circuits. A pair of coupled lines and its equivalent circuit are shown in Figure 5.60. Here we see that the equivalent

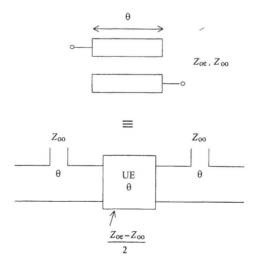

Figure 5.60 Equivalent circuit of a parallel coupled-line pair

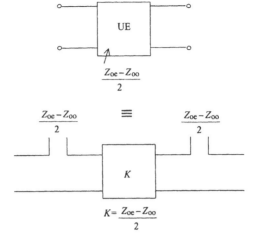

Figure 5.61 Equivalent circuit of a unit element

circuit consists of series open circuited stubs separated by a UE. Z_{oe} and Z_{oo} are the even- and odd-mode characteristic impedances of the coupled-line pair. Now a UE may be decomposed into a pair of open circuited stubs separated by an inverter as shown in Figure 5.61. Combining this with the equivalent circuit in Figure 5.60 we obtain a final equivalent circuit consisting of series open circuit stubs separated by inverters, shown in Figure 5.62. A cascade of $N - 1$ coupled-line pairs results in a circuit consisting of N series stubs separated by inverters (Figure 5.63). The filter can be designed by applying the Richards highpass transformation to the series inductor/inverter coupled prototype. Impedance scaling and introduction of redundant transformer elements can be carried out in a similar way to the methods described for interdigital and combline filters. Details of this procedure are provided in References 5, 6 and 7.

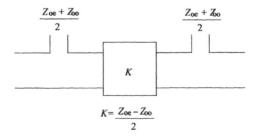

Figure 5.62 Equivalent circuit of a coupled-line pair

TEM transmission line filters 197

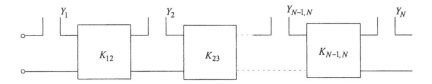

Figure 5.63   Equivalent circuit of the parallel coupled-line filter

5.8 Narrowband coaxial resonator filters

In cellular radio base station applications the bandwidths required are narrow, typically 3.5 per cent, and the low loss specifications require resonator Q factors of up to 5000. Consequently the filters require large ground plane spacings and combline realisations would result in unacceptably large inter-resonator spacings. Consequently a coaxial resonator approach is taken where individual combline resonators are constructed in separate cavities with apertures providing the required weak couplings between resonators. Furthermore, these devices usually require asymmetric generalised Chebyshev characteristics with real frequency transmission zeros located on one side of

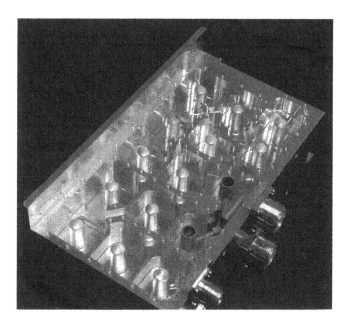

Figure 5.64   A coaxial resonator filter
(photograph courtesy of Filtronic plc)

198 Theory and design of microwave filters

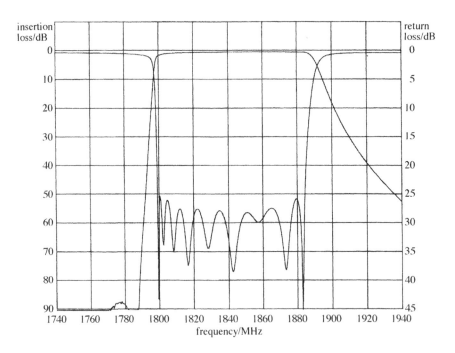

Figure 5.65 Measured performance of a coaxial resonator filter
(courtesy of Filtronic plc)

the passband. A single real frequency transmission zero may be realised by coupling around three resonators as described in Chapter 3. A photograph of a typical filter is shown in Figure 5.64. Its measured performance is shown in Figure 5.65.

5.9 Summary

This chapter has been concerned with the theory and design of filters consisting of interconnections of TEM transmission lines. Initially the use of the Richards transformation to convert lumped prototype networks into distributed quasi-lowpass and quasi-highpass filters is described. These filters consist of interconnections of open and short circuit stubs separated by inverters which may be approximated by quarter wave sections of line. Next the design of lowpass filters consisting entirely of a cascade of commensurate UEs of transmission line is described. This is illustrated by an example and is followed by the design of highly selective lowpass and highpass distributed filters with generalised Chebyshev characteristics. The lowpass design is illustrated by an example.

Next the theory of coupled transmission lines is developed in terms of the admittance matrix for a system of N coupled lines. This is followed by detailed design procedures of two types of coupled-line filter, the interdigital and the combline filter. Design examples of both these filters are presented including information on the physical realisation. The parallel coupled-line filter, which is convenient to realise in microstrip, is also described. Finally the use of iris-coupled coaxial resonator filters is described and illustrated with a real device.

5.10 References

1 RICHARDS, P.I.: 'Resistor transmission line networks', *Proceedings of the IRE*, 1948, **30**, pp. 217–20
2 RICHARDS, P.I.: 'General impedance function theory', *Quarterly of Applied Mathematics*, 1948, **6**, pp. 21–29
3 BAHER, H.: 'Synthesis of electrical networks' (Wiley, New York, 1984) pp. 133–53
4 RHODES, J.D.: 'Theory of electrical filters' (Wiley, New York, 1976) pp. 134–49
5 TRINOGGA, L.A., GUO, K., and HUNTER, I.C.: 'Practical microstrip circuit design' (Ellis Horwood, Chichester, 1991) pp. 168–85
6 EDWARDS, T.: 'Foundations for microstrip circuit design' (Wiley, New York, 1992, 2nd edn.)
7 ALSAYAB, S.A.: 'A novel class of generalised Chebyshev lowpass prototype for suspended substrate stripline filters', *IEEE Transactions on Microwave Theory and Techniques*, 1982, **MTT-30** (9), pp. 1341–47
8 GETSINGER, W.J.: 'Coupled rectangular bars between parallel plates', *IEEE Transactions on Microwave Theory and Techniques*, 1962, **10** (1), pp. 65–72
9 MATTHAEI, G., YOUNG, L., and JONES, E.M.T.: 'Microwave filters, impedance matching networks and coupling structures' (Artech House, Norwood, MA, 1980) pp. 163–229
10 DEAN, J.E., and RHODES, J.D.: 'MIC broadband filters and multiplexers', Proceedings of the 9th European Microwave Conference, 1979
11 ZYSMAN, G.I., and JOHNSON, A.K.: 'Coupled transmission line networks in an inhomogeneous dielectric medium', *IEEE Transactions on Microwave Theory and Techniques*, 1969, **MTT-17** (10), pp. 753–59
12 DEAN, J.E., and RHODES, J.D.: 'Design of MIC broadband multiplexers', *Microwave Theory and Techniques 5*, International Microwave Symposium 1980, Digest 80.1, pp. 147–49
13 SCANLAN, J.O.: 'Theory of microwave coupled-line networks', *Proceedings of the IEEE*, 1980, **68** (2), pp. 209–31
14 WENZEL, R.J.: 'Exact theory of interdigital bandpass filters and related coupled structures', *IEEE Transactions on Microwave Theory and Techniques*, 1965, **MTT-13** (5), pp. 559–75

15 MATTHAEI, G.L.: 'Comb-line bandpass filters of narrow or moderate bandwidth', *Microwave Journal*, 1963, **1**, pp. 82–91
16 HUNTER, I.C., and RHODES, J.D.: 'Electronically tunable microwave bandpass filters', *IEEE Transactions on Microwave Theory and Techniques*, 1982, **MTT-30** (9), pp. 1354–60
17 COHN, S.B.: 'Parallel coupled transmission line resonator filters', *IRE Transactions on Microwave Theory and Techniques*, 1958, **46**, pp. 223–33

Chapter 6
Waveguide filters

6.1 Introduction

A waveguide is a structure which directs the propagation of an electromagnetic wave in a particular direction by confining the wave energy. Waveguides normally consist of hollow metallic pipes with uniform cross-section. The use of dielectric rods as waveguides is also common and these will be discussed in Chapter 7. Waveguide resonators are useful elements in filter design as they generally have much higher Q factors than coaxial or other TEM resonators.

There are distinct differences between waveguides and TEM transmission lines. A transmission line has a minimum of two conductors and supports the TEM mode of propagation, which has zero cut-off frequency. There is no minimum size for the cross-section of a TEM line in order for signal propagation to occur, other than that determined by dissipation losses. On the other hand, a waveguide has only one conductor consisting of the boundary of the pipe. The waveguide has a distinct cut-off frequency above which electromagnetic energy will propagate and below which it is attenuated. The cut-off frequency of the waveguide is determined by its cross-sectional dimensions. For example, a rectangular cross-section waveguide must have a width at least greater than one-half of the free space wavelength for propagation to occur at a particular frequency.

Furthermore, propagation in waveguides occurs with distinct field patterns, or modes. Any waveguide can support an infinite number of modes each of which have their own cut-off frequency. Also, both the characteristic impedance and the propagation constant of a waveguide are functions of frequency.

In this chapter we will examine the design techniques for filters comprising interconnections of waveguides. Initially the basic theory of rectangular and circular cross-section waveguides will be described. This is essential for a proper understanding of the modal fields, and in order to develop expressions for cut-off frequency and the resonant frequency and Q of waveguide resonators. Design techniques for various waveguide filters will then be developed.

6.2 Basic theory of waveguides

A waveguide normally consists of a hollow conducting pipe of arbitrary cross-section (Figure 6.1). In the ideal case both the conductor and the dielectric filling the waveguide are assumed lossless.

Analysis of the possible field structures within the guide is accomplished by solution of Maxwells equations, which for sinusoidal excitation are

$$\nabla X E = -j\omega\mu H \tag{6.1}$$

$$\nabla X H = j\omega\varepsilon E \tag{6.2}$$

($\exp(j\omega t)$ dependence assumed) and for a source-free region

$$\nabla \cdot D = \varepsilon \nabla \cdot E = 0 \tag{6.3}$$

$$\nabla \cdot B = \mu \nabla \cdot H = 0 \tag{6.4}$$

Taking the curl of (6.1) and substituting (6.2) we obtain

$$\nabla X \nabla X E = \omega^2 \mu\varepsilon E \tag{6.5}$$

or

$$-\nabla^2 E + \nabla(\nabla \cdot E) = \omega^2 \mu\varepsilon E \tag{6.6}$$

and from (6.3), for source-free regions, we obtain the Helmholtz equations

$$\nabla^2 E = -k^2 E \tag{6.7}$$

$$\nabla^2 H = -k^2 H \tag{6.8}$$

where

$$k^2 = \omega^2 \mu\varepsilon \tag{6.9}$$

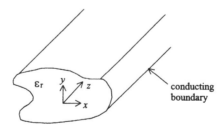

Figure 6.1 Uniform waveguide with arbitrary conducting boundary

If we assume that the direction of propagation is along the z axis then the fields can be expressed in terms of the propagation constant γ:

$$E(x, y, z) = f(x, y)\exp(-\gamma z) \tag{6.10}$$

where for a lossless waveguide $\gamma = \alpha$ implies an exponentially decaying or cut-off wave and $\gamma = j\beta$ implies a propagating wave with sinusoidal variation along the z axis. The Helmholtz equations can be expressed as

$$\nabla_t^2 E = -(\gamma^2 + k^2)E \tag{6.11}$$

$$\nabla_t^2 H = -(\gamma^2 + k^2)H \tag{6.12}$$

where

$$\nabla_t^2 = \frac{\partial^2}{\partial x^2} + \frac{\partial^2}{\partial y^2} \tag{6.13}$$

The E and H fields can be obtained by solving (6.11) and (6.12) with the appropriate boundary conditions, which in this case are that the tangential E field should be zero on the surface of the conducting pipe. General expressions for the fields in waveguides of arbitrary cross-section are difficult to obtain. Fortunately most practical waveguides have simple rectangular or circular cross-sections. Initially we will examine the rectangular waveguide shown in Figure 6.2. TEM modes cannot exist in the waveguide and the simplest modes are those with purely transverse E fields (TE or H modes) or purely transverse H fields (TM or E modes).

6.2.1 TE modes

For TE modes, E_z is equal to zero and H_z is finite; the Helmholtz equation is

$$\frac{\partial^2 H_z}{\partial x^2} + \frac{\partial^2 H_z}{\partial y^2} = -k_c^2 H_z \tag{6.14}$$

where

$$k_c^2 = \gamma^2 + k^2 \tag{6.15}$$

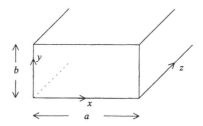

Figure 6.2 Rectangular waveguide

204 Theory and design of microwave filters

Solution of (6.14) by separation of variables yields

$$H_z = [A\sin(k_x x) + B\cos(k_x x)][C\sin(k_y y) + D\cos(k_y y)] \quad (6.16)$$

where

$$k_c^2 = k_x^2 + k_y^2 \quad (6.17)$$

A, B, C, D, k_x and k_y can be found by applying the boundary conditions of the waveguide to the E fields. These may be found from Maxwell's equations as follows:

$$\nabla X E = \begin{bmatrix} i & j & k \\ \dfrac{\partial}{\partial x} & \dfrac{\partial}{\partial y} & \dfrac{\partial}{\partial z} \\ E_x & E_y & E_z \end{bmatrix} = -j\omega\mu H \quad (6.18)$$

and

$$\nabla X H = \begin{bmatrix} i & j & k \\ \dfrac{\partial}{\partial x} & \dfrac{\partial}{\partial y} & \dfrac{\partial}{\partial z} \\ H_x & H_y & H_z \end{bmatrix} = -j\omega\varepsilon E \quad (6.19)$$

By expanding these equations and setting $E_z = 0$ for TE waves we can express the other field components as

$$E_x = \frac{-j\omega\mu}{\gamma^2 + k^2} \frac{\partial H_z}{\partial y} \quad (6.20)$$

$$E_y = \frac{j\omega\mu}{\gamma^2 + k^2} \frac{\partial H_z}{\partial x} \quad (6.21)$$

$$H_x = \frac{-\gamma}{\gamma^2 + k^2} \frac{\partial H_z}{\partial x} \quad (6.22)$$

$$H_y = \frac{-\gamma}{\gamma^2 + k^2} \frac{\partial H_z}{\partial y} \quad (6.23)$$

Hence substituting for H_z in (6.20) and (6.21) we obtain

$$E_x = \frac{-j\omega\mu k_y}{k_c^2}[A\sin(k_x x) + B\cos(k_x x)][C\cos(k_y y) - D\sin(k_y y)] \quad (6.24)$$

and

$$E_y = \frac{j\omega\mu k_y}{k_c^2}[A\cos(k_x x) - B\sin(k_x x)][C\sin(k_y y) - D\cos(k_y y)] \quad (6.25)$$

Now the boundary conditions are that the tangential E field should be zero on the surface of the conductor. Thus

$$E_x = 0\big|_{y=0} \Rightarrow C = 0 \tag{6.26}$$

$$E_y = 0\big|_{x=0} \Rightarrow A = 0 \tag{6.27}$$

Hence

$$H_z = H \cos(k_x x) \cos(k_y y) \tag{6.28}$$

where H is an arbitrary constant. In addition

$$E_x = 0\big|_{y=b} \Rightarrow k_x a = m\pi \tag{6.29}$$

$$E_y = 0\big|_{x=a} \Rightarrow k_y b = n\pi \tag{6.30}$$

and

$$H_z = H \cos\left(\frac{m\pi x}{a}\right) \cos\left(\frac{n\pi y}{b}\right) \tag{6.31}$$

where m and n are integers. The other field components are (for $\gamma = j\beta$)

$$E_x = \frac{j\omega\mu}{k_c^2} k_y H \cos\left(\frac{m\pi x}{a}\right) \sin\left(\frac{n\pi y}{b}\right) \tag{6.32}$$

$$E_y = \frac{-j\omega\mu k_x}{k_c^2} H \sin\left(\frac{m\pi x}{a}\right) \cos\left(\frac{n\pi y}{b}\right) \tag{6.33}$$

$$H_x = \frac{j\beta k_x H}{k_c^2} \sin\left(\frac{m\pi x}{a}\right) \cos\left(\frac{n\pi y}{b}\right) \tag{6.34}$$

$$H_y = \frac{j\beta k_x H}{k_c^2} \cos\left(\frac{m\pi x}{a}\right) \sin\left(\frac{n\pi y}{b}\right) \tag{6.35}$$

In these equations m and n are the mode numbers, representing the number of half wave variations in the field in the x and y directions. There are a doubly infinite set of modes, depending on the value of m and n. These are called the TE_{mn} modes.

Now from (6.15), (6.17), (6.29) and (6.30)

$$k_c^2 = \gamma^2 + k^2 = \left(\frac{m\pi}{a}\right)^2 + \left(\frac{n\pi}{b}\right)^2 \tag{6.36}$$

where $k = \omega(\mu\varepsilon)^{1/2}$. For a propagating mode $\gamma = j\beta$ and

$$\beta = \left[\omega^2 \mu\varepsilon - \left(\frac{m\pi}{a}\right)^2 - \left(\frac{n\pi}{b}\right)^2\right]^{1/2} \tag{6.37}$$

At the cut-off frequency of the guide $\beta = 0$ and

$$\omega_c = \frac{1}{(\mu\varepsilon)^{1/2}} \left[\left(\frac{m\pi}{a}\right)^2 + \left(\frac{n\pi}{b}\right)^2\right]^{1/2} \tag{6.38}$$

From (6.38) we can see that the lower the mode number then the lower the cut-off frequency. From (6.32) to (6.35) we see that either m or n can be zero, but not both, and assuming $a > b$ the lowest cut-off mode is the TE_{10} mode. In this case the fields are

$$H_z = H \cos\left(\frac{\pi x}{a}\right) \tag{6.39}$$

$$E_y = \frac{-j\omega\mu a}{\pi} H \sin\left(\frac{\pi x}{a}\right) \tag{6.40}$$

$$H_x = \frac{ja\beta}{\pi} H \sin\left(\frac{\pi x}{a}\right) \tag{6.41}$$

(E_x and H_y are both zero). The cut-off frequency of this mode is given by

$$\omega_c = \frac{\pi \nu}{a} \tag{6.42}$$

where $\nu = 1/(\mu\varepsilon)^{1/2}$ is the velocity of light in the dielectric medium, and since $\nu = f\lambda$ then

$$a = \frac{\lambda c}{2} \tag{6.43}$$

In other words the a dimension is half the free space wavelength at the cut-off frequency.

The E and H fields for the TE_{10} mode are shown in Figure 6.3.

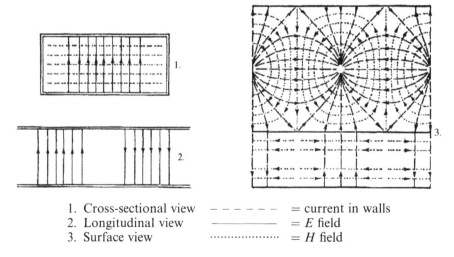

1. Cross-sectional view — — — — — — = current in walls
2. Longitudinal view ——————— = E field
3. Surface view .... = H field

Figure 6.3 Electric and magnetic fields for the TE_{10} mode in a rectangular waveguide

The characteristic impedance of this mode can be defined as the ratio of the transverse E and H fields, i.e.

$$Z_{TE} = \frac{E_y}{H_x} = \frac{\omega\mu}{\beta} \tag{6.44}$$

and the propagation constant β is given by

$$\beta = \left[\omega^2\mu\varepsilon - \left(\frac{\pi}{a}\right)^2\right]^{1/2} \tag{6.45}$$

and since

$$\frac{\pi}{a} = \frac{\omega_c}{\nu} \tag{6.46}$$

$$\beta = \frac{\omega}{\nu}\left[1 - \left(\frac{\omega}{\omega_c}\right)^2\right]^{1/2} \tag{6.47}$$

β is related to the wavelength of propagation in the guide, the guide wavelength λ_g, by

$$\beta = \frac{2\pi}{\lambda_g} \tag{6.48}$$

Hence

$$\lambda_g = \frac{2\pi}{\beta} = \frac{\lambda_0}{[1 - (\omega_c/\omega)^2]^{1/2}} \tag{6.49}$$

Here we see that the propagation constant and guide wavelength are functions of frequency. As we approach cut-off β tends to zero and λ_g tends to infinity. As we approach infinite frequency they both tend towards their free space value.

From (6.46) the wave impedance is given by

$$Z_{TE} = \frac{\omega\mu}{\beta} = \frac{\omega\mu\lambda_g}{2\pi} = \frac{\eta}{[1 - (\omega_c/\omega)^2]^{1/2}} \tag{6.50}$$

and

$$\eta = (\mu/\varepsilon)^{1/2} \tag{6.51}$$

where η is the characteristic impedance of free space.

Alternatively

$$Z_{TE} = \frac{\eta\lambda_g}{\lambda_0} \tag{6.52}$$

where λ_0 is the free space wavelength and

$$\lambda_0 = \frac{c}{f} \tag{6.53}$$

208 Theory and design of microwave filters

We can compute the group velocity of the waveguide from

$$v_g = \frac{d\omega}{d\beta} = v\left[1 - \left(\frac{\omega_c}{\omega}\right)^2\right]^{1/2} \tag{6.54}$$

We see that v_g approaches zero as ω approaches ω_c. This can cause phase distortion of modulated waves if the signal frequency approaches too close to the cut-off frequency.

6.2.2 TM modes

TM modes have zero H_z and finite E_z. In this case their behaviour is described by the solution of

$$\nabla_t^2 E_z = \frac{\partial^2 E_z}{\partial x^2} + \frac{\partial^2 E_z}{\partial y^2} = -k_c^2 E_z \tag{6.55}$$

This equation can be solved in a similar way to the TE wave equation yielding

$$E_z = E \sin\left(\frac{m\pi x}{a}\right) \sin\left(\frac{n\pi y}{b}\right) \tag{6.56}$$

In this case in order for the field to exist neither of the mode numbers can be zero and the lowest cut-off mode is the TM_{11} mode. The field pattern for this mode is shown in Figure 6.4.

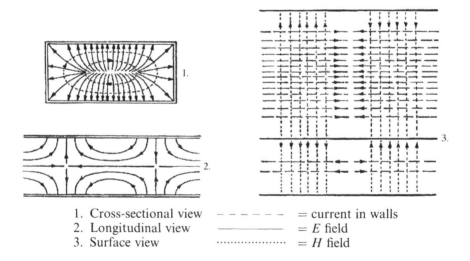

1. Cross-sectional view
2. Longitudinal view
3. Surface view

– – – – – – = current in walls
―――――― = E field
·············· = H field

Figure 6.4 Electric and magnetic fields for the TM_{11} mode in a rectangular waveguide

6.2.3 Relative cut-off frequencies of modes

The cut-off frequency of any mode in a rectangular waveguide is given by

$$\omega_c = \nu \left[\left(\frac{m\pi}{a}\right)^2 + \left(\frac{n\pi}{b}\right)^2\right]^{1/2} \tag{6.57}$$

For a typical aspect ratio of $a = 2b$ and $\varepsilon = \varepsilon_0$, ω_c is given by

$$\omega_c = \frac{c\pi}{a}(m^2 + 4n^2)^{1/2} \tag{6.58}$$

Assuming that the cut-off frequency of the TE_{10} mode in a particular waveguide is 1 GHz, i.e. $a = 15$ cm, then the next propagating modes are the TE_{01} and TE_{20} modes with a cut-off frequency of 2 GHz. These are followed by the TE_{11} and TM_{11} modes, both with a cut-off frequency of 2.236 GHz. The TE_{10} mode is often called the dominant mode as the waveguide can be operated in this mode over a broad spurious free bandwidth. Normally the waveguide would be operated at least 25 per cent above cut-off to avoid phase distortion.

6.2.4 Rectangular waveguide resonators

A waveguide can be formed into a resonant circuit by placing short circuited boundary conditions one half guide wavelength apart to form a box, as shown in Figure 6.5. If the mode of propagation in the waveguide is the TE_{10} mode propagating along z then the E field must be zero at $z = 0$ and $z = \ell$. Thus ℓ must be one half guide wavelength. Therefore

$$\ell = \frac{\lambda_g}{2} = \frac{\lambda_0}{2[1 - (\omega_c/\omega)^2]^{1/2}} = \frac{\lambda_0}{2[1 - (\lambda_0/2a)^2]^{1/2}} \tag{6.59}$$

The resonant frequency is

$$f_0 = \frac{c}{\lambda_0} = \frac{c(a^2 + \ell^2)^{1/2}}{2a\ell} \tag{6.60}$$

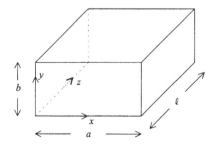

Figure 6.5 Rectangular waveguide resonator

This is independent of b as there is no field variation along the y axis for TE_{m0} modes. This resonant mode is called the TE_{101} mode.

The effects of finite losses in the conducting walls determine the unloaded Q of the resonator. This can be calculated by forming a volume integral of E_y to determine the stored energy, and dividing it by the dissipated energy due to currents in the walls of the resonator. The result, quoted here without proof, is [1]

$$Q_{u01} = \frac{\lambda}{\delta} \frac{abl}{2} \frac{\left(\frac{1}{a^2} + \frac{1}{l^2}\right)^{3/2}}{\frac{l}{a^2}(a+2b) + \frac{a}{l^2}(l+2b)} \qquad (6.61)$$

where λ/δ is the ratio of free space wavelength to skin depth at the resonant frequency. For silver

$$\frac{\lambda}{\delta} = \frac{1.479 \times 10^5}{\sqrt{f}} \qquad (6.62)$$

For brass

$$\frac{\lambda}{\delta} = \frac{7.462 \times 10^4}{\sqrt{f}} \qquad (6.63)$$

where f is the frequency in gigahertz.

6.2.5 Numerical example

A rectangular waveguide has an a dimension of 2 cm and a b dimension of 1 cm. Calculate the length l for a resonant frequency of 10 GHz and calculate the unloaded Q of the resonator, assuming it is silver-plated.

First we compute

$$l = \frac{\lambda_0}{2[1-(\lambda_0/2a)^2]^{1/2}} \qquad (6.64)$$

where

$$\lambda_0 = \frac{c}{f} = \frac{3 \times 10^8}{10^{10}} = 0.03 \qquad (6.65)$$

Hence

$$l = \frac{0.03}{2[1-(0.03/0.04)^2]^{1/2}} = 0.0226 \qquad (6.66)$$

That is,

$$l = 2.26 \text{ cm} \qquad (6.67)$$

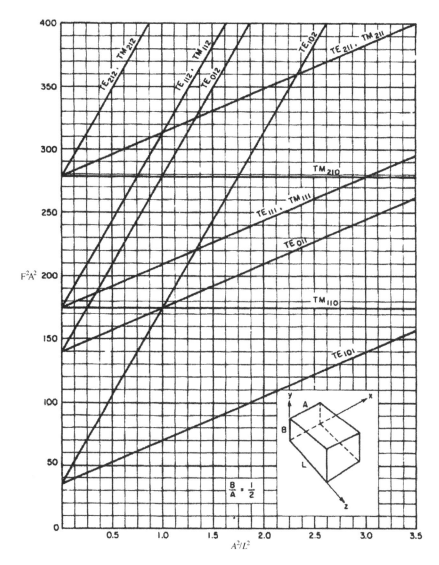

Figure 6.6 Mode chart for a rectangular waveguide resonator with $b = 2a$
(reprinted with permission from Matthaei, G., Young, L., and Jones, E.M.T.: 'Microwave filters, impedance matching networks and coupling structures' (Artech House, Norwood, MA, 1980); www.artechhouse.com)

Now

$$\frac{\lambda}{\delta} = \frac{1.479 \times 10^5}{\sqrt{(10)}} = 46\,770 \tag{6.68}$$

and from (6.61)

$$Q_{u\ 101} = 8009 \tag{6.69}$$

Note that although the b dimension does not affect the resonant frequency it does affect the Q. This is analogous to the ground plane spacing in TEM resonators.

6.2.6 Spurious resonances

As we have already found there are an infinite number of possible propagating modes in a waveguide, with an infinite number of cut-off frequencies. Consequently a waveguide resonator has an infinite number of resonant frequencies. It is useful to be able to predict these frequencies in order to gain insight into the spurious performance of a filter. The easiest way to predict these is to use a mode chart as shown in Figure 6.6. The chart enables a graphical method of predicting the resonant frequencies of a rectangular waveguide resonator with $b = 2a$. Here a, b and ℓ are measured in inches with f in gigahertz. Taking the previous example with $a = 2\,\text{cm}$ (0.787") and $\ell = 2.26\,\text{cm}$ (0.889") then $(a/\ell)^2 = 0.783$ and the chart gives the expected resonant frequency of 10 GHz for the TE_{101} mode. Moving vertically up the chart until we intersect the line for the first spurious (TE_{102}) mode, we obtain a resonant frequency of 15.65 GHz.

6.2.7 Circular waveguides

Circular waveguides are often used in filters because of the very high Q factors which can be obtained from the TE_{0N} modes. Furthermore they are often used in dual-mode configuration where two orthogonal degenerate modes (e.g. TE_{111}) exist in a single resonator. The analysis of circular waveguides is best done in a cylindrical coordinate system as shown in Figure 6.7. Again TEM modes cannot exist in circular waveguides; TE and TM modes will be treated separately.

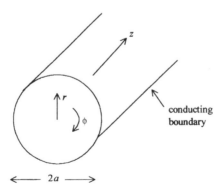

Figure 6.7   Circular waveguide

6.2.8 TE modes

The differential Helmholtz equation for H_z is [2]

$$\nabla_t^2 H_z = \frac{1}{r}\frac{\partial}{\partial r}\left(r\frac{\partial H_z}{\partial r}\right) + \frac{1}{r^2}\frac{\partial^2 H_z}{\partial \phi^2} = -k_c^2 H_z \tag{6.70}$$

where

$$k_c^2 = \gamma^2 + k^2 \tag{6.71}$$

The solution for this equation is

$$H_z(r, \phi) = [AJ_n(k_c r) + BN_n(k_c r)][C\cos(n\phi) + D\sin(n\phi)] \tag{6.72}$$

where J_n and N_n are nth-order Bessel functions of the first and second kind, respectively [3]. In fact the second kind of Bessel function has a singularity at $r = 0$ and thus cannot be a solution in regions which include the axis. In addition we will choose the orientation such that we only take the $\cos(n\phi)$ solution. Thus

$$H_z(r, \phi) = HJ_n(k_c r)\cos(n\phi) \tag{6.73}$$

The other field components can be found from Maxwell's equations in cylindrical coordinates with $E_z = 0$ and $\gamma = j\beta$, giving

$$E_r = \frac{-j\omega\mu}{rk_c^2}\frac{\partial H_z}{\partial \phi} \tag{6.74}$$

$$E_\phi = \frac{j\omega\mu}{k_c^2}\frac{\partial H_z}{\partial r} \tag{6.75}$$

$$H_r = \frac{-j\beta}{k_c^2}\frac{\partial H_z}{\partial r} \tag{6.76}$$

$$H_\phi = \frac{-j\beta}{k_c^2 r}\frac{\partial H_z}{\partial \phi} \tag{6.77}$$

Thus from (6.73)–(6.77)

$$E_r = Z_{TE} H_\phi = \frac{j\omega\mu n}{k_c^2 r} HJ_n(k_c r)\sin(n\phi) \tag{6.78}$$

$$E_\phi = -Z_{TE} H_r = \frac{j\omega\mu}{k_c} HJ_n'(k_c r)\cos(n\phi) \tag{6.79}$$

and

$$Z_{TE} = \frac{\omega\mu}{\beta} \tag{6.80}$$

The boundary condition for the waveguide is that the tangential E fields must be zero at the surface of the cylindrical conductor. Thus

$$E_\phi\big|_{r=a} = 0 \qquad \text{for all } \phi \tag{6.81}$$

Therefore

$$J_n'(k_c a) = 0 \tag{6.82}$$

Now
$$k_c^2 = \gamma^2 + k^2 = k^2 - \beta^2 \tag{6.83}$$

At cut-off $\beta = 0$ and k_c is thus the wavenumber at cut-off, i.e.

$$k_c = \frac{2\pi}{\lambda_c} = \frac{\omega_c}{\nu} \tag{6.84}$$

and

$$k_c a = \frac{2\pi a}{\lambda_c} = \frac{\omega_c a}{\nu} = p'(n, \ell) \tag{6.85}$$

where $p'(n, \ell)$ are the doubly infinite set of zeros of the derivatives of Bessel functions of the first kind. These zeros are given for the first three Bessel functions in Table 6.1.

The lowest zero of the derivatives which gives finite fields is the first zero of J_1' with a value of 1.841. This corresponds to the lowest cut-off frequency of all the TE modes, the TE_{11} mode. The cut-off frequency of this mode is given by

$$\frac{2\pi f_c}{\nu} = 1.841 \tag{6.86}$$

and if ν is the velocity of light in a vacuum then

$$a = \frac{8.79 \times 10^7}{f_c} \tag{6.87}$$

Thus for a cut-off frequency of 1 GHz, $a = 8.79$ cm.

In general the mode numbers are designated $TE_{n,\ell}$ where n denotes the angular variation in ϕ and ℓ is the variation in radial position r determined by the number of the zero. The next lowest cut-off frequency is for the TE_{01} mode with $p_{01} = 3.832$. For $a = 8.79$ cm this has a cut-off frequency of 2.08 GHz.

The field components for the TE_{01} mode are given by

$$H_z = HJ_0\left(\frac{3.832 r}{a}\right) \tag{6.88}$$

$$E_\phi = -Z_{TE} H_r = \frac{j\omega\mu}{K_c} HJ_0'\left(\frac{3.832 r}{a}\right) \tag{6.89}$$

$$E_r = H_\phi = 0 \tag{6.90}$$

These fields have no variation with ϕ as shown in Figure 6.8.

Table 6.1 Zeros of Bessel functions and their derivatives

J_0	J_1	J_2	J_0'	J_1'	J_2'
2.405	3.832	5.136	0	1.841	3.054
5.520	7.016	8.417	3.832	5.331	6.706
8.654	10.173	11.620	7.016	8.536	9.909

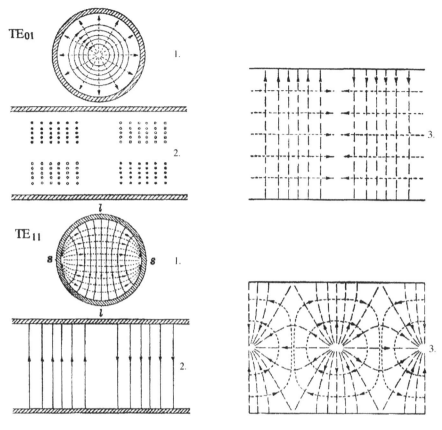

1. Cross-sectional view
2. Longitudinal view through plane $\ell-\ell$
3. Surface view through $s-s$

Figure 6.8 Field patterns for TE_{01} and TE_{11} modes in a circular waveguide

6.2.9 TM modes

For TM waves the solution of the Helmholtz equation is

$$E_z = E J_n(k_c r) \cos(n\phi) \tag{6.91}$$

$$E_r = Z_{TM} H_\phi = \frac{-j\beta}{k_c} E J_n'(k_c r) \cos(n\phi) \tag{6.92}$$

$$E_\phi = Z_{TM} H_r = \frac{-j\beta n}{k_c r} E J_n'(k_c r) \sin(n\phi) \tag{6.93}$$

216 *Theory and design of microwave filters*

where

$$Z_{TM} = \frac{\beta}{\omega\varepsilon} \qquad (6.94)$$

The boundary conditions require E_z and E_ϕ to be zero at $r = a$. Thus

$$k_c a = \frac{\omega_c a}{\nu} = \frac{2\pi a}{\lambda_c} = p_{n,\ell} \qquad (6.95)$$

where $p_{n,\ell}$ are the zeros of the Bessel function of order n. The lowest order mode is the TM_{01} mode with $p_{01} = 2.405$, giving a cut-off frequency (for $a = 8.78$ cm)

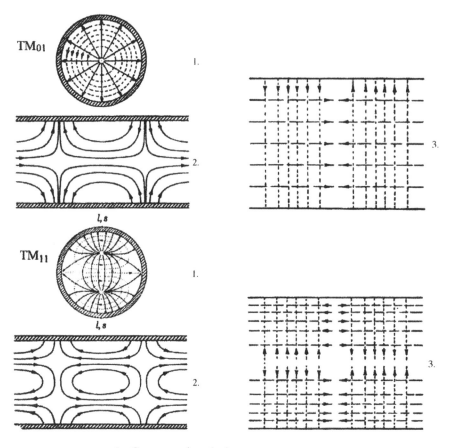

1. Cross-sectional view
2. Longitudinal view through plane $\ell-\ell$
3. Surface view through $s-s$

Figure 6.9 Field patterns for TM_{01} and TM_{11} modes in circular waveguides

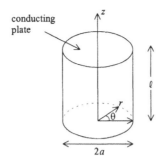

Figure 6.10 Circular waveguide resonator

of 1.306 GHz. The next lowest order mode is the TM_{11} mode with $p_{11} = 3.832$. This is degenerate with the TE_{01} mode. Field patterns for these two modes are shown in Figure 6.9.

6.2.10 Circular waveguide resonators

Resonators can be formed from circular waveguides by forming a cylindrical box with conducting plates across the surface of the waveguide as shown in Figure 6.10. In this case the modes are designated $TE_{\ell,m,n}$ or $TM_{\ell,m,n}$ where ℓ, m and n are the mode numbers for the number of variations along ϕ, r and z respectively.

The resonant frequencies for the various TE modes are given by

$$(fa)^2 = 224.6 \left[\left(\frac{p'_{\ell m}}{\pi}\right)^2 + n\left(\frac{a}{\ell}\right)^2 \right] \tag{6.96}$$

For TM modes simply substitute $p_{\ell m}$ for $p'_{\ell m}$ in (6.96).

More conveniently a mode chart for circular cylinder resonators is shown in Figure 6.11. The theoretical Q factors for various TE modes are shown in Figure 6.12.

6.2.11 Numerical example

Design a circular cylindrical resonator for operation in the TE_{011} mode with a resonant frequency of 10 GHz. Optimise the resonator for a reasonable compromise between unloaded Q and spurious-free performance.

From the mode chart a value of $(a/\ell)^2 = 0.5$ is reasonable for spurious-free performance. Note that the TE_{111} mode is degenerate with the TE_{011} mode and care must be taken not to excite this mode. From Figure 6.12 with $a/\ell = 1/\sqrt{2}$ we obtain

$$6.95 \times 10^{-6} \sqrt{f} \, Q_u \approx 0.64 \tag{6.97}$$

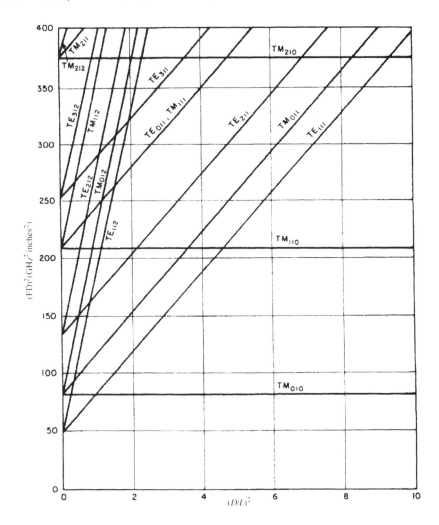

Figure 6.11 Mode chart for circular cylinder waveguide resonators
(Source: Montgomery, C.G.: 'Technique of microwave measurements' (McGraw Hill, New York, 1947))

Therefore

$$Q_u \approx 29\,000 \tag{6.98}$$

The dimensions are most accurately calculated using the formula rather than the mode chart with $p'_{\ell m} = 3.832$, $n = 1$ and $f = 10$. Hence

$$a^2 = \frac{224.6}{100}\left[\left(\frac{3.832}{\pi}\right)^2 + 0.5\right] \tag{6.99}$$

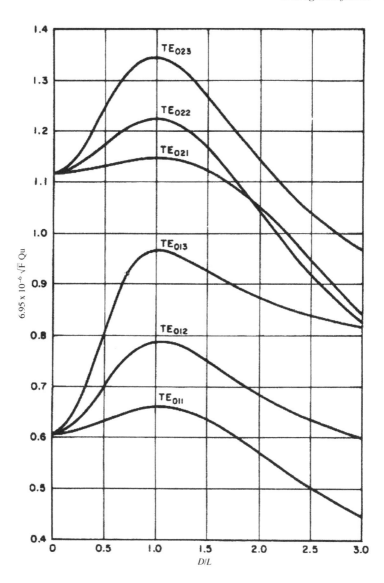

Figure 6.12 Theoretical unloaded Q for TE modes in circular cylinder waveguide resonators
(Source: Montgomery, C.G.: 'Technique of microwave measurements' (McGraw-Hill, New York, 1947))

Therefore $a = 2.113$ cm and $\ell = 2.988$ cm.

Moving up the mode chart the TE_{311} mode and TE_{112} modes both resonate at 10.83 GHz. In addition the TM_{110} mode resonates at 8.71 GHz.

6.3 Design of waveguide bandpass filters

Waveguide bandpass filters can be constructed from uniform lengths of waveguide loaded with shunt discontinuities. A particular example of a rectangular waveguide with posts connected across the broad wall of the guide is shown in Figure 6.13.

The principle of operation of the filter is that the posts act as shunt inductive discontinuities (with associated reference planes) and the sections of waveguide between posts are half wave resonators [4]. It will be shown that an inductive post embedded in a waveguide can behave as an impedance inverter over relatively broad bandwidths. Thus the physical structure has an equivalent circuit consisting of bandpass resonators separated by inverters, which is suitable for a bandpass filter. It now remains to develop a design theory.

From a theoretical viewpoint it is unimportant whether we are working with rectangular or circular guides. We can choose the appropriate guide and mode of operation from considerations such as physical size and Q.

A section of waveguide may be defined by its transfer matrix which is that of a transmission line with frequency-dependent propagation constant and characteristic impedance, e.g.

$$Z_0 = \frac{\eta \lambda_g}{\lambda_0} = \frac{\eta}{[1 - (\omega_c/\omega)^2]^{1/2}} \tag{6.100}$$

and

$$\beta = \frac{2\pi}{\lambda_g} = \frac{\omega}{\nu}\left[1 - \left(\omega_c/\omega\right)^2\right]^{1/2} \tag{6.101}$$

The characteristic impedance is unimportant, as we can normalise all the

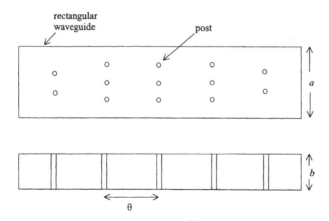

Figure 6.13 Rectangular waveguide bandpass filter

elements in the filter with respect to terminating impedances equivalent to an infinite uniform guide. Thus we can define the transfer matrix for a length of guide as

$$[T] = \begin{pmatrix} \cos(\theta) & j\sin(\theta) \\ j\sin(\theta) & \cos(\theta) \end{pmatrix} \tag{6.102}$$

where

$$\theta = \beta\ell = \frac{2\pi\ell}{\lambda_g} \tag{6.103}$$

or

$$\theta = \frac{\pi\lambda_{g0}}{\lambda_g} \tag{6.104}$$

Here λ_{g0} is proportional to the length of guide. In this particular case λ_{g0} has been chosen to be the guide wavelength when the guide is half a wavelength long, i.e.

$$\ell = \frac{\lambda_{g0}}{2} \tag{6.105}$$

Now we would like to introduce discontinuities into the guide to form inverters. Series discontinuities in waveguides are difficult to produce so we will only consider shunt elements. A shunt inductive discontinuity can be introduced by a vane in the side of a rectangular waveguide [5], although this is not good for suppressing spurious modes. This type of discontinuity can also be introduced

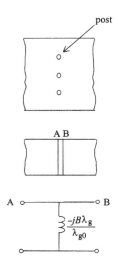

Figure 6.14 Shunt inductive discontinuity in a rectangular waveguide

222   Theory and design of microwave filters

by inserting posts across the broad wall of a rectangular guide operating in the dominant TE_{10} mode, or across the middle of a circular guide operating in the TE_{11} mode. Other types of inductive iris may be formed by holes in plates across the guide [6]. The equivalent circuit of the shunt discontinuity consists of a shunt inductor with appropriate reference planes (Figure 6.14).

The transfer matrix of the shunt inductor is

$$[T] = \begin{bmatrix} 1 & 0 \\ \dfrac{-jB\lambda_g}{\lambda_{g0}} & 1 \end{bmatrix} \tag{6.106}$$

The reference planes A and B are normally within the diameter of the post. The number of posts and their diameters determine the parameter B and the reference plane locations. B will be determined by the particular design which itself then determines the post structure.

We now modify the inductive iris section by symmetrically embedding it in a uniform section of waveguide of electrical length

$$\psi = \frac{\psi_0 \lambda_{g0}}{\lambda_g} \tag{6.107}$$

as shown in Figure 6.15.

The transfer matrix of the new section is

$$\begin{bmatrix} \cos(\psi/2) & j\sin(\psi/2) \\ j\sin(\psi/2) & \cos(\psi/2) \end{bmatrix} \begin{bmatrix} 1 & 0 \\ \dfrac{jB\lambda_g}{\lambda_{g0}} & 1 \end{bmatrix} \begin{bmatrix} \cos(\psi/2) & j\sin(\psi/2) \\ j\sin(\psi/2) & \cos(\psi/2) \end{bmatrix}$$

$$= \begin{bmatrix} \cos(\psi/2) + \dfrac{B\lambda_g}{\lambda_{g0}} \sin(\psi/2) & j\sin(\psi/2) \\ j\sin(\psi/2) - \dfrac{jB\lambda_g}{\lambda_{g0}} \cos(\psi/2) & \cos(\psi/2) \end{bmatrix} \begin{bmatrix} \cos(\psi/2) & j\sin(\psi/2) \\ j\sin(\psi/2) & \cos(\psi/2) \end{bmatrix}$$

$$= \begin{bmatrix} \cos^2(\psi/2) - \sin^2(\psi/2) + \dfrac{B\lambda_g}{\lambda_{g0}} \sin(\psi/2)\cos(\psi/2) & j\left[2\cos(\psi/2)\sin(\psi/2) + \dfrac{B\lambda_g}{\lambda_{g0}} \sin^2(\psi/2) \right] \\ j\left[2\cos(\psi/2)\sin(\psi/2) - \dfrac{B\lambda_g}{\lambda_{g0}} \cos^2(\psi/2) \right] & \cos^2(\psi/2) - \sin^2(\psi/2) + \dfrac{B\lambda_g}{\lambda_{g0}} \sin(\psi/2)\cos(\psi/2) \end{bmatrix}$$

$$= \begin{bmatrix} \cos(\psi) + \dfrac{B}{2}\dfrac{\lambda_g}{\lambda_{g0}} \sin(\psi) & j\left[\sin(\psi) + \dfrac{B\lambda_g}{\lambda_{g0}} \sin^2(\psi/2) \right] \\ j\left[\sin(\psi) - \dfrac{B\lambda_g}{\lambda_{g0}} \cos^2(\psi/2) \right] & \cos(\psi) + \dfrac{B}{2}\dfrac{\lambda_g}{\lambda_{g0}} \sin(\psi) \end{bmatrix} \tag{6.108}$$

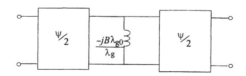

Figure 6.15   Shunt inductive iris embedded in a section of waveguide

Now we equate this to an inverter with the transfer matrix

$$[T] = \begin{bmatrix} 0 & \dfrac{-j\lambda_{g0}}{K\lambda_g} \\ \dfrac{-jK\lambda_g}{\lambda_{g0}} & 0 \end{bmatrix} \quad (6.109)$$

where K is the characteristic admittance when $\lambda_g = \lambda_{g0}$. Hence

$$\cos(\psi_0) + \frac{B}{2}\sin(\psi_0) = 0 \quad (6.110)$$

That is,

$$\psi_0 = -\tan^{-1}\left(\frac{2}{B}\right) \quad (6.111)$$

Now subtracting the C parameter from the B parameter in each matrix and equating at λ_{g0} we obtain

$$K - \frac{1}{K} = B[\sin^2(\psi/2) + \cos^2(\psi/2)] = B \quad (6.112)$$

The susceptance B of the inductive iris is positive. Therefore

$$1/K < K \text{ and } K > 1 \quad (6.113)$$

From (6.111) with B positive, ψ_0 must be negative and the line cannot be realised in isolation. In reality we shall be connecting the impedance inverter to lengths of waveguide and the negative lengths can be absorbed into these. Thus we can say that a shunt inductive iris can be represented by an inverter with reference planes defined by (6.111) and the physical reference planes in Figure 6.14.

It is useful to understand how well the iris approximates to an inverter over broad bandwidths since the design equations are only strictly correct at λ_{g0}. Examining the A parameter in (6.108) we have

$$A = \cos\left(\frac{\psi_0 \lambda_{g0}}{\lambda_g}\right) + \frac{B}{2}\frac{\lambda_g}{\lambda_{g0}}\sin\left(\frac{\psi_0 \lambda_{g0}}{\lambda_g}\right) \quad (6.114)$$

and from (6.111)

$$\frac{B}{2} = -\frac{1}{\tan(\psi_0)} \quad (6.115)$$

Hence

$$A = \cos\left(\frac{\psi_0 \lambda_{g0}}{\lambda_g}\right) - \frac{\lambda_g}{\lambda_{g0}\tan(\psi_0)}\sin\left(\frac{\psi_0 \lambda_{g0}}{\lambda_g}\right) \quad (6.116)$$

224 *Theory and design of microwave filters*

Differentiating A with respect to λ_g we obtain

$$\frac{dA}{d\lambda_g} = \frac{\psi_0 \lambda_{g0}}{\lambda_g^2} \sin\left(\frac{\psi_0 \lambda_{g0}}{\lambda_g}\right) - \frac{1}{\lambda_{g0}\tan(\psi_0)}\sin\left(\frac{\psi_0 \lambda_{g0}}{\lambda_g}\right)$$
$$+ \frac{\lambda_g}{\lambda_{g0}\tan(\psi_0)}\cos\left(\frac{\psi_0 \lambda_{g0}}{\lambda_g}\right)\left(\frac{\psi_0 \lambda_{g0}}{\lambda_g^2}\right) \quad (6.117)$$

and if ψ_0 is relatively small

$$\frac{dA}{d\lambda_g} \approx \frac{\psi_0^2 \lambda_{g0}^2}{\lambda_g} - \frac{1}{\lambda_g} + \frac{1}{\lambda_g} = \frac{\psi_0^2 \lambda_{g0}^2}{\lambda_g} \quad (6.118)$$

which is small for ψ_0 small; hence A is approximately zero over a relatively broad band.

Furthermore, examining the B parameter in (6.108) we obtain

$$B = \sin\left(\frac{\psi_0 \lambda_{g0}}{\lambda_g}\right) - \frac{2\lambda_g}{\tan(\psi_0)\lambda_{g0}}\sin^2\left(\frac{\psi_0 \lambda_{g0}}{2\lambda_g}\right) \quad (6.119)$$

Again differentiating with respect to λ_g we obtain

$$\frac{dB}{d\lambda_g} = -\frac{\psi_0 \lambda_{g0}}{\lambda_g^2}\cos\left(\frac{\psi_0 \lambda_{g0}}{\lambda_g}\right) - \frac{2}{\tan(\psi_0)\lambda_{g0}}\sin^2\left(\frac{\psi_0 \lambda_{g0}}{2\lambda_g}\right)$$
$$+ \frac{4\lambda_g}{\tan(\psi_0)\lambda_{g0}}\sin\left(\frac{\psi_0 \lambda_{g0}}{2\lambda_g}\right)\cos\left(\frac{\psi_0 \lambda_{g0}}{2\lambda_g}\right)\frac{\psi_0 \lambda_{g0}}{2\lambda_g^2} \quad (6.120)$$

and for ψ_0 small

$$\frac{dB}{d\lambda_g} = -\frac{\psi_0 \lambda_{g0}}{2\lambda_g^2} \quad (6.121)$$

Now the differential of the B parameter in (6.109) is given by

$$\frac{dB}{d\lambda_g} = \frac{\lambda_{g0}}{K\lambda_g^2} \quad (6.122)$$

and since ψ_0 is negative the differentials of the two B parameters have the same functional behaviour with respect to λ_g. Hence the inductive iris embedded in a waveguide is a good approximation to an inverter over broad bandwidths.

Inverters can also be formed from a shunt capacitive iris with the transfer matrix

$$[T] = \begin{bmatrix} 1 & 0 \\ \frac{jB\lambda_{g0}}{\lambda_g} & 1 \end{bmatrix} \quad (6.123)$$

The inverter can be formed again by embedding the iris in a section of waveguide. In this case the A parameter is

$$A = \cos\left(\frac{\psi_0 \lambda_{g0}}{\lambda_g}\right) - \frac{\lambda_{g0}}{\tan(\psi_0)\lambda_g}\sin\left(\frac{\psi_0 \lambda_{g0}}{\lambda_g}\right) \quad (6.124)$$

Differentiating with respect to λ_g and evaluating for small ψ we obtain

$$\frac{dA}{d\lambda_g} = \frac{\psi_0^2 \lambda_{g0}^2}{\lambda_g^3} + \frac{2\lambda_{g0}^2}{\lambda_g^3} \quad (6.125)$$

The second term in this expression is significant, even for small ψ, and hence the A parameter quickly deviates from zero when λ_g deviates from λ_0. Thus the inverter approximation is only valid for narrow bandwidths.

A design procedure for all-pole type waveguide bandpass filters will now be developed. Having established that a single inductive discontinuity embedded in a waveguide is a good approximation to an inverter then the waveguide bandpass filter in Figure 6.14 is equivalent to a cascade of UEs separated by inverters. Apart from the λ_{g0}/λ_g frequency dependence, the inverters only change the impedance level in the network and hence the entire network is equivalent to a cascade of UEs as shown in Figure 6.16.

The optimum equiripple bandpass response for a cascade of UEs is given by

$$|S_{12}(j\omega)|^2 = \frac{1}{1 + \varepsilon^2 T_N^2[\alpha \sin(\theta)]} \quad (6.126)$$

where

$$\theta = \frac{\pi \lambda_{g0}}{\lambda_g} \quad (6.127)$$

The response must be modified to take into account the frequency dependence of the inverters giving

$$|S_{12}(j\omega)|^2 = \frac{1}{1 + \varepsilon^2 T_N^2[\alpha(\lambda_g/\lambda_{g0})\sin(\pi(\lambda_{g0}/\lambda_g))]} \quad (6.128)$$

Given the degree N, the ripple level ε, and the two band-edge frequencies ω_1 and ω_2 we can find the two guide wavelengths λ_{g1} and λ_{g2} for the particular waveguide used. For example for a rectangular waveguide it was shown that

$$\lambda_g = \frac{\lambda_0}{[1 - (\omega_c/\omega)^2]^{1/2}} \quad (6.129)$$

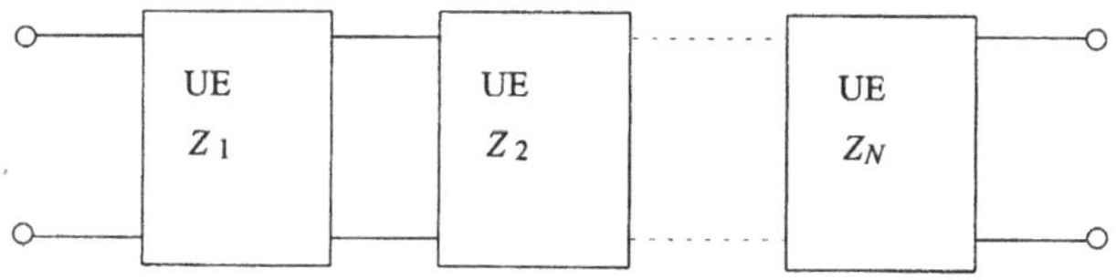

Figure 6.16   Equivalent circuit of a waveguide bandpass filter

Now equating $\omega = \pm 1$ in the lowpass prototype to λ_{g1} and λ_{g2} in the waveguide filter we have

$$\alpha \frac{\lambda_{g1}}{\lambda_{g0}} \sin\left(\frac{\pi \lambda_{g0}}{\lambda_{g1}}\right) = 1 \tag{6.130}$$

$$\alpha \frac{\lambda_{g2}}{\lambda_{g0}} \sin\left(\frac{\pi \lambda_{g0}}{\lambda_{g2}}\right) = -1 \tag{6.131}$$

Hence

$$\lambda_{g1} \sin\left(\frac{\pi \lambda_{g0}}{\lambda_{g1}}\right) + \lambda_{g2} \sin\left(\frac{\pi \lambda_{g0}}{\lambda_{g2}}\right) = 0 \tag{6.132}$$

This equation can be solved using the Newton–Raphson technique as follows. Let

$$F(\lambda_{g0}) = \lambda_{g1} \sin\left(\frac{\pi \lambda_{g0}}{\lambda_{g1}}\right) + \lambda_{g2} \sin\left(\frac{\pi \lambda_{g0}}{\lambda_{g2}}\right) \tag{6.133}$$

Then

$$F'(\lambda_{g0}) = \pi \cos\left(\frac{\pi \lambda_{g0}}{\lambda_{g1}}\right) + \pi \cos\left(\frac{\pi \lambda_{g0}}{\lambda_{g2}}\right) \tag{6.134}$$

We can make an initial approximation to λ_{g0} of

$$\lambda_{g0} \approx \frac{\lambda_{g1} + \lambda_{g2}}{2} \tag{6.135}$$

This is then modified to

$$\lambda_{g0} = \frac{\lambda_{g1} + \lambda_{g2}}{2} - \frac{F(\lambda_{g0})}{F'(\lambda_{g0})} \tag{6.136}$$

That is,

$$\lambda_{g0} = \frac{\lambda_{g1} + \lambda_{g2}}{2} + \frac{1}{\pi} \left[\frac{\lambda_{g1} \cos\left(\frac{\pi}{2} \frac{\lambda_{g2}}{\lambda_{g1}}\right) + \lambda_{g2} \cos\left(\frac{\pi}{2} \frac{\lambda_{g1}}{\lambda_{g2}}\right)}{\sin\left(\frac{\pi}{2} \frac{\lambda_{g2}}{\lambda_{g1}}\right) + \sin\left(\frac{\pi}{2} \frac{\lambda_{g1}}{\lambda_{g2}}\right)} \right] \tag{6.137}$$

Then from (6.130)

$$\alpha = \left[\frac{\lambda_{g1}}{\lambda_{g0}} \sin\left(\frac{\pi \lambda_{g0}}{\lambda_{g1}}\right) \right]^{-1} \tag{6.138}$$

The element values of a cascade of UEs with equiripple response were presented in Chapter 5. These may be modified to account for introducing impedance inverters as follows.

The element values are then

$$K'_{r,r+1} = \frac{K_{r,r+1}}{(Z_r Z_{r+1})^{1/2}} \tag{6.139}$$

$$Z_r = \frac{2\alpha \sin\left[\frac{(2r-1)\pi}{2N}\right]}{\eta}$$

$$- \frac{1}{4\eta\alpha}\left\{\frac{\eta^2 + \sin^2(r\pi/N)}{\sin[(2r+1)\pi/2_N]} + \frac{\eta^2 + \sin^2[(r-1)\pi/N]}{\sin[(2r-3)\pi/2_N]}\right\} \tag{6.140}$$

for $r = 1, \ldots, N$, and

$$K_{r,r+1} = \frac{[\eta^2 + \sin^2(r\pi/N)]^{1/2}}{\eta} \tag{6.141}$$

for $r = 0, \ldots, N$. Here

$$\eta = \sinh\left[\frac{1}{N}\sinh^{-1}\left(\frac{1}{\varepsilon}\right)\right] \tag{6.142}$$

Note that we have introduced unity impedance inverters at the input and output, which are needed to define the first and last UEs. Finally we would like to realise the filter in a uniform guide and thus we scale the internal impedance level of the filter to obtain

$$K'_{r,r+1} = \frac{K_{r,r+1}}{(Z_r Z_{r+1})^{1/2}} \quad (r = 1, \ldots, N) \tag{6.143}$$

with

$$Z_0 = Z_{N+1} = 1 \tag{6.144}$$

The design process may now be summarised as follows.

First we assume that the band-edge frequencies f_1 and f_2 are known. The ripple level ε is determined by the required return loss. Given the frequency band the waveguide size can be selected from standard sizes. The cut-off frequency of the waveguide is then determined, and λ_{g1} and λ_{g2} can be found from (6.129), λ_{g0} from (6.137) and α from (6.138).

The degree of the filter can be determined by analysis of the insertion loss function

$$L = 10\log_{10}\left\{1 + \varepsilon^2 T_N^2\left[\alpha \frac{\lambda_g}{\lambda_{g0}} \sin\left(\pi \frac{\lambda_{g0}}{\lambda_g}\right)\right]\right\} \tag{6.145}$$

with N chosen to meet the most severe specification on insertion loss. Z_r and $K_{r,r+1}$ can then be found from (6.140)–(6.142).

From (6.112) and (6.143) we obtain the susceptances of the inductive irises

$$B_{r,r+1} = \frac{(Z_r Z_{r+1})^{1/2}}{K_{r,r+1}} - \frac{K_{r,r+1}}{(Z_r Z_{r+1})^{1/2}} \tag{6.146}$$

and the negative lengths of guide are subtracted from a half wavelength to obtain the actual lengths of guide between the irises, ψ_r, giving

$$\psi_r = \pi - \frac{1}{2}\left[\cot^{-1}\left(\frac{B_{r-1,r}}{2}\right) + \cot^{-1}\left(\frac{B_{r,r+1}}{2}\right)\right] \quad (6.147)$$

where the electrical length of π corresponds to a physical distance $\lambda_{g0}/2$.

Finally the electrical parameters for the irises must be converted into actual physical dimensions. The simplest method of producing shunt inductive irises is to use circular posts of fixed diameter shorted across the broad wall of the waveguide. A number of posts can be used depending on the required susceptance. Normally they are located symmetrically across the guide in order to suppress higher order mode propagation through the waveguide structure. With a fixed number of posts a fine adjustment in susceptance can be made by adjusting the distances between posts.

The data for post susceptances can be obtained experimentally. For example, a single cavity filter can be constructed using two identical irises and the measured insertion loss can then be used to deduce the susceptances and the reference planes for the irises.

6.3.1 Design example

We shall design a rectangular waveguide filter to meet the following specification:

Passband	8.5–9.5 GHz
Return loss	≥ 20 dB
Stopband insertion loss	≥ 25 dB at 10.5 GHz
	≥ 40 dB at 8 GHz

Thus $f_1 = 8.5$ GHz, $f_2 = 9.5$ GHz and $\varepsilon = 0.1$.

A suitable waveguide is WG16 with internal dimensions of 22.86 mm × 10.16 mm. This has a TE$_{10}$ mode cut-off frequency of 6.56 GHz. λ_{g1} and λ_{g2} are determined from

$$\lambda_g = \frac{\lambda_0}{[1 - (\omega_c/\omega)^2]^{1/2}} \quad (6.148)$$

with $\lambda_{01} = 35.29$ mm and $\lambda_{02} = 31.58$ mm. Thus $\lambda_{g1} = 55.49$ mm and $\lambda_{g2} = 43.66$ mm. From (6.137) we obtain $\lambda_{g0} = 49.611$ mm and $\alpha = 2.7367$.

The required degree is determined by analysis of (6.145) with various values of N. For $N = 5$ we have

$$T_5(x) = 16x^5 - 20x^3 + 5x \quad (6.149)$$

Substituting for $T_5(x)$ in (6.148) gives the frequency response shown in Figure 6.17 and the response meets the desired specifications. Here we see that the frequency response of the filter is more selective on the low frequency side of the passband. This is due to the transmission zeros introduced by the

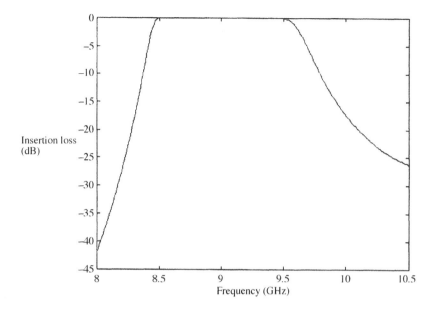

Figure 6.17   Transfer function of a waveguide bandpass filter

finite cut-off frequency of the waveguide and also by the relatively wide passband bandwidth.

Having determined $N = 5$ the prototype element values (normalised to 1 Ω) are determined from (6.140) to (6.143) giving

$$Z_1 = Z_5 = 2.71338$$
$$Z_2 = Z_4 = 6.42334 \tag{6.150}$$
$$Z_3 = 8.13821$$

$$K_{01} = K_{56} = 1$$
$$K_{12} = K_{45} = 1.36144 \tag{6.151}$$
$$K_{23} = K_{34} = 1.79848$$

The susceptances of the inductive irises are then determined from (6.146).

$$B_{01} = B_{56} = 1.0402$$
$$B_{12} = B_{23} = 2.7404 \tag{6.152}$$
$$B_{23} = B_{34} = 3.7714$$

The phase lengths of the guide between the irises are determined from (6.147).

$$\psi_1 = \psi_5 = 2.2807$$
$$\psi_2 = \psi_4 = 2.5825 \qquad (6.153)$$
$$\psi_3 = 2.654$$

The phase lengths are in radians and the physical lengths are given by

$$\ell_r = \frac{\psi_r}{\pi} \frac{\lambda_{g0}}{2} \qquad (6.154)$$

Thus

$$\ell_1 = \ell_5 = 18.01 \text{ mm}$$
$$\ell_2 = \ell_4 = 20.39 \text{ mm} \qquad (6.155)$$
$$\ell_3 = 20.96 \text{ mm}$$

The final circuit of the filter is shown in Figure 6.18.

6.4 The generalised direct-coupled cavity waveguide filter

Certain applications such as satellite communications require very severe filtering functions. Low passband loss may be combined with extreme selectivity and group delay linearity requirements. In such cases, conventional all-pole transfer functions may not be suitable and filters with generalised Chebyshev characteristics are required.

Classical cascade synthesis procedures enable transmission zeros to be independently realised by cascades of two-port networks such as the Brune section. However, when the transmission zeros are on the real axis or in the complex plane it is usually more convenient to synthesise cross-coupled ladder network prototypes as described in Chapter 3. Certain transfer functions are then realisable by the generalised direct-coupled cavity waveguide filter described in this section [7].

The symmetrical generalised direct-coupled cavity waveguide filter is shown in Figure 6.19. It consists of two identical shunt inductive-iris coupled waveguide structures where adjacent cavities in the two halves are cross-coupled through apertures in the common narrow wall. The equivalent circuit for the

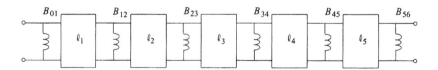

Figure 6.18 Final circuit for the waveguide bandpass filter design

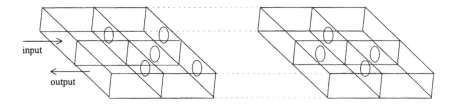

Figure 6.19 The generalised direct-coupled cavity waveguide filter

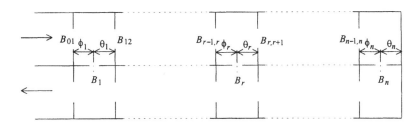

Figure 6.20 Midband susceptances and electrical lengths defining the generalised direct-coupled cavity waveguide filter

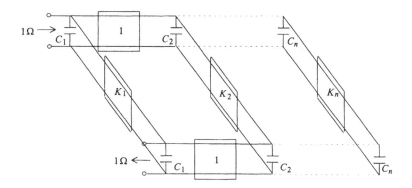

Figure 6.21 Cross-coupled array lowpass prototype

filter will be derived and equated to the cross-coupled lowpass prototype filter at midband. Formulae for the iris susceptances and the electrical lengths of the cavities, shown in Figure 6.20, then follow directly.

The cross-coupled lowpass prototype network is shown in Figure 6.21. Note that for an Nth-degree filter $n = N/2$ in this prototype. The even-mode equivalent circuit of this prototype network is shown in Figure 6.22. The odd-mode equivalent circuit is simply the complex conjugate of the even-mode circuit.

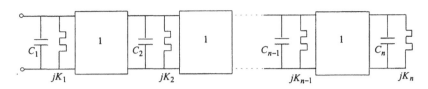

Figure 6.22 The even-mode equivalent circuit of the cross-coupled array lowpass prototype

The equivalent circuit of a typical section of the generalised direct-coupled cavity waveguide filter is shown in Figure 6.23. In this equivalent circuit the main lines consist of UEs with unity impedance and electrical lengths ϕ_r and θ_r. These are separated by shunt inductive susceptances which are normalised to λ_{g0} in a similar way to the conventional waveguide filter. The equivalent circuit of the cross-coupling arm is interesting as it consists of a cascade of two frequency-dependent inverters separated by a shunt inductive iris. The iris susceptance is normalised to the admittance of the inverters and λ_{g0}. The shunt susceptance is similar in form to the susceptances in the main branch as the input admittance of the shunt inverter terminated in the susceptance is given by

$$Y(j\omega) = \frac{K^2}{-j(\lambda_g \lambda_{g0}/4 a^2 B_r)} = \frac{j\lambda_g B_r}{\lambda_{g0}} \tag{6.156}$$

The reason that the inverters exist in the cross-coupling branch is because the waveguide is a two-dimensional structure and one has to consider the spatial

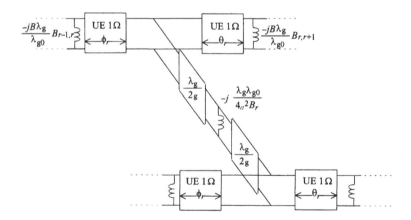

Figure 6.23 Equivalent circuit for a typical section of the generalised direct-coupled cavity waveguide filter

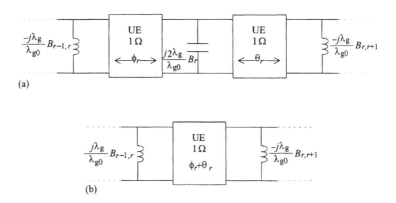

Figure 6.24 Even and odd-mode equivalent circuits of a typical section of the generalised direct-coupled cavity filter

distance from the centre to the edge of the waveguide. Note that the equivalent circuit of the waveguide filter is for a single mode of propagation which can only be assumed correct if the cross-coupling apertures are small, which is true for narrowband designs.

The even- and odd-mode equivalent circuits for the typical filter section may be derived by applying open and short circuits along the line of symmetry in Figure 6.23. These are shown in Figure 6.24. The even- and odd-mode equivalent circuits of the filter will now be transformed into an equivalent form to the lowpass prototype. First, in a similar way to the conventional filter, we consider a shunt inductor symmetrically located in a unit impedance guide (Figure 6.25) of phase length Φ where

$$\Phi = \frac{\Phi_0 \lambda_{g0}}{\lambda_g} \tag{6.157}$$

The transfer matrix of this section is

$$\begin{bmatrix} \cos\left(\frac{\Phi}{2}\right) & j\sin\left(\frac{\Phi}{2}\right) \\ j\sin\left(\frac{\Phi}{2}\right) & \cos\left(\frac{\Phi}{2}\right) \end{bmatrix} \begin{bmatrix} 1 & 0 \\ \frac{-jB\lambda_g}{\lambda_{g0}} & 1 \end{bmatrix} \begin{bmatrix} \cos\left(\frac{\Phi}{2}\right) & j\sin\left(\frac{\Phi}{2}\right) \\ j\sin\left(\frac{\Phi}{2}\right) & \cos\left(\frac{\Phi}{2}\right) \end{bmatrix}$$

$$= \begin{bmatrix} \cos(\Phi) + \dfrac{\lambda_g B \sin(\Phi)}{2\lambda_{g0}} & j\left\{\sin(\Phi) + \dfrac{B\lambda_g}{2\lambda_{g0}}[1-\cos(\Phi)]\right\} \\ j\left\{\sin(\Phi) - \dfrac{B\lambda_g}{2\lambda_{g0}}[1+\cos(\Phi)]\right\} & \cos(\Phi) + \dfrac{B\lambda_g \sin(\Phi)}{2\lambda_{g0}} \end{bmatrix}$$

(6.158)

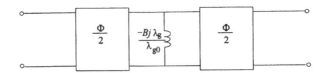

Figure 6.25   Shunt inductive iris embedded in a length of guide

Equating this at midband to an inverter of admittance K yields

$$\Phi = -\cot^{-1}\left(\frac{B}{2}\right) \tag{6.159}$$

$$B = K - \frac{1}{K} \tag{6.160}$$

The section may be represented over relatively broad bands by a frequency-dependent inverter with the transfer matrix

$$[T] = \begin{bmatrix} 0 & \dfrac{j\lambda_{g0}}{K\lambda_g} \\ \dfrac{jK\lambda_g}{\lambda_{g0}} & 0 \end{bmatrix} \tag{6.161}$$

Justification of the validity of this representation over broad bandwidths was given in the previous section.

Now consider the transfer matrix of a shunt susceptance $2B\lambda_g/\lambda_{g0}$ of length θ where

$$\theta = \frac{\theta_0 \lambda_{g0}}{\lambda_g} \quad (\theta \approx \pi) \tag{6.162}$$

$$[T] = \begin{bmatrix} \cos\left(\dfrac{\theta}{2}\right) & j\sin\left(\dfrac{\theta}{2}\right) \\ j\sin\left(\dfrac{\theta}{2}\right) & \cos\left(\dfrac{\theta}{2}\right) \end{bmatrix} \begin{bmatrix} 1 & 0 \\ \dfrac{-2jB\lambda_g}{\lambda_{g0}} & 1 \end{bmatrix} \begin{bmatrix} \cos\left(\dfrac{\theta}{2}\right) & j\sin\left(\dfrac{\theta}{2}\right) \\ j\sin\left(\dfrac{\theta}{2}\right) & \cos\left(\dfrac{\theta}{2}\right) \end{bmatrix}$$

$$= \begin{bmatrix} \cos(\theta) - \dfrac{B\lambda_g}{\lambda_{g0}}\sin(\theta) & j\left\{\sin(\theta) - \dfrac{B\lambda_g}{\lambda_{g0}}[1-\cos(\theta)]\right\} \\ j\left\{\sin(\theta) + \dfrac{B\lambda_g}{\lambda_{g0}}[1+\cos(\theta)]\right\} & \cos(\theta) - \dfrac{B\lambda_g}{\lambda_{g0}}\sin(\theta) \end{bmatrix}$$

$$\tag{6.163}$$

Now consider the pi network shown in Figure 6.26. This has a transfer

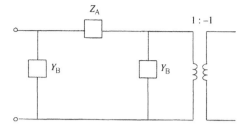

Figure 6.26 Pi section equivalent for the susceptance loaded line

matrix

$$[T] = \begin{bmatrix} 1 & 0 \\ Y_B & 1 \end{bmatrix} \begin{bmatrix} 1 & Z_A \\ 0 & 1 \end{bmatrix} \begin{bmatrix} 1 & 0 \\ Y_B & 1 \end{bmatrix} \begin{bmatrix} -1 & 0 \\ 0 & -1 \end{bmatrix}$$

$$= \begin{bmatrix} -(1 + Z_A Y_B) & -Z_A \\ -Y_B(2 + Y_B Z_A) & -(1 - Y_B Z_A) \end{bmatrix} \quad (6.164)$$

By equating the B parameters of the two transfer matrices we obtain

$$Z_A = -j\left[\sin(\theta) - \frac{B\lambda_g}{\lambda_{g0}}[1 - \cos(\theta)]\right] \quad (6.165)$$

and equating the A parameters

$$1 + Z_A Y_B = \frac{B\lambda_g}{\lambda_{g0}}\sin(\theta) - \cos(\theta) \quad (6.166)$$

From (6.165) and (6.166)

$$Y_B = -j\cot\left(\frac{\theta}{2}\right) \quad (6.167)$$

Now applying these two results to the even-mode and odd-mode equivalent circuits of the waveguide filter we obtain the equivalent circuit shown in Figure 6.27. Here from (6.165)

$$Z_{re} = -j\left\{\sin\left(\frac{\theta_r \lambda_{g0}}{\lambda_g}\right) - \frac{B_r \lambda_g}{\lambda_{g0}}\left[1 - \cos\left(\frac{\theta_r \lambda_{g0}}{\lambda_g}\right)\right]\right\} \quad (6.168)$$

In the odd-mode case the embedded shunt susceptance is zero and we obtain

$$Z_{ro} = -j\sin\left(\frac{\theta_r \lambda_{g0}}{\lambda_g}\right) \quad (6.169)$$

Now scaling through the λ_{g0}/λ_g frequency dependence of the inverters in

236 *Theory and design of microwave filters*

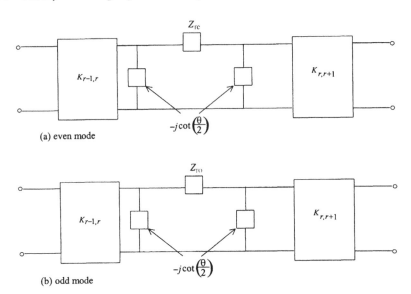

Figure 6.27 Even- and odd-mode equivalent circuits of a basic section of generalised direct-coupled cavity filter

(6.161) we obtain

$$Z_{re} = \frac{-j\lambda_g}{\lambda_{g0}} \left\{ \sin\left(\frac{\theta_r \lambda_{g0}}{\lambda_g}\right) - \frac{B_r \lambda_g}{\lambda_{g0}} \left[1 - \cos\left(\frac{\theta_r \lambda_{g0}}{\lambda_g}\right)\right] \right\} \quad (6.170)$$

and

$$Z_{ro} = \frac{-j\lambda_g}{\lambda_{g0}} \sin\left(\frac{\theta_r \lambda_{g0}}{\lambda_g}\right) \quad (6.171)$$

The shunt elements then become

$$\frac{-j\lambda_g}{\lambda_{g0}} \cot\left(\frac{\theta}{2}\right) \quad (6.172)$$

For narrow bandwidths the inverters are large relative to the shunt elements and they may be neglected.

The even-mode equivalent circuit of the lowpass prototype shown in Figure 6.24 may now be transformed into the network shown in Figure 6.28. This is achieved by introducing a redundant inverter at the input and scaling the network admittance level at each internal node in order to obtain equal-valued series inductive elements. Again the odd-mode circuit is simply the complex conjugate of the even-mode circuit. We can now see the similarity between the lowpass prototype and the equivalent circuit of the waveguide filter.

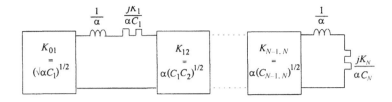

Figure 6.28 Transformed (dual) lowpass prototype even-mode circuit with equal-valued inductors

Now applying the bandpass transformation

$$\omega \to -\frac{\alpha \lambda_g}{\lambda_{g0}} \sin\left(\frac{\pi \lambda_{g0}}{\lambda_g}\right) \tag{6.173}$$

to the lowpass prototype, the series elements in Figure 6.28 become

$$\frac{jK_r}{\alpha C_r} + \frac{j\omega}{\alpha} \Rightarrow \frac{jK_r}{\alpha C_r} \frac{-j\lambda_g}{\lambda_{g0}} \sin\left(\frac{\pi \lambda_{g0}}{\lambda_g}\right) \tag{6.174}$$

and equating with the waveguide filter in Figure 6.27, at midband we obtain in the odd-mode case

$$\frac{-\lambda_g}{\lambda_{g0}} \sin\left(\frac{\pi \lambda_{g0}}{\lambda_g}\right) + \frac{K_r}{\alpha C_r} = \frac{-\lambda_g}{\lambda_{g0}} \sin\left(\frac{\theta_r \lambda_{g0}}{\lambda_g}\right) \tag{6.175}$$

and in the even-mode case

$$\frac{-\lambda_g}{\lambda_{g0}} \sin\left(\frac{\pi \lambda_{g0}}{\lambda_g}\right) - \frac{K_r}{\alpha C_r} = \frac{-\lambda_g}{\lambda_{g0}} \left\{ \sin\left(\frac{\theta_r \lambda_{g0}}{\lambda_g}\right) - \frac{B_r \lambda_g}{\lambda_{g0}} \left[1 - \cos\left(\frac{\theta_r \lambda_{g0}}{\lambda_g}\right)\right]\right\} \tag{6.176}$$

Evaluating at $\lambda_g = \lambda_{g0}$ we obtain

$$\theta_r = \pi - \sin^{-1}\left(\frac{K_r}{\alpha C_r}\right) \tag{6.177}$$

and

$$B_r = \frac{K_r}{\alpha C_r} \tag{6.178}$$

Now over a narrowband around $\lambda_g = \lambda_{g0}$ the bandpass frequency transformation (6.173) reduces to

$$\omega \to \alpha \pi \left(1 - \frac{\lambda_g}{\lambda_{g0}}\right) \tag{6.179}$$

Let λ_{s1} and λ_{s2} be the guide wavelengths at the band-edge frequencies f_1 and f_2.

If the lowpass prototype cuts off at $\omega = \pm 1$ then

$$-1 = \alpha\pi\left(1 - \frac{\lambda_{g1}}{\lambda_{g0}}\right) \tag{6.180}$$

$$+1 = \alpha\pi\left(1 - \frac{\lambda_{g2}}{\lambda_{g0}}\right) \tag{6.181}$$

Adding (6.180) and (6.181) we obtain

$$\frac{\lambda_{g1}}{\lambda_{g0}} + \frac{\lambda_{g2}}{\lambda_{g0}} = 2 \tag{6.182}$$

Therefore

$$\lambda_{g0} = \frac{\lambda_{g1} + \lambda_{g2}}{2} \tag{6.183}$$

Subtracting we obtain

$$2 = \alpha\pi\left(\frac{\lambda_{g1}}{\lambda_{g0}} - \frac{\lambda_{g2}}{\lambda_{g0}}\right) \tag{6.184}$$

Hence from (6.183) and (6.182)

$$\alpha = \frac{\lambda_{g1} + \lambda_{g2}}{\pi(\lambda_{g1} - \lambda_{g2})} \tag{6.185}$$

To summarise the design equations we have

$$B_{r,r+1} = \alpha(C_r C_{r+1})^{1/2} - \frac{1}{\alpha(C_r C_{r+1})^{1/2}} \tag{6.186}$$

$$B_r = \frac{K_r}{\alpha C_r} \tag{6.187}$$

$$\Phi_r = \frac{\pi}{2} - \frac{1}{2}\left[\cot^{-1}\left(\frac{B_{r-1,r}}{2}\right) + \sin^{-1}(B_r)\right] \tag{6.188}$$

and

$$\Phi_r = \frac{\pi}{2} - \frac{1}{2}\left[\cot^{-1}\left(\frac{B_{r,r+1}}{2}\right) + \sin^{-1}(B_r)\right] \tag{6.189}$$

where

$$C_0 = \frac{1}{\alpha} \qquad C_{N+1} = \infty \tag{6.190}$$

and as in the case of the conventional waveguide filter an electrical length of π radians corresponds to λ_{g0}.

The generalised direct-coupled cavity filter has certain limitations as only positive couplings may be realised. Thus the locations of transmission zeros are restricted and cannot be on the imaginary axis. However, it also forms a useful building block when combined with the extracted pole filter as described in the next section.

6.5 Extracted pole waveguide filters

The generalised direct-coupled cavity filter described in the previous section has certain limitations. The structure is ideal for linear phase filters with monotonic stopbands where all the couplings are of the same sign. However, there are difficulties in realising filters with real frequency transmission zeros as these require both signs of couplings, which are difficult to achieve in simple waveguide structures. Furthermore, consider the case where real frequency transmission zeros are required to be symmetrically located on either side of the passband. In the fourth-degree case a pair of zeros would be associated with a single cross-coupling around all four resonators. Thus the pair of transmission zeros are not independently tunable. This can cause problems in manufacturing associated with sensitivity. It is more desirable to be able to extract individual transmission zeros above or below the passband in an independent manner.

This may be achieved by cross-coupled networks in which the cross-coupling is realised by a coupling around three resonators, as described in Chapter 3. This is a useful technique for coaxial resonator and dielectric resonator filters. It is not always suitable for waveguide filters because of the difficulties in coupling around three resonators which requires coupling from both narrow and broad walls of the waveguide. More importantly, when the transmission zeros are located very close to the passband-edge the values of the required couplings may be physically unrealisable.

It is often more desirable to synthesise the transmission zeros using bandstop resonators or extracted poles. Each bandstop resonator corresponds to a transmission zero (and one return loss ripple) and thus independent tuning is achieved. The optimum solution for complex transfer functions is to use a combination of extracted poles and cross-couplings as required. A technique for achieving this will be described in this section. One of the main advantages of this method is that the entire structure may be realised with positive couplings enabling a simple realisation for TE_{011} mode waveguide filters [8].

The only real restriction is that the synthesis procedure is limited to transfer functions with a symmetrical frequency response. The networks thus have complex conjugate symmetry. In other words a bandstop resonator at one end of the filter producing a transmission zero on one side of the passband always has its associated complex conjugate at the other end of the filter producing the transmission zero on the other side of the passband.

Complex conjugate symmetry means that at any stage in the synthesis procedure the transfer matrix will always be of the form

$$[T] = \frac{1}{jF}\begin{bmatrix} A_1 + jA_2 & B \\ C & A_1 - jA_2 \end{bmatrix} \tag{6.191}$$

where A_1 is odd and F, A_2, B and C are even polynomials in p. Furthermore, by reciprocity

$$A_1^2 + A_2^2 - BC = F^2 \qquad (6.192)$$

The synthesis procedure will be developed along with an example, in this case a degree 6 equiripple prototype with 20 dB return loss, two transmission zeros at infinity, a pair at $p = \pm j1.414$ and a pair on the real axis at $p = \pm 0.953076$.

Generation of the polynomial for $S_{12}(p)$ and synthesis of the cross-coupled lowpass prototype is described in Chapter 3. The network may be synthesised as a cross-coupled ladder as shown in Figure 6.29. Here

$$\begin{aligned} C_1 &= 1.00367 \\ C_2 &= 1.43354 \\ C_3 &= 1.938932 \end{aligned} \qquad (6.193)$$

$$\begin{aligned} K_1 &= -0.078796 \\ K_2 &= -0.000237 \\ K_3 &= 1.184116 \end{aligned} \qquad (6.194)$$

To proceed with the extracted pole synthesis we must first form the transfer matrix of the filter in order that it can be re-synthesised in the correct form. The even-mode admittance is

$$\begin{aligned} Y_e &= C_1 p + jK_1 + \frac{1}{C_2 p + jK_2 + 1/(C_3 p + jK_3)} \\ &= \frac{2.78056p^3 + j1.47862p^2 + 3.0733p + j1.1053}{2.77954p^2 + j1.69702p + 1.00028} \end{aligned} \qquad (6.195)$$

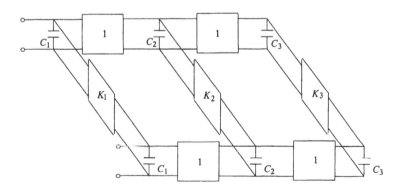

Figure 6.29 Cross-coupled ladder realisation of a degree 6 generalised Chebyshev filter

The odd-mode admittance $Y_o = Y_e^*$ and the transfer matrix is given by

$$[T] = \frac{1}{jF}\begin{bmatrix} A & B \\ C & A \end{bmatrix} = \frac{1}{Y_e - Y_o}\begin{bmatrix} Y_e + Y_o & 2 \\ 2Y_e Y_o & Y_e + Y_o \end{bmatrix} \quad (6.196)$$

Thus

$$F = 1.10561 - 0.664187p^2 - 0.608764p^4 \quad (6.197)$$

and F contains the real frequency transmission zero. Thus

$$F = (p^2 + 2)(0.552973 - 0.608764p^2) \quad (6.198)$$

and

$$A = 4.94987p + 13.8329p^3 + 7.72866p^5 \quad (6.199)$$

$$B = 1.00056 + 8.4405p^2 + 7.72582p^4 \quad (6.200)$$

$$C = 1.22168 + 12.7138p^2 + 19.2773p^4 + 7.7315p^6 \quad (6.201)$$

(Here $A_1 = A$ and $A_2 = 0$.)

The synthesis procedure starts by extracting a unity impedance phase shifter of phase length ψ_1 from the output of the network and its conjugate $-\psi_1$ from the input. The remaining transfer matrix is obtained by pre- and post-multiplying the transfer matrix by the inverse transfer matrix, yielding

$$[T] = \frac{1}{jF}\begin{bmatrix} \cos(\psi_1) & j\sin(\psi_1) \\ j\sin(\psi_1) & \cos(\psi_1) \end{bmatrix}\begin{bmatrix} A_1 + jA_2 & B \\ C & A_1 - jA_2 \end{bmatrix}\begin{bmatrix} \cos(\psi_1) & -j\sin(\psi_1) \\ -j\sin(\psi_1) & \cos(\psi_1) \end{bmatrix}$$

$$= \frac{1}{jF}\begin{bmatrix} A_1 + j\left[A_2\cos(2\psi_1) + \dfrac{C-B}{2}\sin(2\psi_1)\right] & B\cos^2(\psi_1) + C\sin^2(\psi) + A_2\sin(2\psi_1) \\ C\cos^2(\psi_1) + B\sin^2(\psi_1) - A_2\sin(2\psi_1) & A_1 - j\left[A_2\cos(2\psi_1) + \dfrac{C-B}{2}\sin(2\psi_1)\right] \end{bmatrix}$$

$$(6.202)$$

We now choose the value of ψ_1 such that the B parameter has a factor $p^2 + \omega_1^2$. This enables the transmission zeros to be extracted by shunt resonators. Hence

$$B\cos^2(\psi_1) + C\sin^2(\psi_1) + A_2\sin(2\psi_1) = 0\big|_{p=\pm j\omega_1} \quad (6.203)$$

or

$$\frac{B}{C} + \tan^2(\psi_1) + \frac{2A_2}{C}\tan(\psi_1) = 0\big|_{p=\pm j\omega_1} \quad (6.204)$$

Hence

$$t_1 = \tan(\psi_1) = \frac{-A_2 \pm (A_2^2 - BC)^{1/2}}{C} \quad (6.205)$$

Now from (6.192)

$$A_2^2 - BC = F^2 - A_1^2 \quad (6.206)$$

242 Theory and design of microwave filters

and F contains the transmission zeros as factors. Thus

$$A_2^2 - BC\big|_{p=\pm j\omega_1} = -A_1^2 \tag{6.207}$$

Thus from (6.205) and (6.207) a solution for t_1 is

$$t_1 = \frac{-A_2 + jA_1}{C}\bigg|_{p=j\omega_1} = \frac{B}{-A_2 - jA_1}\bigg|_{p=j\omega_1} \tag{6.208}$$

This results in the new A parameter possessing a factor $p + j\omega_1$ and the new D parameter possessing a factor $p - j\omega_1$. The transfer matrix is then given by

$$[T] = \frac{1}{jF} \begin{bmatrix} j(p+j\omega_1)A_1' + jA_2') & B'(p^2+\omega_1^2) \\ C' & -j(p-\omega_1)(A_1' - jA_2') \end{bmatrix} \tag{6.209}$$

where

$$A_1' + jA_2' = \frac{A_2 \cos(2\psi_1) + [(C-B)/2]\sin(2\psi_1) - jA_1}{p + j\omega_1} \tag{6.210}$$

$$B' = \frac{B\cos^2(\psi_1) + C\sin^2(\psi_1) + A_2 \sin(2\psi_1)}{p^2 + \omega_1^2} \tag{6.211}$$

$$C' = C\cos^2(\psi_1) + B\sin^2(\psi_1) - A_2 \sin(2\psi_1) \tag{6.212}$$

Now from (6.208) in the example

$$t_1 = \frac{j(4.94987p + 13.8329p^3 + 7.72866p^5)}{1.22168 + 12.7138p^2 + 19.2773p^4 + 7.7315p^6}\bigg|_{p=j\sqrt{2}} \tag{6.213}$$

and

$$\psi_1 = -52.3531° \tag{6.214}$$

The matrix is then evaluated from (6.202) giving

$$[T] = \frac{1}{jF} \begin{bmatrix} A_1' + jA_2' & B' \\ C' & A_1' - jA_2' \end{bmatrix} = \frac{1}{jF} \begin{bmatrix} A_1 + jA_2 & B \\ C & A_1 - jA_2 \end{bmatrix} \tag{6.215}$$

with

$$A_1 = 4.94987p + 13.8329p^3 + 7.72866p^5 \tag{6.216}$$

$$A_2 = 0.106939 + 2.66665p^2 + 5.58652p^4 + 3.73911p^6 \tag{6.217}$$

$$B = 1.13919 + 11.1196p^2 + 14.9678p^4 + 4.84712p^6 \tag{6.218}$$

$$C = 1.08306 + 10.0347p^2 + 12.0353p^4 + 2.88438p^6 \tag{6.219}$$

and factorising the matrix into the form in (6.210)–(6.212) we obtain

$$A_1' + jA_2' = 0.756286j + 3.5541p - 1.05194jp^2 + 9.03779p^3$$
$$- 2.44156jp^4 + 3.73911p^5 \tag{6.220}$$

$$B' = 0.569767 + 5.27649p^2 + 4.84712p^4 \tag{6.221}$$

$$C' = 1.08306 + 10.0347p^2 + 12.0353p^4 + 2.88438p^6 \tag{6.222}$$

$$F = (p^2 + 1.999396)(0.552973 - 0.608704p^2) \tag{6.223}$$

We may now extract shunt resonators of admittance

$$\frac{b_1}{p + j\omega_1} \quad \text{and} \quad \frac{b_1}{p - j\omega_1} \tag{6.224}$$

from the input and output, respectively. Again this is done by pre- and post-multiplying by their inverse transfer matrices yielding

$$[T] = \frac{1}{jF} \begin{bmatrix} 1 & 0 \\ \frac{-b_1}{p+j\omega_1} & 1 \end{bmatrix} \begin{bmatrix} j(p+j\omega_1)(A'_1 + jA'_2) & B'(p^2+\omega_1^2) \\ C' & -j(p+j\omega_1)(A'_1 - jA'_2) \end{bmatrix} \begin{bmatrix} 1 & 0 \\ \frac{-b_1}{p-j\omega_1} & 1 \end{bmatrix}$$

$$= \frac{1}{jF} \begin{bmatrix} (p+j\omega_1)(A'_2 + jA'_1 - b_1 B') & B'(p^2+\omega_1^2) \\ C' + 2b_1 A'_2 + b_1^2 B' & (p+j\omega_1)(-b_1 B' - A'_2 + jA'_1) \end{bmatrix}$$

$$= \frac{1}{jF''} \begin{bmatrix} A''_1 + jA''_2 & B'' \\ C'' & A''_1 + jA''_2 \end{bmatrix} \tag{6.225}$$

where

$$A''_1 + jA''_2 = -\frac{A'_2 - b_1 B' + jA'_1}{p - j\omega_1} \tag{6.226}$$

$$B'' = B' \tag{6.227}$$

$$C'' = \frac{C' + b_1^2 B' + 2b_1 A'_2}{p^2 + \omega_1^2} \tag{6.228}$$

$$F'' = \frac{F}{p^2 + \omega_1^2} \tag{6.229}$$

b_1 can be calculated by forcing the D parameter to be zero at $p = j\omega_1$ yielding

$$b_1 = \frac{A'_2 + jA'_1}{B'} \tag{6.230}$$

This completes the synthesis cycle to extract a pair of transmission zeros at $p = \pm j\omega_1$. The cycle is shown in Figure 6.30. This process may be repeated until the desired number of real frequency transmission zeros have been extracted.

In the case of the example we have

$$b_1 = 0.871402 \tag{6.231}$$

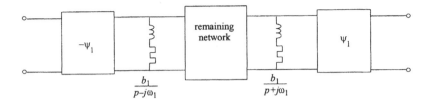

Figure 6.30   Synthesis cycle for extracting a pair of j-axis transmission zeros

and the remaining transfer matrix is given by

$$B'' = 0.569767 + 5.27649p^2 + 4.84712p^4 \tag{6.232}$$

$$C'' = 0.824005 + 5.69375p^2 + 2.88438p^4 \tag{6.233}$$

$$A_1'' + jA_2'' = 0.404614 - 2.22736jp + 4.083p^2$$
$$- 3.50487jp^3 + 3.73911p^4 \tag{6.234}$$

$$F'' = 0.552973 - 0.608764p^2 \tag{6.235}$$

The remaining network may contain transmission zeros which are not on the $j\omega$ axis and may be synthesised as a cross-coupled array. Assuming that at least two of the remaining transmission zeros are at infinity then we first extract phase shifters in order for this pair of zeros to be extracted as shunt capacitors. The value of the phase shifters is determined such that the remaining B parameter is of degree 2 lower than the C parameter. This can be determined from (6.208). However, in our example the original matrix was real; thus we simply extract $-\psi_1$. Extracting this phase shifter we obtain a new matrix

$$[T] = \frac{1}{jF}\begin{bmatrix} A_1 + jA_2 & B \\ C & A_1 - jA_2 \end{bmatrix} \tag{6.236}$$

where B is two degrees lower than C. In the example we obtain

$$A_1 + jA_2 = 2.22736p + 3.50487p^3 - j0.22567 - j1.23832p^2 \tag{6.237}$$

$$B = 0.337797 + 1.58884p^2 \tag{6.238}$$

$$C = 1.05597 + 9.38141p^2 + 7.7315p^4 \tag{6.239}$$

$$F = 0.552973 - 0.608764p^2 \tag{6.240}$$

We now extract the pair of transmission zeros at infinity by removing a capacitor C_1 in parallel with a frequency-invariant reactance B_1 from the

Waveguide filters 245

input of the network and similarly $C_1 p - jB_1$ from the output of the network. The new transfer matrix is then

$$[T] = \frac{1}{jF} \begin{bmatrix} 1 & 0 \\ -C_1 p - jB_1 & 1 \end{bmatrix} \begin{bmatrix} A_1 + jA_2 & B \\ C & A_1 - jA_2 \end{bmatrix} \begin{bmatrix} 1 & 0 \\ -C_1 p + jB_1 & 1 \end{bmatrix}$$

$$= \frac{1}{jF} \begin{bmatrix} A_1 - C_1 Bp + j(A_2 + B_1 B) & B \\ C - 2A_1 C_1 p + 2A_2 B_1 + C_1^2 p^2 B + B_1^2 B & A_1 - C_1 Bp - j(A_2 + B_1 B) \end{bmatrix}$$

$$= \frac{1}{jF} \begin{bmatrix} A_1' + jA_2' & B' \\ C' & A_1' - jA_2' \end{bmatrix} \quad (6.241)$$

where

$$A_1' = A_1 - C_1 pB \quad (6.242)$$
$$A_2' = A_2 + B_1 B \quad (6.243)$$
$$B' = B \quad (6.244)$$
$$C' = C - 2A_1 C_1 p + 2A_2 B_1 + C_1^2 p^2 B + B_1^2 B \quad (6.245)$$

and for the transmission zeros at infinity to have been successfully extracted

$$A_1' = 0 \big|_{p = \infty} \Rightarrow A_1 - C_1 pB = 0 \big|_{p = \infty} \quad (6.246)$$

Therefore

$$C_1 = \frac{A_1}{Bp} \bigg|_{p = \infty} \quad (6.247)$$

$$A_2' = 0 \big|_{p = \infty} \Rightarrow A_2 + B_1 B = 0 \big|_{p = \infty} \quad (6.248)$$

Therefore

$$B_1 = \frac{-A_2}{B} \bigg|_{p = \infty} \quad (6.249)$$

In the example we obtain

$$A_1' = 1.4822p \quad (6.250)$$
$$A_2' = 0.0376023 \quad (6.251)$$
$$B' = 0.337797 + 1.5884p^2 \quad (6.252)$$
$$C' = 0.909396 + 0.233249p^2 \quad (6.253)$$
$$F = 0.552973 - 0.608764p^2 \quad (6.254)$$
$$C_1 = 2.20593 \qquad B_1 = 0.779383 \quad (6.255)$$

The synthesis procedure may now be continued by extracting an inverter of admittance K_{12} in parallel with the remaining network such that the remaining

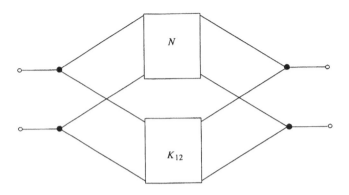

Figure 6.31 Parallel extraction of an inverter

network has a double ordered transmission zero at infinity. This is equivalent to moving the transmission zeros onto the $j\omega$ axis so that ladder network extraction may be continued (Figure 6.31).

By conversion to Y matrices and back to transfer matrices the remaining two-port transfer matrix is

$$\frac{1}{j(F - K_{12}B')} \begin{bmatrix} A'_1 + jA'_2 & B' \\ C' - 2FK_{12} + K_{12}^2 B' & A'_1 - jA'_2 \end{bmatrix} \qquad (6.256)$$

and for transmission zeros at infinity

$$K_{12} = \frac{F}{B'} \bigg|_{p=\infty} \qquad (6.257)$$

Now we continue by extracting unity impedance inverters from both ends of the network to leave

$$[T] = \frac{1}{jF''} \begin{bmatrix} A''_1 + jA''_2 & B'' \\ C'' & A''_1 - jA''_2 \end{bmatrix} \qquad (6.258)$$

where

$$F'' = -(F - K_{12}B') \qquad (6.259)$$
$$A''_1 = A'_1 \qquad (6.260)$$
$$A''_2 = -A'_2 \qquad (6.261)$$
$$C'' = B' \qquad (6.262)$$
$$B'' = C' - 2FK_{12} + K_{12}^2 B' \qquad (6.263)$$

The synthesis cycle may then be repeated until the complete network has been synthesised.

From (6.250)–(6.255) and (6.257) we obtain for the example

$$K_{12} = -0.38315 \tag{6.264}$$

and the remaining transfer matrix is given by

$$F'' = -0.6824 \tag{6.265}$$

$$A_1'' = 1.4822p \tag{6.266}$$

$$A_2'' = -0.037602 \tag{6.267}$$

$$C'' = 0.337797 + 1.5884p^2 \tag{6.268}$$

$$B'' = 1.38273 \tag{6.269}$$

Finally a shunt capacitor in parallel with a frequency-invariant reactance may be extracted from the input and its complex conjugate at the output with element values

$$C_2 = 1.07194 \tag{6.270}$$

$$B_2 = 0.027194 \tag{6.271}$$

$$K_{23} = -0.49352 \tag{6.272}$$

The complete network is shown in Figure 6.32. Note that K_{12} and K_{23} can be changed to positive sign giving all positive couplings in the prototype. This merely introduces a constant $180°$ phase shift into the transfer function of the filter. The simulated response of the prototype network is shown in Figure 6.33.

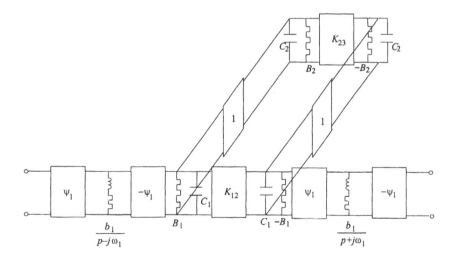

Figure 6.32   The final network for the sixth-degree extracted pole/cross-coupled filter

248 Theory and design of microwave filters

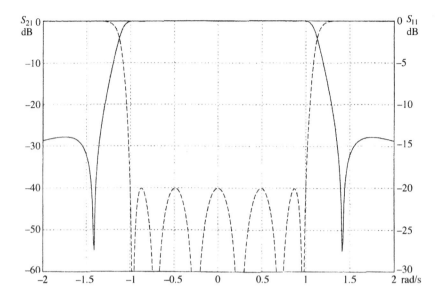

Figure 6.33 Simulated response of a lowpass prototype extracted pole/cross-coupled filter

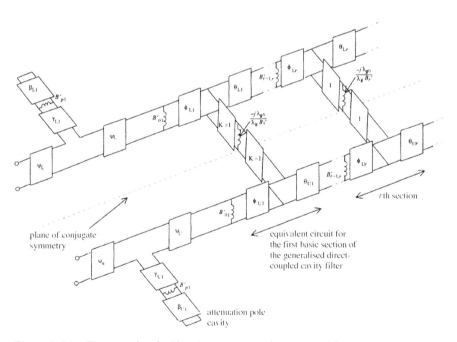

Figure 6.34 Extracted pole filter in a rectangular waveguide

6.5.1 Realisation in waveguide

A procedure will now be developed for transforming the extracted pole prototype shown in Figure 6.32 into the waveguide structure shown in Figure 6.34. The realisation starts with synthesis of the cross-coupled part of the filter, followed by the pole cavities and finally by the phase lengths between the pole cavities and the cross-coupled part of the filter.

The synthesis of the cross-coupled section follows closely the synthesis of the generalised direct-coupled cavity filter described in the previous section. This procedure gave the following design formulae:

$$B'_r = K_{r,r+1}/\alpha C_r \tag{6.273}$$

$$B'_{r-1,r} = \alpha(C_r C_{r-1})^{1/2} - \frac{1}{\alpha(C_r C_{r-1})^{1/2}} \tag{6.274}$$

$$\phi_{Lr} = \phi_{Ur} = \frac{\pi}{2} - \frac{1}{2}\left[\cot^{-1}\left(\frac{B'_{r-1,r}}{2}\right) + \sin^{-1}(B'_r)\right] \tag{6.275}$$

$$\theta_{Lr} = \theta_{Ur} = \frac{\pi}{2} - \frac{1}{2}\left[\cot^{-1}\left(\frac{B'_{r,r+1}}{2}\right) + \sin^{-1}(B'_r)\right] \tag{6.276}$$

$$C_0 = \frac{1}{\alpha} \quad C_{n+1} = \infty \quad r = 1, 2, \ldots, n \tag{6.277}$$

$$\alpha = \frac{\lambda_{g1} + \lambda_{g2}}{\pi(\lambda_{g1} - \lambda_{g2})} \tag{6.278}$$

where, as in the previous sections, λ_{g1} and λ_{g2} are the guide wavelengths at the band-edges of the filter.

We can now form the even-mode network for the cross-coupled part of the prototype shown in Figure 6.32. This is obtained by placing an open circuit plane along the line of complex conjugate symmetry through K_{12} and K_{23}. The even-mode circuit is shown in Figure 6.35. The $\pm B_r$ in Figure 6.35 refers to each arm of the complex conjugate symmetric array. The odd-mode networks are similar with $K_{r,r+1}$ replaced by $-K_{r,r+1}$.

Now comparing this even-mode circuit with the one for the symmetrical waveguide structure in Figure 6.22 we see that the only difference is the inclusion of the frequency-invariant susceptances B_r in parallel with the other

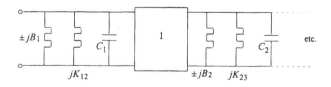

Figure 6.35   Even-mode network for the complex-conjugate symmetric array

two susceptances at each node. These do not change sign between the even and odd modes of the network and so may be realised by adding small sections of waveguide at each node. The formulae for the phase lengths of the two halves of each cavity thus become modified to

$$\phi_{Ur} = \frac{\pi}{2} - \frac{1}{2}\left[\cot^{-1}\left(\frac{B'_{r,1,r}}{2}\right) + \sin^{-1}(B'_r + B'_{Cr})\right] \quad (6.279)$$

$$\theta_{Ur} = \frac{\pi}{2} - \frac{1}{2}\left[\cot^{-1}\left(\frac{B'_{r,r+1}}{2}\right) + \sin^{-1}(B'_r + B'_{Cr})\right] \quad (6.280)$$

$$\phi_{Lr} = \frac{\pi}{2} - \frac{1}{2}\left[\cot^{-1}\left(\frac{B'_{r,1,r}}{2}\right) + \sin^{-1}(B'_r - B'_{Cr})\right] \quad (6.281)$$

$$\theta_{Lr} = \frac{\pi}{2} - \frac{1}{2}\left[\cot^{-1}\left(\frac{B'_{r,r+1}}{2}\right) + \sin^{-1}(B'_r - B'_{Cr})\right] \quad (6.282)$$

where

$$B'_{Cr} = B_r/\alpha C_r \quad (6.283)$$

The resonators of the two halves of the network are thus slightly different in length. However, the length difference is small and may be taken up by tuning screws.

The pole cavity pairs are synthesised by assuming that each pair forms a single-section complex conjugate cross-coupled array with no cross-coupling. This can be seen by examining the left-hand end of Figure 6.34. The above design formulae can then be used with $n = 1$ and $B'_r = 0$; hence

$$B'_{p1} = (\alpha C_{p1})^{1/2} - \frac{1}{(\alpha C_{p1})^{1/2}} \quad (6.284)$$

$$B'_{Cp1} = \frac{B_{p1}}{\alpha C_{p1}} \quad (6.285)$$

$$B'_r = 0 \quad (6.286)$$

$$\beta_{U1} = \phi_{Up1} + \theta_{Up1} = \pi - \frac{1}{2}\cot^{-1}\left(\frac{B'_{p1}}{2}\right) - \sin^{-1}(B'_{Cp1}) \quad (6.287)$$

$$\beta_{L1} = \phi_{Lp1} + \theta_{Lp1} = \pi - \frac{1}{2}\cot^{-1}\left(\frac{B'_{p1}}{2}\right) + \sin^{-1}(B'_{Cp1}) \quad (6.288)$$

The synthesis of the pole cavities in the two branches is illustrated in Figure 6.36. Each shunt resonator is transformed into its series dual. By equating the two circuits we obtain

$$B_{p1} = \frac{\pm \omega_1}{b_1} \quad (6.289)$$

$$C_{p1} = \frac{1}{b_1} \quad (6.290)$$

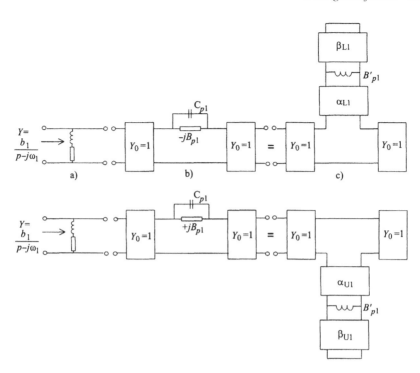

Figure 6.36 Synthesis of the pole cavities: (a) lowpass prototype; (b) dual circuit; (c) waveguide circuit

The shunt reactances are realised as irises and the phase lengths β_{U1} and β_{L1} as short circuited lengths of waveguide approximately half a wavelength long using (6.287) and (6.288). To compensate for the short negative lengths of transmission line associated with the iris susceptances lengths of waveguide γ_U and γ_L approximating $\lambda/2$ in length are included between the irises and the main waveguide, with $\gamma_{U1} = \beta_{U1}$ and $\gamma_{L1} = \beta_{L1}$.

The lengths of waveguide between the cross-coupled part of the filter and the pole cavities are modified to take into account the phase lengths ψ in the prototype, the short negative lengths of line associated with the input and output susceptances of the body of the filter, and the inverter associated with the pole cavity. Thus the phase length ψ between the front pole cavity and the cross-coupled part of the filter in the prototype is modified to

$$\psi \to \psi + \frac{\pi}{2} - \frac{1}{2}\cot^{-1}\left(\frac{B_{01}^1}{2}\right) \qquad (6.291)$$

6.5.2 Design example

We shall use the sixth-degree prototype previously synthesised to design a filter with a centre frequency of 10 GHz and a bandwidth of 40 MHz. In order to simplify the analysis of the circuit we shall assume that the cut-off frequency of the waveguide is zero. This reduces the equivalent circuit to a dispersionless transmission line circuit enabling most circuit analysis packages to analyse the network.

Thus the guide wavelengths are now given by

$$\lambda_g = \frac{\lambda_0}{[1-(\omega_c/\omega)]^{1/2}}\bigg|_{\omega_c=0} = \lambda_0 \qquad (6.292)$$

With band-edges at 9.98 GHz and 10.02 GHz we obtain $\alpha = 159.1549$. The susceptances in the cross-coupled part of the circuit are

$$B'_1 = \frac{K_{12}}{\alpha C_1} = 1.0913 \times 10^{-3} \qquad (6.293)$$

$$B'_2 = \frac{K_{23}}{\alpha C_2} = 2.8927 \times 10^{-3} \qquad (6.294)$$

$$B'_{01} = (\alpha C_1)^{1/2} - \frac{1}{(\alpha C_1)^{1/2}} = 18.6838 \qquad (6.295)$$

$$B'_{12} = \alpha(C_1 C_2)^{1/2} - \frac{1}{\alpha(C_1 C_2)^{1/2}} = 244.733 \qquad (6.296)$$

$$B'_{C1} = \frac{B_1}{\alpha C_1} = 2.21993 \times 10^{-3} \qquad (6.297)$$

$$B'_{C2} = \frac{B_2}{\alpha C_2} = 1.59397 \times 10^{-4} \qquad (6.298)$$

The phase lengths of the cross-coupled part are then given by

$$\phi_{U1} = \frac{\pi}{2} - \frac{1}{2}\left[\cot^{-1}\left(\frac{B'_{01}}{2}\right) + \sin^{-1}(B'_1 + B'_{C1})\right] = 86.8502° \qquad (6.299)$$

$$\phi_{L1} = \frac{\pi}{2} - \frac{1}{2}\left[\cot^{-1}\left(\frac{B'_{01}}{2}\right) + \sin^{-1}(B'_1 - B'_{C1})\right] = 86.9773° \qquad (6.300)$$

$$\theta_{U1} = \frac{\pi}{2} - \frac{1}{2}\left[\cot^{-1}\left(\frac{B'_{12}}{2}\right) + \sin^{-1}(B'_1 + B'_{C1})\right] = 89.6767° \qquad (6.301)$$

$$\theta_{L1} = \frac{\pi}{2} - \frac{1}{2}\left[\cot^{-1}\left(\frac{B'_{12}}{2}\right) + \sin^{-1}(B'_1 - B'_{C1})\right] = 89.7982° \qquad (6.302)$$

$$\psi_{U2} = \frac{\pi}{2} - \frac{1}{2}\left[\cot^{-1}\left(\frac{B'_{12}}{2}\right) + \sin^{-1}(B'_2 - B'_{C2})\right] = 89.6784° \qquad (6.303)$$

$$\psi_{L2} = \frac{\pi}{2} - \frac{1}{2}\left[\cot^{-1}\left(\frac{B'_{12}}{2}\right) + \sin^{-1}(B'_2 - B'_{C2})\right] = 89.6876° \quad (6.304)$$

$$\theta_{U2} = \frac{\pi}{2} - \frac{1}{2}\sin^{-1}(B'_2 + B'_{C2}) = 89.9126° \quad (6.305)$$

$$\theta_{L2} = \frac{\pi}{2} - \frac{1}{2}\sin^{-1}(B'_2 - B'_{C2}) = 89.9217° \quad (6.306)$$

Each of these phases is the phase length at the midband of the filter.

Next the susceptance coupling into the pole cavities is given by

$$B'_{p1} = (\alpha C_{p1})^{1/2} - \frac{1}{(\alpha C_{p1})^{1/2}} \quad (6.307)$$

where

$$C_{p1} = \frac{1}{b_1} = 1.147576 \quad (6.308)$$

Thus

$$B'_{p1} = 13.4405 \quad (6.309)$$

The phase lengths of the pole cavities are given by

$$\beta_{U1} = \pi - \frac{1}{2}\cot^{-1}\left(\frac{B'_{p1}}{2}\right) - \sin^{-1}(B'_{Cp1}) \quad (6.310)$$

$$\beta_{L1} = \pi - \frac{1}{2}\cot^{-1}\left(\frac{B'_{p1}}{2}\right) + \sin^{-1}(B'_{Cp1}) \quad (6.311)$$

with

$$B'_{Cp1} = \frac{\omega_1}{\alpha} = 8.8844 \times 10^{-3} \quad (6.312)$$

Thus

$$\beta_{U1} = 175.259° \qquad B_{L1} = 176.277° \quad (6.313)$$

The phase lengths between the pole cavities and the cross-coupled part of the filter are given by

$$\psi_U = -52.3531° + 90° - \frac{1}{2}\cot^{-1}\left(\frac{B'_{01}}{2}\right) = 34.592° \quad (6.314)$$

$$\psi_L = 52.3531° + 90° - \frac{1}{2}\cot^{-1}\left(\frac{B'_{01}}{2}\right) = 139.298° \quad (6.315)$$

The simulated frequency response of the complete equivalent circuit of the filter is shown in Figure 6.37. In a true waveguide design dispersion would cause a slight asymmetry in the frequency response although this would hardly be apparent in a narrowband design. To do the design example for a true

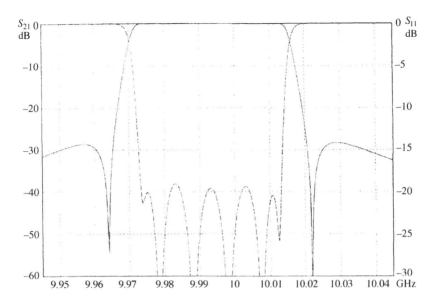

Figure 6.37 Simulated frequency response of an extracted pole waveguide filter

waveguide structure we decide on a waveguide size which determines its cut-off frequency; this then determines a and the design equations are used as above.

6.5.3 Realisation in TE_{011} mode cavities

Although the previous section concentrates on a rectangular waveguide realisation, the extracted pole filter is ideal for realisation with the higher Q TE_{011} mode cavities. The original work by Atia and Williams [9] showed that general transfer functions were realisable with this structure but they required both positive and negative couplings. This required physically offset cavities which are complicated to manufacture. The extracted pole filter only requires positive couplings and thus complex TE_{011} mode filters can be realised with a simple physical structure. It is worth noting, however, that whatever circuit synthesis technique is used the TE_{011} mode is degenerate with the TE_{111} mode. For a useful filter realisation this mode must be suppressed so that no energy is transferred to or absorbed in its resonance. The degeneracy of this mode with the TE_{011} mode can easily be split by introducing perturbations into the cavity. One method is to use a non-contacting tuning plunger in the cavity. Since the TE_{011} mode has no longitudinal current in the cavity walls the plunger will shift the frequency at the TE_{111} mode without affecting the TE_{011} mode. Alternatively a set of posts can be placed in the end plates which push the resonant frequencies of the two modes apart. In reality the effect of a filter tuning screw in the end plate has a similar effect.

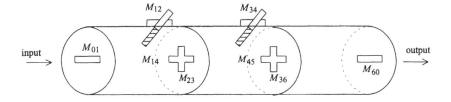

Figure 6.38 Degree 6 dual TE_{111} mode cylindrical waveguide filter

6.6 Dual-mode waveguide filters

High performance waveguide filters with high Q cavities may take up a significant physical volume. This is disadvantageous in many telecommunications and space applications. One method of size reduction is to exploit the existence of multiple degenerate modes in waveguide cavities. This was first reported by Lin in 1951, for air cavities [10]. A complete theory for dual (two-mode) TE_{11n} mode waveguide bandpass filters was first reported by Atia and Williams in 1971 [11]. Since then further developments have been reported by Rhodes and Zabalawi [12] and Cameron and Rhodes [13]. Some of the most important results will be described in this section.

Consider the waveguide structure shown in Figure 6.38. In this structure each waveguide cavity supports two orthogonally polarised degenerate TE_{111} mode resonances. Thus a $2n$th-degree filter is realisable with n cavities, giving a significant size reduction. The modes in each cavity are coupled together by a tuning screw or other discontinuity which is oriented at 45° to the input iris. The two horizontally and vertically polarised modes in each cavity are coupled to the corresponding modes in adjacent cavities by a cruciform iris. The complete structure is known as a dual-mode in-line waveguide filter. For obvious reasons of isolation the input and output ports are at opposite ends of the structure.

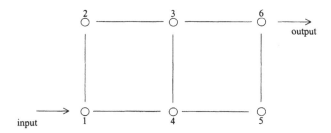

Figure 6.39 Equivalent circuit of a sixth-degree dual-mode in-line waveguide filter

The equivalent circuit of the sixth-degree filter is shown in Figure 6.39. In this diagram the circles represent nodes and the lines represent inverters. In the bandpass case it is assumed that resonant circuits are connected from the nodes to ground. Alternatively the diagram may also represent a lowpass prototype with shunt capacitors to ground at each node. The lowpass prototype for the sixth-degree dual-mode filter is shown in Figure 6.40. Here we see that at $\omega = \infty$ the shunt capacitors short circuit to ground and since there are a minimum of three inverters between input and output then there must be a minimum of four transmission zeros at infinity. By analysis of different degrees of network we see that for a symmetrical even-degree network of degree $N = 2n$ with transfer function

$$|S_{12}(j\omega)|^2 = \frac{1}{1 + F_{2n}^2(\omega)} \tag{6.316}$$

the minimum number of transmission zeros at infinity is $2m$, where

$$2m = n = \frac{N}{2} \text{ for } n \text{ even} \tag{6.317}$$

$$2m = n + 1 = \frac{N}{2} + 1 \text{ for } n \text{ odd} \tag{6.318}$$

Thus a twelfth-degree filter ($N = 12$, $n = 6$) has a minimum of six transmission zeros at infinity. This is one of the main limitations of the dual-mode in-line structure in that there are more transmission zeros at infinity than for the cross-coupled array.

In order to design the dual-mode filter we must synthesise a lowpass prototype network of the appropriate form. Once this has been done the rest of the design is a relatively standard waveguide filter design problem. For complex filter transfer functions it is useful to start with the cross-coupled array which has already been described.

A sixth-degree symmetrical cross-coupled array prototype network is shown in Figure 6.41. In this particular example there is only a single inverter between

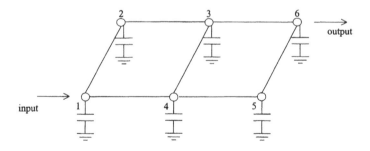

Figure 6.40   Lowpass prototype for a sixth-degree dual-mode in-line filter

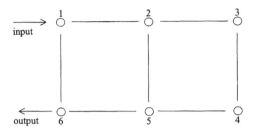

Figure 6.41 Degree 6 cross-coupled array prototype filter

input and output; thus the network only has two transmission zeros at infinity and is not suitable for the design of a dual-mode filter. For it to be so the inverter between nodes 1 and 6 must be eliminated. Furthermore, the dual-mode realisation of this network would result in the input and output being in the same physical cavity which is impractical. Thus the fundamental design problems are to choose a cross-coupled prototype with the correct number of transmission zeros at infinity and to transform this network into one suitable for dual-mode in-line realisation. The starting point for the general $2n$th-degree cross-coupled array with $2m$ transmission zeros at infinity is shown in Figure 6.42.

Now the assumption is that the lowpass prototype is symmetrical. Thus it can be defined by its even- and odd-mode subnetworks and its even- and odd-mode admittances Y_e and Y_o; Y_e is a reactance function with complex coefficients and Y_o is its complex conjugate. The even-mode network for Figure 6.44 is shown in Figure 6.43. The odd-mode network would be obtained by replacing K_{rr} by $-K_{rr}$.

In order to transform the network into the in-line prototype form it is first necessary to scale all the internal nodes of the network to make all the capacitors equal to unity. This can also be accomplished for the first capacitor

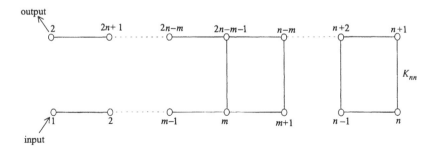

Figure 6.42 General cross-coupled array prototype network suitable for dual-mode in-line realisation

258 Theory and design of microwave filters

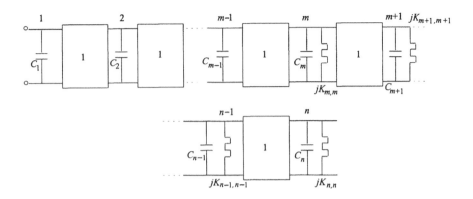

Figure 6.43 Even-mode network for the cross-coupled array prototype

by introducing an extra inverter at the input of the network. The scaled network is shown in Figure 6.44.

The nodal admittance matrix for the scaled even-mode network is then given by

$$[Y] = \begin{bmatrix} 0 & jK_{01} & 0 & 0 & \cdots & & & \\ jK_{01} & p+jK_{11} & jK_{12} & 0 & & & & \\ 0 & jK_{12} & p+jK_{22} & jK_{23} & & & & \\ & & & & \ddots & & & \\ 1 & & & & & & & \\ 1 & & & & & p+jK_{n-1,n-1} & jK_{n-1,n} \\ 1 & & & & & jK_{n-1,n} & p+jK_{n-1,n} \end{bmatrix}$$

(6.319)

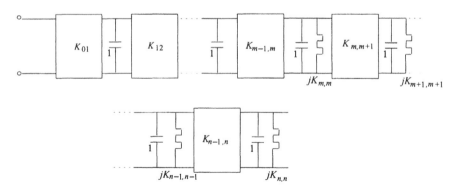

Figure 6.44 Scaled even-mode network with unity capacitors

Now K_{01} only acts as an impedance transformer and can be removed, leaving the internal matrix

$$[Y] = p[I] + j[K] \tag{6.320}$$

where $[I]$ is the $n \times n$ identity matrix and $[K]$ is the $n \times n$ coupling matrix

$$[K] = \begin{bmatrix} K_{11} & K_{12} & 0 & 0 & \cdots & & \\ K_{12} & K_{22} & K_{23} & 0 & & & \\ 0 & K_{23} & K_{33} & K_{34} & & & \\ 0 & & & & \ddots & & \\ \vdots & & & & & K_{n-1,n-1} & K_{n-1,n} \\ & & & & & K_{n-1,n} & K_{n,n} \end{bmatrix} \tag{6.321}$$

Because of the condition on the minimum number of transmission zeros at infinity in the dual-mode in-line filter

$$K_{rr} = 0 \qquad \text{for } r = 1, \ldots, m-1 \tag{6.322}$$

Transformations can now be applied to the nodal matrix and provided they do not affect the first row and column they will not affect the even-mode admittance of the network. The capacitors only exist between nodes and ground; thus the complex frequency variable p only exists on the main diagonal of the nodal matrix. An infinite number of new matrices may be generated all with the same transfer function, and if the matrix is post-multiplied by a matrix $[T]$ it must be pre-multiplied by $[T]^{-1}$. This will not affect the capacitors and we have the new matrix

$$[Y] = [T]^{-1} p[I][T] + j[T]^{-1}[K][T]$$
$$= p[I] + j[T]^{-1}[K][T] \tag{6.323}$$

The new coupling matrix is given by

$$[M] = [T]^{-1}[K][T] \tag{6.324}$$

$[T]$ consists of rotational or similarity transformations where

$$[T] = \prod_{r=1}^{q} [P]_r \tag{6.325}$$

$[P]_r$ is the matrix for a single transformation containing a single rotation of the ith row and column with respect to the jth row and column, with elements

$$P_{\ell,\ell} = 1 \qquad (\ell \neq i, \ell \neq j) \tag{6.326}$$

$$P_{ii} = P_{jj} = \cos(\theta_r) = C_r \tag{6.327}$$

$$P_{i,j} = -P_{j,i} = \sin(\theta_r) = S_r \tag{6.328}$$

$$P_{\ell,s} = 0 \qquad (\ell, s \neq i, j) \tag{6.329}$$

260 Theory and design of microwave filters

For example

$$[P]_r = \begin{bmatrix} 1 & 0 & 0 & & \cdots & \\ 0 & \cos(\theta_r) & \sin(\theta_r) & 0 & & \\ 0 & -\sin(\theta_r) & \cos(\theta_r) & 0 & & \\ 0 & 0 & 0 & 1 & & \\ \vdots & & & & 1 & \\ \vdots & & & & & \ddots \end{bmatrix} \quad (6.330)$$

The principle of the rotational transformations is to progressively apply them and in the process annihilate couplings until the coupling matrix of the cross-coupled array is transformed into that for the dual-mode in-line filter. Unfortunately there does not appear to be any definite pattern to the transformations and each degree must be considered individually. For $N = 4$ the two types of network are of the same physical form and the first meaningful case is for $N = 6$.

For the sixth-degree case the forms of the cross-coupled and in-line circuits are as shown in Figure 6.45. The even-mode coupling matrix for the cross-coupled network is given by inspection of Figure 6.45(a):

$$[K] = \begin{bmatrix} 0 & K_{12} & 0 \\ K_{12} & K_{22} & K_{23} \\ 0 & K_{23} & K_{33} \end{bmatrix} \quad (6.331)$$

It is easy to see the form of the coupling matrix for the in-line filter if we redraw it with the nodes in the same position as the cross-coupled filter (Figure 6.46). The inverters between nodes 1 and 4 and 3 and 6 pass diagonally through the line of symmetry. Applying positive potentials on the nodes 1 and 6 for the even mode, we can represent the even-mode case as in Figure 6.46(b). The coupling matrix transformation is thus

$$\begin{bmatrix} 0 & K_{12} & 0 \\ K_{12} & K_{22} & K_{23} \\ 0 & K_{23} & K_{33} \end{bmatrix} \rightarrow \begin{bmatrix} 0 & M_{12} & M_{13} \\ M_{12} & 0 & M_{23} \\ M_{13} & M_{23} & M_{33} \end{bmatrix} \quad (6.332)$$

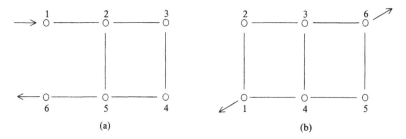

Figure 6.45 Cross-coupled and in-line prototype network of degree 6

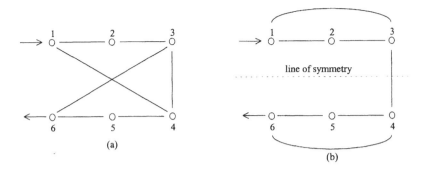

Figure 6.46 In-line prototype network of degree 6: (a) complete circuit; (b) even-mode circuit

The only rotation which can be applied is

$$[P] = \begin{bmatrix} 1 & 0 & 0 \\ 0 & C_1 & S_1 \\ 0 & -S_1 & C_1 \end{bmatrix} \tag{6.333}$$

From (6.324) we have

$[M] = [T]^{-1}[K][T]$

$$= \begin{bmatrix} 1 & 0 & 0 \\ 0 & C_1 & -S_1 \\ 0 & S_1 & C_1 \end{bmatrix} \begin{bmatrix} 0 & K_{12} & 0 \\ K_{12} & K_{22} & K_{23} \\ 0 & K_{23} & K_{33} \end{bmatrix} \begin{bmatrix} 1 & 0 & 0 \\ 0 & C_1 & S_1 \\ 0 & -S_1 & C_1 \end{bmatrix}$$

$$= \begin{bmatrix} 0 & C_1 K_{12} & S_1 K_{12} \\ C_1 K_{12} & C_1^2 K_{22} - 2S_1 C_1 K_{23} + S_1^2 K_{33} & (C_1^2 - S_1^2)K_{23} - S_1 C_1(K_{33} - K_{22}) \\ S_1 K_{12} & (C_1^2 - S_1^2)K_{23} - S_1 C_1(K_{33} - K_{22}) & S_1^2 K_{22} + C_1^2 K_{33} + 2S_1 C_1 K_{23} \end{bmatrix}$$

(6.334)

For this to be in the same form as (6.332) M_{22} must be zero. Thus

$$C_1^2 K_{22} - 2S_1 C_1 K_{23} + S_1^2 K_{33} = 0 \tag{6.335}$$

or

$$K_{33} t_1^2 - 2 K_{23} t_1 + K_{22} = 0 \tag{6.336}$$

where

$$t_1 = \tan(\theta_1) \tag{6.337}$$

Thus

$$t_1 = \frac{K_{23} \pm (K_{23}^2 - K_{22} K_{33})^{1/2}}{K_{33}} \tag{6.338}$$

and for the inverters to be real we have the realisability condition

$$K_{23}^2 \geq K_{22} K_{33} \tag{6.339}$$

which is normally true for cross-coupled networks.

The elements of (6.334) may now be simplified by substituting for t_1 from (6.338) to obtain, for example,

$$\begin{aligned} M_{33} &= S_1^2 K_{22} + C_1^2 K_{33} + 2 S_1 C_1 K_{23} \\ &= S_1^2 (K_{22} - K_{33}) + K_{33} + 2 S_1 C_1 K_{23} \end{aligned} \tag{6.340}$$

and from (6.335)

$$2 S_1 C_1 K_{23} = C_1^2 K_{22} + S_1^2 K_{33} \tag{6.341}$$

Therefore

$$\begin{aligned} M_{33} &= S_1^2 (K_{22} - K_{33}) + K_{33} + C_1^2 K_{22} + S_1^2 K_{33} \\ &= K_{22} + K_{33} \end{aligned} \tag{6.342}$$

The complete coupling matrix is

$$\begin{bmatrix} 0 & C_1 K_{12} & S_1 K_{12} \\ C_1 K_{12} & 0 & K_{23} - K_{33} t_1 \\ S_1 K_{12} & K_{23} - K_{33} t_1 & K_{22} - K_{33} \end{bmatrix} \tag{6.343}$$

The required transformations for the eighth-degree case are shown in Figure 6.47 and the matrix transformation is

$$\begin{bmatrix} 0 & K_{12} & 0 & 0 \\ K_{12} & K_{22} & K_{23} & 0 \\ 0 & K_{23} & K_{33} & K_{34} \\ 0 & 0 & K_{34} & K_{44} \end{bmatrix} \Rightarrow \begin{bmatrix} 0 & M_{12} & 0 & M_{14} \\ M_{12} & 0 & K_{23} & 0 \\ 0 & M_{23} & M_{33} & M_{34} \\ M_{14} & 0 & M_{34} & M_{44} \end{bmatrix} \tag{6.344}$$

To obtain the correct form for the final matrix, two transformations $(3, 4)$ and $(2, 4)$ are used to zero the elements $(2, 2)$ and $(2, 4)$. After a little manipulation it

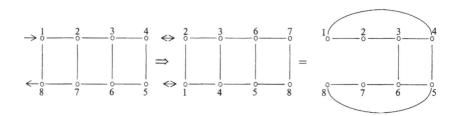

Figure 6.47 Eighth-degree cross-coupled and dual-mode in-line filters

may be shown that the final matrix is given by

$$M_{12} = C_2 K_{12} \tag{6.345}$$

$$M_{14} = S_2 K_{12} \tag{6.346}$$

$$M_{23} = S_2 \left(K_{34} + \frac{K_{44}}{t_1} \right) \tag{6.347}$$

$$M_{34} = C_2 \left(\frac{K_{22}}{S_2^2 t_1} - K_{34} - \frac{K_{44}}{t_1} \right) \tag{6.348}$$

$$M_{33} = K_{33} + K_{44} - \frac{K_{22}}{t_2^2} \tag{6.349}$$

$$M_{44} = \frac{K_{22}}{S_2^2} \tag{6.350}$$

where

$$t_2 = \frac{K_{22}}{S_1 K_{23}} \tag{6.351}$$

and

$$t_1 = \frac{K_{22} K_{34} \pm [K_{22}^2 K_{34}^2 + K_{22} K_{44}(K_{23}^2 - K_{22} K_{33})]^{1/2}}{K_{23}^2 - K_{22} K_{23}} \tag{6.352}$$

The realisability condition is

$$K_{22}^2 K_{34}^2 + K_{22} K_{44}(K_{23}^2 - K_{22} K_{33}) \geq 0 \tag{6.353}$$

The realisability condition restricts the possible locations of transmission zeros in the complex plane. For an eighth-degree filter with four transmission zeros at infinity, circuit analysis shows that the numerator of $S_{12}(p)$ is given by

$$K_{22} p^4 + p^2 [K_{22}(K_{33}^2 + 2K_{34}^2 + K_{44}^2) - K_{33} K_{23}^2]$$
$$+ [K_{34}^2 - K_{33} K_{44}][K_{22} K_{34}^2 + K_{44}(K_{23}^2 - K_{22} K_{23})] \tag{6.354}$$

For most filter characteristics $K_{34}^2 - K_{33} K_{44}$ is greater than zero. Thus if (6.353) is not satisfied the numerator is of the form

$$p^4 + Ap - B = (p^2 + X)(p^2 - Y) \qquad X, Y \geq 0 \tag{6.355}$$

Thus the realisability condition is not satisfied for transmission zeros occurring as a pair on the real axis and a pair on the imaginary axis.

Explicit solutions have also been derived for the tenth- and twelfth-degree cases. Most filter characteristics can be realised in these cases.

6.6.1 Numerical example

As an example we will consider a degree 6 linear phase filter with four transmission zeros at infinity. The lowpass prototype is a generalised Chebyshev filter

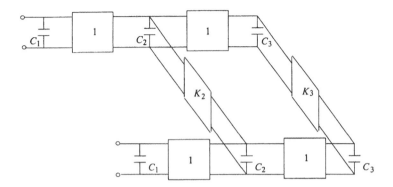

Figure 6.48   Cross-coupled lowpass prototype linear phase filter

with linear phase at the points of perfect transmission (Figure 6.48) [14]. The element values are

$$C_1 = 0.9822, \quad K_1 = 0$$
$$C_2 = 1.3912, \quad K_2 = 0.1744 \tag{6.356}$$
$$C_3 = 1.9185, \quad K_3 = 0.9265$$

After adding a unity impedance inverter at the input, the even-mode circuit is as given in Figure 6.49. The admittance matrix with appropriate row and column scaling factors to make the capacitors unity is

$$
\begin{array}{c}
 \downarrow \frac{1}{\sqrt{C_1}} \quad \downarrow \frac{1}{\sqrt{C_2}} \quad \downarrow \frac{1}{\sqrt{C_3}} \\
\begin{array}{c} \\ \frac{1}{\sqrt{C_1}} \rightarrow \\ \frac{1}{\sqrt{C_2}} \rightarrow \\ \frac{1}{\sqrt{C_3}} \rightarrow \end{array}
\begin{bmatrix}
0 & j & 0 & 0 \\
j & C_1 p & j & 0 \\
0 & j & C_2 p + j K_2 & j \\
0 & 0 & j & C_3 p + j K
\end{bmatrix}
\end{array}
$$

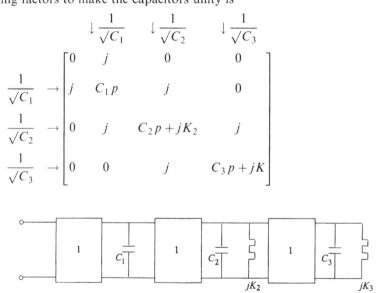

Figure 6.49   Even-mode equivalent circuit of a sixth-degree linear phase filter

$$= \begin{bmatrix} 0 & \dfrac{j}{(C_1)^{1/2}} & 0 & 0 \\ \dfrac{j}{(C_1)^{1/2}} & p & \dfrac{j}{(C_1 C_2)^{1/2}} & 0 \\ 0 & \dfrac{j}{(C_1 C_2)^{1/2}} & p + \dfrac{jK_2}{C_2} & \dfrac{j}{(C_2 C_3)^{1/2}} \\ 0 & 0 & \dfrac{j}{(C_2 C_3)^{1/2}} & p + \dfrac{jK_3}{C_3} \end{bmatrix}$$

$$= \begin{bmatrix} 0 & j1.009 & 0 & 0 \\ j1.009 & p & j0.85547 & 0 \\ 0 & j0.85547 & p + j0.12536 & j0.6121 \\ 0 & 0 & j0.6121 & p + j0.4829 \end{bmatrix} \quad (6.357)$$

After ignoring the input inverter the elements in the coupling matrix are

$$\begin{aligned} K_{12} &= 0.85547 \\ K_{22} &= 0.12536 \\ K_{23} &= 0.6121 \\ K_{33} &= 0.4829 \end{aligned} \quad (6.358)$$

From (6.338)

$$t_1 = \dfrac{K_{23} \pm (K_{23}^2 - K_{22} K_{33})^{1/2}}{K_{33}} = 0.1069 \text{ or } 2.4282 \quad (6.359)$$

and the realisability criterion is satisfied.

Taking the smallest value of t_1 and applying (6.343) we obtain

$$\begin{aligned} M_{12} &= C_1 K_{12} = 0.8506 \\ M_{13} &= S_1 K_{12} = 0.0910 \\ M_{23} &= K_{23} - K_{33} t_1 = 0.5605 \\ M_{33} &= K_{22} + K_{33} = 0.6083 \end{aligned} \quad (6.360)$$

Now from Figure 6.48 M_{13} above represents M_{14} in the in-line prototype. The input coupling M_{01} is obtained from (6.357) and the final in-line filter with its element values is as shown in Figure 6.50. The simulated response of a bandpass version of this filter is shown in Figure 6.51.

6.6.2 Asymmetric realisations for dual-mode filters

Symmetrical dual-mode in-line filters have certain limitations. There are certain transmission zero locations for which the filters are not physically realisable, particularly in the eighth degree case. Second, the methods cannot be used for

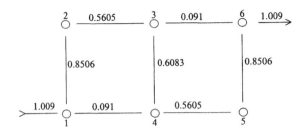

Figure 6.50 Sixth-degree in-line linear phase filter with element values

electrically asymmetric characteristics. Finally there is no known solution for fourteenth-degree filters. By removing the necessity for physical symmetry a more general procedure, operating on the entire coupling matrix of the filter, overcomes these limitations. The only real restriction is on the minimum number of transmission zeros at infinity, which is the same as for the symmetric filter. A systematic procedure for the rotational matrix transformations has been developed. In this case a series of rotations is applied where the angle θ_r of the rth rotation is desired from the elements of the coupling matrix from the previous rotation. Table 6.2 shows the positions and the angles of rotation for degrees 6–14. A photograph of a typical dual-mode device is shown in Figure 6.52.

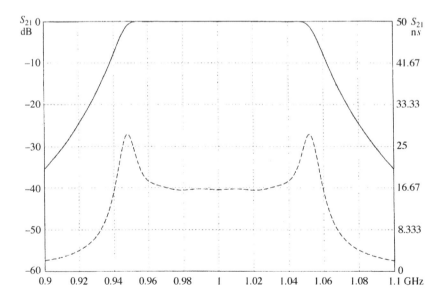

Figure 6.51 Simulated frequency response of a dual-mode in-line linear phase filter

Table 6.2   Pivotal positions and rotation angles for general asymmetric in-line prototype networks

Order N	Rotation number r	Pivot [i, j]	$\theta_r = \tan^{-1}(kM_{u1,u2}/M_{v1,v2})$				
			U1	u2	v1	v2	k
6	1	[2, 4]	2	5	4	5	+1
8	1	[4, 6]	3	6	3	4	−1
	2	[2, 4]	2	7	4	7	+1
	3	[3, 5]	2	5	2	3	−1
	4	[5, 7]	4	7	4	5	−1
10	1	[4, 6]	4	7	6	7	+1
	2	[6, 8]	3	8	3	6	−1
	3	[7, 9]	6	9	6	7	−1
12	1	[5, 9]	4	9	4	5	−1
	2	[3, 5]	3	10	5	10	+1
	3	[2, 4]	2	5	4	5	+1
	4	[6, 8]	3	8	3	6	−1
	5	[7, 9]	6	9	6	7	−1
	6	[8, 10]	5	10	5	8	−1
	7	[9, 11]	8	11	8	9	−1
14	1	[6, 10]	5	10	5	6	−1
	2	[4, 6]	4	11	6	11	+1
	3	[7, 9]	4	9	4	7	−1
	4	[8, 10]	7	10	7	8	−1
	5	[9, 11]	6	11	6	9	−1
	6	[10, 12]	9	12	9	10	−1
	7	[5, 7]	4	7	4	5	−1
	8	[7, 9]	6	9	6	7	−1
	9	[9, 11]	8	11	8	9	−1
	10	[11, 13]	10	13	10	11	−1

Source: Cameron, R., Rhodes, J.D.: 'Asymmetric realisations for dual mode bandpass filters', *IEEE Transactions on Microwave Theory and Techniques*, 1981, **29** (1); © 1981 IEEE.

6.7 Summary

This chapter has been concerned with the design of waveguide filters to realise various transfer functions. Initially a review of the basic theory of rectangular and circular waveguides and waveguide resonators is presented. Next a design procedure for waveguide bandpass filters with all-pole transfer functions is developed, and supported with an example. More complex transfer functions require either cross-coupled or extracted pole filters. The former enable realisation of transfer functions with real-axis transmission zeros, i.e. prototypes with all positive couplings. The development of design procedures for these generalised waveguide filters is presented. The restriction of transmission zero locations

268 Theory and design of microwave filters

Figure 6.52   Typical dual-mode device
              (courtesy of Filtronic plc)

in the real axis is removed by the use of extracted pole waveguide filters, the design theory of which is developed and again supported by an example. Finally techniques for the design of dual-mode filters are presented. It is important to note that the extracted pole and dual-mode techniques are relevant to the dielectric resonator filters described in the next chapter.

6.8 References

1. RAMO, S., WHINNERY, J.R., and VAN DUZER, T.V.: 'Fields and waves in communication electronics' (Wiley, New York, 1993) pp. 493–94
2. BALANIS, C.A.: 'Advanced engineering electromagnetics' (Wiley, New York, 1989) pp. 116–21
3. ABRAMOWITZ, M., and STEGUN, I.A.: 'Handbook of mathematical functions' (Dover, New York, 1988) pp. 355–434
4. MATTHAEI, G., YOUNG, L., and JONES, E.M.T.: 'Microwave filters, impedance matching networks and coupling structures' (Artech House, Norwood, MA, 1980) pp. 450–64
5. MARCUVITZ, N.: 'Waveguide handbook' (IEE, Stevenage, 1986) pp. 249–79
6. SAAD, T.S.: 'Microwave engineers handbook' (Artech House, Norwood, MA, 1971)

7 RHODES, J.D.: 'The generalised direct-coupled cavity linear phase filter', *IEEE Transactions on Microwave Theory and Techniques*, 1970, **18** (6) pp. 308–13
8 RHODES, J.D., and CAMERON, R.J.: 'General extracted pole synthesis technique with applications to low loss TE_{011} mode filters', *IEEE Transactions on Microwave Theory and Techniques*, 1980, **28** (9), pp. 1018–28
9 ATIA, A.E., and WILLIAMS, A.E.: 'General TE_{011} mode waveguide bandpass filters', *IEEE Transactions on Microwave Theory and Techniques*, 1976, **24** (10), pp. 640–48
10 LIN, W.: 'Microwave filters employing a single cavity excited in more than one mode', *Journal of Applied Physics*, 1951, **22** (8), pp. 989–1001
11 ATIA, A.E., and WILLIAMS, A.E.: 'Dual mode canonical waveguide filters', *IEEE Transactions on Microwave Theory and Techniques*, 1977, **25** (12), pp. 1021–26
12 RHODES, J.D., and ZABALAWI, I.H.: 'Synthsis of symmetric dual-mode in-line prototype networks', *International Journal of Circuit Theory and Applications*, 1980, **8** (2), pp. 145–60
13 CAMERON, R.J., and RHODES, J.D.: 'Asymmetric realisations for dual mode bandpass filters', *IEEE Transactions on Microwave Theory and Techniques*, 1981, **29** (1), pp. 51–58
14 RHODES, J.D., and ZABALAWI, I.H.: 'Design of selective linear-phase filters with equiripple amplitude characteristics', *IEEE Transactions on Circuits and Systems*, 1978, **25** (12), pp. 989–1000

Chapter 7
Dielectric resonator filters

7.1 Introduction

The applications of dielectric materials at radio frequencies were first proposed by Rayleigh in 1897 [1] who established the waveguiding properties of a dielectric rod. In 1938, Richtmayer [2] proposed the use of dielectrics as resonators and studied spherical and ring resonators. The first designs for dielectric resonator filters were described by Cohn in 1968 [3], although material properties were too poor at that time for many real filtering applications. More recently the properties of dielectric materials have improved dramatically. Along with the advent of satellite and cellular communications this has resulted in an explosion in the applications for, and published material on, dielectric resonator filters.

A dielectric resonator consists of a cylindrical, cubic or other shaped piece of high dielectric constant material, known as a puck. In conventional operation the puck is held by a supporting structure of low dielectric constant inside a conducting enclosure, which does not contact the puck (Figure 7.1). Typically the relative permittivity of the puck is between 20 and 80 and the puck is remote from the enclosure with $b \geq a$. At the resonant frequency most of the electromagnetic energy is stored within the dielectric. The enclosure stops radiation and because it is remote the resonant frequency is largely controlled by the dimensions and permittivity of the puck. The fields outside the puck are evanescent and decay rapidly with distance away from the puck. The remoteness of the enclosure ensures that the unloaded Q_u factor is dominated by the loss tangent of the dielectric. Very low loss dielectrics are now available, enabling Q_u factors of 50 000 or more. Dielectric resonators can thus be thought of as 'super-insulators'. Very temperature stable dielectrics now available enable resonators to be constructed with extremely low temperature coefficients of resonant frequency. Properties of typical dielectric materials are listed in Table 7.1 [4].

272 Theory and design of microwave filters

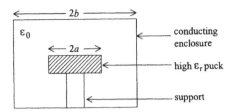

Figure 7.1 Cross-section of a typical dielectric resonator structure

The most important properties of a dielectric resonator are its field pattern, Q factor, resonant frequency and spurious-free bandwidth. These depend on the material used, the shape of the resonator and the particular resonant mode used. The fundamental properties of single-, dual- and triple-mode resonators and their application in filter design will be presented in this chapter.

7.2 Dielectric rod waveguides and the $TE_{01\delta}$ mode

The most commonly used resonator structure uses a cylindrical puck operating in the $TE_{01\delta}$ mode, originally described by Cohn. A simple model can be used to describe the properties of this mode. The dielectric puck is assumed to be a section of dielectrically loaded circular waveguide with magnetic wall boundary conditions on its lateral surface. Energy is allowed to leak out of the flat surfaces of the puck. The construction of this resonator is shown in Figure 7.2. Here the dielectric puck of permittivity ε_r, radius a and height ℓ is centrally located in a cavity of height $\ell + 2\ell_1$.

The cylindrical puck may be considered as a truncated section of dielectric rod waveguide as shown in Figure 7.3. Expressions for the various field components of the modes in a dielectric rod waveguide may be obtained by solving the

Table 7.1 Properties of typical dielectric material

Material	ε_r	Q_u at (F) GHz	Temperature coefficient of resonant frequency (ppm/°)
Barium zinc tantalate	29	48 000 (2)	−2 to +4
Zirconium tin titanate	35	16 000 (2)	−1 to +8
Calcium titanate − neodymium aluminate	45	30 000 (1)	−7 to +8
Calcium titanate − barium tungstate	55	25 000 (0.8)	+6
Lanthanum zinc titanate	80	5 000 (1)	−1 to +9

Dielectric resonator filters 273

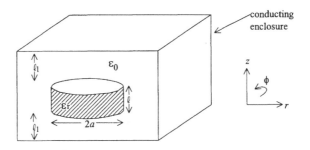

Figure 7.2 Cylindrical dielectric resonator structure

vector Helmholtz equation for the z-directed field components E_z and H_z in cylindrical coordinates.

$$(\nabla_T^2 + k^2) E_z = 0 \tag{7.1}$$

$$(\nabla_T^2 + k^2) H_z = 0 \tag{7.2}$$

where

$$k^2 = \omega^2 \mu \varepsilon \tag{7.3}$$

The analysis is somewhat involved and will not be repeated here. A detailed presentation is given in Reference 5. There are three basic types of mode in a dielectric rod waveguide: transverse electric (TE), transverse magnetic (TM), and hybrid (HE) modes. The purely transverse modes exhibit circularly symmetric field patterns with no ϕ variation. The propagation constant for various modes has been computed as a function of frequency for a dielectric rod enclosed in a metallic waveguide, shown in Figure 7.4.

Plots of the propagation constant versus frequency are shown in Figure 7.5 for all modes up to HE_{36}. Here the y axis indicates the propagation constant, positive numbers indicating propagating waves and negative numbers cut-off waves. Dotted lines indicate cases with complex propagation constants [6]. In this particular structure the cut-off frequencies of the HE_{11} and TE_{01} modes are

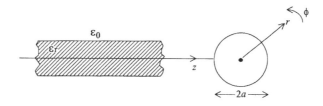

Figure 7.3 Dielectric rod waveguide

274 *Theory and design of microwave filters*

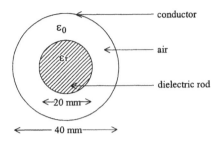

Figure 7.4 *Dielectric rod enclosed in a metallic waveguide*

1.7504 GHz and 1.868 GHz. The TM_{01} mode cuts off at 1.03 GHz. The first two modes are generally of more interest for dielectric resonators as most of their E field is confined to the dielectric, indicating a potentially high resonator Q factor.

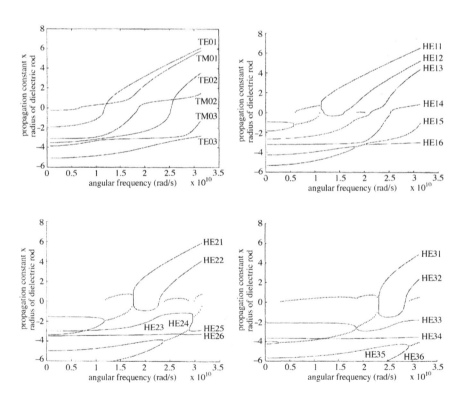

Figure 7.5 *Propagation constants for various modes in a dielectric rod enclosed in a metallic waveguide*

The transverse field components of the TE$_{01}$ and HE$_{11}$ modes are shown in Figure 7.6. Here we see that the TE$_{01}$ mode is circularly symmetric but the HE$_{11}$ mode can support two orthogonally polarised field patterns enabling single- and dual-mode operation respectively.

A simplified model for the TE$_{01}$ mode may be constructed by observing that the tangential magnetic field at the interface between a high permittivity medium and air is approximately zero. Consider the interface shown in Figure 7.7. Assume a plane wave is propagating in region 1 in the z direction. It has an x-directed E field and a y-directed H field; both are tangential to the interface between ε_r and ε_0. At the interface some of the field is reflected and some is transmitted into region 2. Denoting the forward wave in region 1 by A, the reflected wave by B and the transmitted wave into region 2 by C we have in region 1

$$E_1 = E_A + E_B \tag{7.4}$$

$$H_1 = \frac{1}{\eta}(E_A - E_B) \tag{7.5}$$

(here the $\exp(\pm j\beta z)$ propagation is assumed) where

$$\eta_1 = \left(\frac{\mu_0}{\varepsilon_r}\right)^{1/2} \tag{7.6}$$

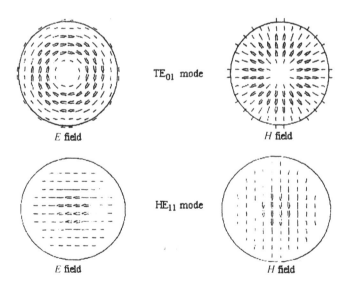

Figure 7.6 Transverse E and H fields for TE$_{01}$ and HE$_{11}$ modes in a dielectric rod waveguide
(reproduced with permission from Kajfez, D., and Guillon, P.: 'Dielectric Resonators' (Artech House, Norwood, MA, 1986); www.artechhouse.com)

276 Theory and design of microwave filters

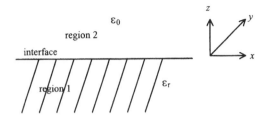

Figure 7.7 Dielectric–air interface

In region 2

$$E_2 = E_C \tag{7.7}$$

$$H_2 = \frac{E_C}{\eta_2} \tag{7.8}$$

where

$$\eta_2 = \left(\frac{\mu_0}{\varepsilon_r}\right)^{1/2} \tag{7.9}$$

At the interface both the tangential electric and magnetic fields must be continuous; thus

$$E_A + E_B = E_C \tag{7.10}$$

and

$$\frac{1}{\eta}(E_A - E_B) = \frac{E_C}{\eta_2} \tag{7.11}$$

Defining the reflection coefficient τ as the ratio of E_B to E_A we obtain

$$\tau = \frac{E_B}{E_A} = \frac{\sqrt{\varepsilon_r} - 1}{\sqrt{\varepsilon_r} + 1} \tag{7.12}$$

Thus as ε_r tends to infinity E_B tends to E_A and there is total reflection from the interface. In this case from (7.5) $H_1 = 0$ and the tangential magnetic field at the interface is zero. This is analogous to an electric conductor where the tangential electrical field would be zero. In this case the interface approximates a (physically unrealisable) ideal magnetic conductor. As an example, from (7.12) for a dielectric constant of 45, $\tau = 0.799$. The magnetic conductor is often referred to as a magnetic wall.

The concept of the ideal magnetic wall is used in the Cohn model [3] to simplify the analysis of resonators operating in the TE_{01} mode. In this model it is assumed that the puck consists of a dielectric rod waveguide with magnetic wall boundary conditions. Energy is allowed to leak out of the flat surfaces of the puck but is assumed to be still confined within the same cross-section. In

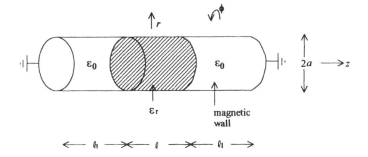

Figure 7.8 The Cohn model for a dielectric resonator

other words the magnetic wall boundary condition is assumed to continue into the air space above and below the puck until the waveguide is terminated in a short circuit (the conducting enclosure). This is shown in Figure 7.8.

From (6.78) to (6.85) the field components for TE modes in a circular waveguide are given by

$$H_z = HJ_n(k_c r)\cos(n\phi)\exp(\pm\gamma z) \quad (7.13)$$

$$E_r = Z_{TE} H_\phi = \frac{j\omega\mu n}{k_c^2 r} HJ_n(k_c r)\sin(n\phi)\exp(\pm\gamma z) \quad (7.14)$$

$$E_\phi = -Z_{TE} H_r = \frac{j\omega\mu}{k_c} HJ'_n(k_c r)\cos(n\phi)\exp(\pm\gamma z) \quad (7.15)$$

with

$$Z_{TE} = \frac{j\omega\mu}{\gamma} \quad (7.16)$$

and γ is the propagation constant.

For the TE_{01} mode these equations simplify to

$$H_z = HJ_0(k_c r) \quad (7.17)$$

$$E_\phi = -Z_{TE} H_r = \frac{j\omega\mu}{k_c} J_1(k_c r) \quad (7.18)$$

$$E_r = H_\phi = 0 \quad (7.19)$$

(The $\exp(\pm j\beta z)$ dependence is assumed.) Now applying an ideal magnetic wall boundary condition at $r = a$ then the tangential magnetic field at $r = a$ is zero. Thus

$$H_z = 0|_{r=a} \quad (7.20)$$

Hence

$$J_0(k_c a) = 0 \quad (7.21)$$

278 Theory and design of microwave filters

or

$$k_c = \frac{2.408}{a} \tag{7.22}$$

and

$$k_c^2 = \gamma^2 + \omega^2\mu\varepsilon \tag{7.23}$$

or

$$\gamma = \left[\omega^2\mu_0\varepsilon_0\varepsilon_r - \left(\frac{2.408}{a}\right)^2\right]^{1/2} \tag{7.24}$$

For propagating modes in a lossless waveguide γ is purely imaginary and

$$\gamma = j\beta \qquad \beta = \left[\omega^2\mu_0\varepsilon_0\varepsilon_r - \left(\frac{2.408}{a}\right)^2\right]^{1/2} \tag{7.25}$$

For cut-off modes γ is real:

$$\gamma = \alpha = \left[\left(\frac{2.408}{a}\right)^2 - \omega^2\mu_0\varepsilon_0\varepsilon_r\right]^{1/2} \tag{7.26}$$

The wave impedance for propagating modes is

$$Z_{TE} = Z_p = \frac{\omega\mu}{\beta} \tag{7.27}$$

and for non-propagating modes

$$Z_{TE} = Z_c = \frac{j\omega\mu_0}{\alpha} \tag{7.28}$$

From (7.18) we observe that the transverse field components E_ψ and H_r have a similar variation across the transverse plane, given by $J_1(k_c r)$. Therefore the dielectric rod waveguide can be described as a single-mode transmission line with propagation constant β and characteristic impedance Z_p. Similarly the air-filled waveguides in the Cohn model can be represented by sections of cut-off waveguide terminated in short circuits. This is shown in Figure 7.9.

The transfer matrix of the propagating guide, looking into the circuit at the line of symmetry, is

$$[T] = \begin{bmatrix} \cos(\beta\ell) & \dfrac{j\omega\mu}{\beta}\sin(\beta\ell) \\ \dfrac{j\beta}{\omega\mu}\sin(\beta\ell) & \cos(\beta\ell) \end{bmatrix} \tag{7.29}$$

This is terminated in an impedance Z_L, the input impedance of the short

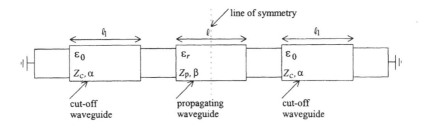

Figure 7.9 Single-mode equivalent circuit of the Cohn model for a dielectric resonator

circuited section of cut-off guide where

$$Z_L = \frac{j\omega\mu_0}{\alpha}\tanh(\alpha\ell_1) \tag{7.30}$$

The circuit is terminated at both ends in short circuits so resonance occurs when the input impedance looking in at the line of symmetry is infinite. The input impedance is

$$Z_{in} = \frac{AZ_L + B}{CZ_L + D} \tag{7.31}$$

Resonance occurs when the denominator of (7.31) is zero, when

$$\frac{\beta}{\alpha}\tan\left(\frac{\beta\ell}{2}\right)\tan(\alpha\ell_1) = 1 \tag{7.32}$$

This is the resonance equation for the $TE_{01\delta}$ mode. Since $\tanh(\alpha\ell_1)$ is positive, $\beta\ell/2$ must be less then 90° – hence the use of δ for a mode number as $\beta\ell$ is less then 180° at resonance. In other words there is less than one half wavelength variation in the transverse fields in the dielectric region at resonance. A lumped element equivalent to the resonance would be to consider the dielectric region as a capacitor and the air-filled region as an inductor.

The input impedance of the resonator is infinite at resonance; consequently the transverse H field, which is analogous to current, is zero at the centre of the resonator. Analysis of the equivalent circuit shows that H_r is a maximum at the flat ends of the dielectric and rolls off to zero at the end conducting plates. Plots of H_r and E_ψ are shown as functions of axial position in Figure 7.10. Analysis of TM modes for the cylindrical resonator gives the resonance equation

$$\frac{\beta}{\alpha}\tan\left(\frac{\beta\ell}{2}\right)\tanh(\alpha\ell_1) = -1 \tag{7.33}$$

Since $\tanh(\alpha\ell_1)$ is positive, $\tan(\beta\ell/2)$ must be negative and there is a $\delta + 1$ variation in the dielectric. The lowest resonant TM mode is thus the $TM_{01\delta+1}$ mode which resonates at a much higher frequency than the $TE_{01\delta}$ mode.

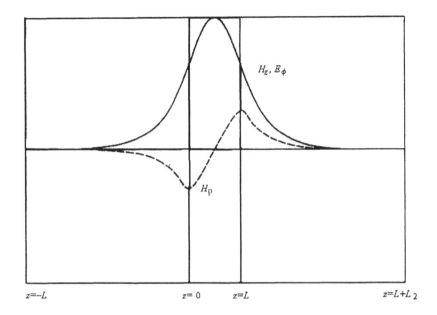

Figure 7.10 Transverse field variation as a function of axial position for the TE$_{01\delta}$ resonator
(reproduced with permission from Kajfez, D., and Guillon, P.: 'Dielectric Resonators' (Artech House, Norwood, MA, 1986); www.artechhouse.com)

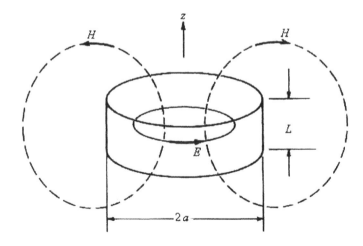

Figure 7.11 Field pattern of the TE$_{01\delta}$ mode

Table 7.2 Resonant frequencies of modes in a dielectric resonator

Mode	Resonant frequency/MHz
$TE_{01\delta}$	990
$HE_{11\delta}$	1298
$HE_{11\delta+1}$	1341
$TM_{01\delta}$	1513
$HE_{21\delta}$	1575

Since the H_z component shows a similar variation with z to E_φ, the H field turns round near the flat surface of the puck; the complete field patterns are shown in Figure 7.11.

As an example we will analyse a resonator using the resonance equation. Consider a puck with $\varepsilon_r = 45$, radius 2.5 cm and height 2 cm located centrally in a cubic conducting enclosure of internal dimension 10 cm. Thus $a = 2.5$ cm, $\ell = 2$ cm and $\ell_1 = 4$ cm. Equation (7.32) is transcendental and must be solved numerically, giving a resonant frequency of 920 MHz. More accurate methods of solving for the resonant frequency are given in References 7 and 8. Alternatively one can use an EM simulator to obtain accurate results. Analysing this example using HFSS we obtained the resonant frequencies of various modes given in Table 7.2.

The ratio of the resonant frequencies of the fundamental mode and the first spurious mode is 1.303 : 1. This ratio is important in filter design as it determines the spurious-free stopband performance. The aspect ratio of the puck, 2.5 : 1, is nearly optimum in this respect. However, the spurious performance may be improved by introducing a hole in the centre of the resonator, forming it into a ring (Figure 7.12) [9, 10]. The $TE_{01\delta}$ mode has zero E field in the centre of the puck where other modes have finite E field. Since the dielectric acts on the E field, removing regions of the puck where the E field of a particular mode is

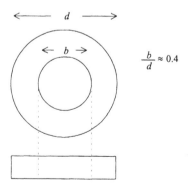

Figure 7.12 Dielectric resonator with improved spurious performance

strong will raise the resonant frequency of that mode. In our example we introduced a 20 mm diameter hole in the puck. This raised the resonant frequency of the first spurious mode to 1.549 GHz while the $TE_{01\delta}$ mode only increased slightly to 1.027 GHz. Thus the spurious ratio increased to 1.508, a useful improvement.

It is interesting to consider the relative amounts of energy stored within the dielectric and in the air cavity. For a resonator with $\varepsilon_r = 38$ it was reported in Reference 11 that 97 per cent of the electric energy and 63 per cent of the magnetic energy were stored in the dielectric puck. Furthermore, the H field outside the puck decays exponentially with distance away from the dielectric. Obviously the H field induces currents in the walls of the conducting enclosure but provided they are far enough away from the puck they will have little effect on the unloaded Q_u. Typically the enclosure diameter should be double the puck diameter for permittivities in the range 36–44.

Couplings between dielectric resonators rely on the magnetic field since there is so little electric field in the air region. One method of coupling is via an aperture in the common wall between two cavities, as shown in Figure 7.13. The coupling bandwidths between resonators may be obtained experimentally using the procedures described in Chapter 4.

A typical example of coupling bandwidth versus aperture depth d is given in Table 7.3. In this case the cavity was an 80 mm cube, and the pucks were 50 mm in diameter and 20 mm high with $\varepsilon_r = 45$. The aperture width w was the same as the width of the cavity.

In cross-coupled filters we often require both positive and negative couplings. These can be achieved by inserting a coaxial resonator vertically in the aperture

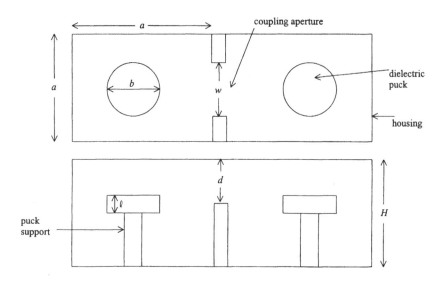

Figure 7.13 *Aperture coupling of dielectric resonators*

Table 7.3   Coupling bandwidth versus aperture depth for coupled $TE_{01\delta}$ resonators

Aperture depth (mm)	Coupling bandwidth (MHz)
30	2
35	17
40	28
50	42
80	77

w. If the resonator is resonant above the $TE_{01\delta}$ resonance we obtain a positive coupling. Alternatively if the resonator is tuned below the $TE_{01\delta}$ resonance then we obtain a negative coupling.

A picture of a typical $TE_{01\delta}$ filter is shown in Figure 7.14 and its measured response is shown in Figure 7.15.

7.3 Dual-mode dielectric resonator filters

The earliest dual-mode dielectric resonator filters were reported in 1982 by Fiedziuszko [12]. A picture of one of these filters is shown in Figure 7.16. The resonant mode in these devices is the $HE_{11\delta}$ which again uses a puck supported in a cavity. The magnetic wall waveguide model, or Cohn model, may again be

Figure 7.14   A $TE_{01\delta}$ mode dielectric resonator filter
     (courtesy of Filtronic plc)

284 Theory and design of microwave filters

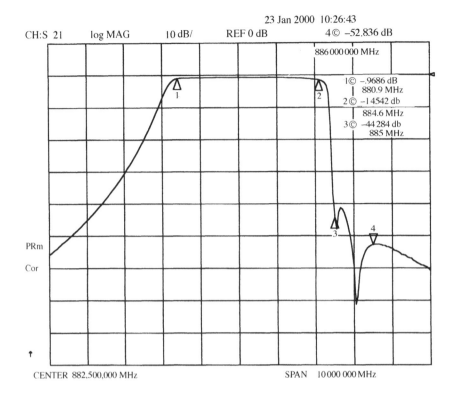

Figure 7.15 Measured response of a $TE_{01\delta}$ mode dielectric resonator filter (courtesy of Filtronic plc)

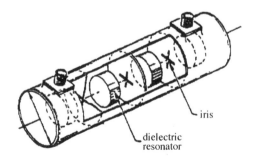

Figure 7.16 Dual-mode dielectric resonator filter

used to obtain an approximate resonance equation giving

$$\frac{\beta}{\alpha}\tan\left(\frac{\beta\ell}{2}\right)\tanh(\alpha\ell_1) = 1 \tag{7.34}$$

where

$$\beta = \left[\omega^2 \mu_0 \varepsilon_0 \varepsilon_r - \left(\frac{3.832}{a}\right)^2\right]^{1/2} \qquad (7.35)$$

$$\alpha = \left[\left(\frac{3.832}{a}\right)^2 \omega^2 \mu_0 \varepsilon_0\right]^{1/2} \qquad (7.36)$$

The $HE_{11\delta}$ mode is not the fundamental mode for a puck centred in a conducting enclosure. Thus for a given resonant frequency the $HE_{11\delta}$ resonator will be larger than a $TE_{01\delta}$ resonator. However, the dual-mode resonance still gives a significant size reduction and typically a volume reduction of 30 per cent can be achieved for a given filter transfer function, compared with $TE_{01\delta}$ designs. The realisation shown in Figure 7.16 is a dual-mode in-line filter. This can be designed using the methods described in Chapter 6. These devices are widely used in communication satellite transponders where size, weight and performance are all of importance. A planar version of the dual HE mode device is reported in Reference 13. An alternative dual-mode $TE_{01\delta}$ resonator consisting of two intersecting cylindrical pucks is reported in Reference 14.

7.3.1 Dual-mode conductor-loaded dielectric resonator filters

In GSM cellular radio base station applications, filter requirements typically need resonators with unloaded Q factors of 5000. These are normally realised using coaxial resonators and are physically quite large. It is desirable to achieve similar Q factors in a much reduced size. The normal configuration for dielectric resonators, with a puck suspended in the middle of a conducting enclosure, gives unloaded Q factors which are only restricted by the loss tangent of the dielectric material. Thus very high Q factors are achieved but the size is large. A typical cavity volume for a single $TE_{01\delta}$ resonator would be 600 cm^3 for a Q_u of 30 000. It is possible to trade off Q_u for volume reduction using the method described in this section. First consider a cylindrical puck suspended in the middle of a conducting enclosure. The order of resonant frequencies is $TE_{01\delta}$ followed by $HE_{11\delta}$. If we now move the puck down towards the base of the housing then the $TE_{01\delta}$ mode goes up in frequency and the $HE_{11\delta}$ goes down in frequency. Eventually they cross over and when the puck is resting on the base of the housing the $HE_{11\delta}$ is the lowest resonance. The two resonant frequencies may still be quite close, however. Also the field pattern of the $HE_{11\delta}$ mode is now distorted by the electric wall touching one of its flat surfaces. Thus we have achieved a fundamental dual-mode resonator although the spurious performance is quite poor. The Q factor is lowered but it is still quite high. Finally we introduce a conductor on the top flat surface of the puck. The effect of the conductor is to push the fundamental dual mode further down in frequency while not significantly affecting the other mode. The resonator is known as a conductor-loaded dielectric resonator [15] and is shown in Figure 7.17.

286 *Theory and design of microwave filters*

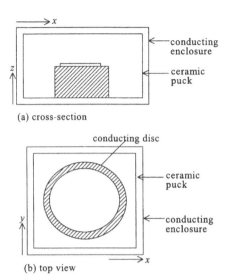

Figure 7.17 Dual-mode conductor-loaded dielectric resonator

It is interesting to consider the fields in the conductor-loaded resonator. First we see a simulation of the magnitude of the E fields of the fundamental mode along the axis of the puck in Figure 7.18. We can see that apart from some fringing fields around the axis of the puck, the field intensity of the fundamental mode is nearly constant along the axis. Now the transverse E field must be zero

Figure 7.18 Magnitude of the E field of the fundamental mode of a dual-mode conductor-loaded dielectric resonator

at the flat surface of the disc and at the base of the housing. Consequently, since there is no variation along z, the transverse E field must be zero everywhere. The resonant mode is thus extremely similar to the TM_{110} mode used to describe the fundamental modes in microstrip patch antennas and ferrite circulators. A simple cavity model can be derived by assuming that there is an electrical wall on the top and bottom of the puck and a magnetic wall around its lateral surface (Figure 7.19).

The field components for the TM_{110} mode are evaluated as follows. First

$$H_z = E_r = E_\phi = 0 \qquad (7.37)$$

Now there is no variation in field along z so resonance must occur at cut-off where $\beta_z = 0$. Now

$$E_z = EJ_1(k_c r)\cos(\phi) \qquad (7.38)$$

$$H_r = \frac{j}{k_c^2}\left(\frac{\omega\varepsilon}{r}\frac{\partial E_z}{\partial \phi} - \frac{\beta \partial H_z}{\partial r}\right) \qquad (7.39)$$

$$H_\phi = \frac{j}{k_c^2}\left(\omega\varepsilon\frac{\partial E_z}{\partial r} + \frac{\beta}{r}\frac{\partial H_z}{\partial r}\right) \qquad (7.40)$$

and with $\beta_z = 0$

$$H_r = \frac{jE\omega\varepsilon}{rk_c^2}J_1(k_c r)\sin(\phi) \qquad (7.41)$$

$$H_\phi = \frac{j\omega\varepsilon}{k_c}EJ_1'(k_c r)\cos(\phi) \qquad (7.42)$$

where

$$k_c^2 = k^2 - \beta^2 = \omega^2\mu\varepsilon \qquad (7.43)$$

Now there is a magnetic wall at $r = a$ so

$$H_\phi = 0|_{r=a} \qquad (7.44)$$

Hence

$$J_1'(k_c a) = 0 \qquad (7.45)$$

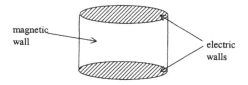

Figure 7.19 Cavity model for the dual TM_{110} mode

or

$$k_c = 1.841 \tag{7.46}$$

and the resonant frequency is given by

$$\omega_c = \frac{1.841c}{a\sqrt{\varepsilon_r}} \tag{7.47}$$

Since the mode has zero field variation along the z axis the resonant frequency is determined by the diameter of the puck using (7.47). However, the resonant frequency and spurious performance of the resonator are also affected by the dimensions of the conducting enclosure. Thus equation (7.47) is not very accurate. As an example a 40 mm diameter puck with $\varepsilon_r = 44$ resonated at 900 MHz while the equation predicts 660 MHz. The field equations do, however, give a good description of the fields in the puck.

The unloaded Q factor and spurious performance of the resonator are determined by the height of the puck. The higher the puck the higher the Q_u, but since the first spurious mode has a half wave variation along the axis of the puck (Figure 7.20) the higher the puck the lower the spurious resonant frequency.

As an example a resonator was constructed with a puck 40 mm in diameter and 24 mm high, with a permittivity of 44 and a loss tangent of 3.3×10^{-5}, in a silver-plated cavity with internal dimensions 65 mm × 65 mm × 40 mm. The silver-plated aluminium disc was 35 mm in diameter and 3 mm thick, to reduce losses due to current flow in the edge of the disc. The fundamental resonant frequency was 930 MHz with a Q_u of 6300. A TEM resonator constructed in the same physical volume would have a Q_u of approximately 5200. Scaling all the dimensions of the resonator to half the above gives a Q_u factor of 4000 at 1.86 GHz.

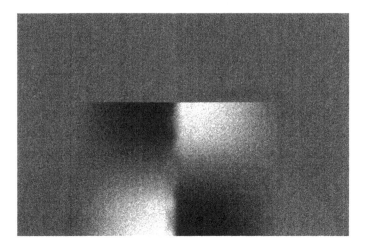

Figure 7.20 H field of the first spurious mode along the axis of the puck

Table 7.4 Resonant frequencies and coupling bandwidths for the dual-mode filter

Resonator	Resonant frequency (MHz)	Coupling bandwidth (MHz)	
1	924.68	0, 1 = 25.34	1, 2 = 30.76
2	951.50	0, 2 = 17.89	2, 3 = 25.20
3	944.85	3, 4 = 20.73	
4	943.99	4, 5 = 20.19	
5	943.85	5, 6 = 20.46	
6	944.59	6, 7 = 24.25	
7	950.69	7, 8 = 31.03	7, 9 = 11.65
8	927.53	8, 9 = 30.08	

The spurious performance of the resonator was limited by the HE_{111} mode which resonated at 370 MHz above the fundamental mode. The spurious performance may be improved by introducing a hole along the axis of the puck. An optimised resonator with a circular hole exhibited a fundamental resonance of 919 MHz with the first spurious mode at 1420 MHz.

This type of resonator is useful for cellular base station filtering applications. As an example a GSM base station filter was designed. This had a 925–960 MHz passband with stopband rejection of 80 dB at 915 MHz. This required an eight-pole generalised Chebyshev filter with two transmission zeros on the low side of the passband. The prototype network was synthesised with cross-couplings from the input node to the second resonator and from the output node to the seventh resonator. Synthesis of this type of asymmetric

Figure 7.21 Dual-mode base station filter
(Reproduced courtesy of Filtronic plc).

290 *Theory and design of microwave filters*

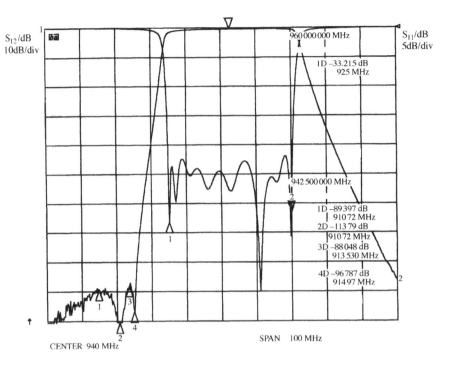

Figure 7.22 *Measured performance of the dual-mode base station filter*

generalised filter has been described in Chapter 3. The resonant frequencies and coupling bandwidths of the bandpass filter were as given in Table 7.4. A photograph of the filter is shown in Figure 7.21. Here we can see that the input and output feeds are inclined relative to the two modes. This has the effect of introducing a cross-coupling by coupling into the first two modes simultaneously. The measured performance of the filter is shown in Figure 7.22.

7.4 Triple-mode dielectric resonator filters

Triply degenerate resonances occur in structures with symmetry in all three dimensions such as spheres and cubes. Initially we shall consider spherical resonators which, because of their very special symmetry, can be analysed exactly. These will then be compared with cubic resonators (which are easier to manufacture) and then a design procedure for triple-mode filters will be described.

7.4.1 Spherical dielectric resonators

Field solutions for spherical resonators are obtained by solving Maxwell's equations in spherical coordinates, r, θ, ϕ. General solutions are given by

Stratton [16]. In our case we will assume axial symmetry. For TE modes the Helmholtz equation in spherical coordinates is given by

$$\frac{\partial^2}{\partial r^2}(rE\phi) + \frac{1}{r^2}\frac{\partial}{\partial \theta}\left\{\frac{1}{\sin(\theta)}\frac{\partial}{\partial \theta}[rE\phi \sin(\theta)]\right\} + K^2(rE\phi) = 0 \quad (7.48)$$

A solution may be obtained by separation of variables if

$$rE\phi = R(r)\theta(\theta) \quad (7.49)$$

where

$$\theta(\theta) = P'_n[\cos(\theta)] \quad (7.50)$$

$P'_n[\cos(\theta)]$ are the nth-degree Legendre polynomials given by

$$P'_0[\cos(\theta)] = 0 \quad (7.51)$$

$$P'_1[\cos(\theta)] = \sin(\theta) \quad (7.52)$$

$$P'_2[\cos(\theta)] = 3\sin(\theta)\cos(\theta) \quad (7.53)$$

etc. and

$$R = \sqrt{r}[A_n J_{n+1/2}(k_r) + B_n N_{n+1/2}(k_r)] \quad (7.54)$$

where $J_{n+1/2}$ and $N_{n+1/2}$ are half-integral-order or spherical Bessel functions of the first and second kind respectively, e.g.

$$J_{1/2}(x) = \left(\frac{2}{\pi x}\right)^{1/2} \sin(x) \quad (7.55)$$

$$N_{1/2}(x) = -\left(\frac{2}{\pi x}\right)^{1/2} \cos(x) \quad (7.56)$$

$$J_{3/2}(x) = \left(\frac{2}{\pi x}\right)^{1/2}\left[\frac{\sin(x)}{x} - \cos(x)\right] \quad (7.57)$$

$$N_{3/2}(x) = \left(\frac{2}{\pi x}\right)^{1/2}\left[\sin(x) + \frac{\cos(x)}{x}\right] \quad (7.58)$$

The field solutions for symmetric TE and TM modes may be given in terms of linear combinations of the J and N functions.

$$Z_{n+1/2}(kr) = J_{n+1/2}(kr) + N_{n+1/2}(kr) \quad (7.59)$$

where $Z_{n+1/2}$ is a spherical Hankel function, and for TE modes

$$E_\phi = \frac{H}{\sqrt{r}} P'_n[\cos(\theta)] Z_{n+1/2}(kr) \quad (7.60)$$

$$H_\theta = \frac{jHP'_n[\cos(\theta)]}{\omega\mu r^{3/2}}[nZ_{n+1/2}(kr) - krZ_{n-1/2}(kr)] \quad (7.61)$$

$$H_r = \frac{-jH_n Z_{n+1/2}(kr)}{\omega\mu r^{3/2}\sin(\theta)}\{\cos(\theta) P'_n[\cos(\theta)] - P'_{n+1}[\cos(\theta)]\} \quad (7.62)$$

These solutions simplify if the origin is included in the structure since the second type of spherical Bessel function has a singularity at the origin and cannot be included in the solution. The lowest order TE mode is the TE_{01} mode, which for solutions including the origin has field components

$$H_r = \frac{-2jH\cos(\theta)}{k^2 r^2}\left[\frac{\sin(kr)}{kr} - \cos(kr)\right] \quad (7.63)$$

$$H_\theta = \frac{jH\sin(\theta)}{k^2 r^2}\left[\frac{(kr)^2 - 1}{kr}\sin(kr) + \cos(kr)\right] \quad (7.64)$$

$$E_\phi = \frac{-H\sin(\theta)}{kr}\left[\frac{\sin(kr)}{kr} - \cos(kr)\right] \quad (7.65)$$

Now consider the spherical dielectric resonator shown in Figure 7.23. The resonator structure consists of a spherical puck of radius a and permittivity ε_r enclosed in an air-filled conducting enclosure of radius b. To a first degree of approximation we can assume that the surface of the dielectric sphere can be represented by a perfect magnetic conductor. Thus

$$H_\theta = 0|_{r=a} \quad (7.66)$$

and from (7.64)

$$\tan(ka) = \frac{ka}{1 - (ka)^2} \quad (7.67)$$

Solving this numerically we find $ka \approx 2.74$ and the resonant frequency is approximated by

$$F_0 = \frac{1.31 \times 10^8}{a\sqrt{\varepsilon_r}} \quad (7.68)$$

As an example, for $\varepsilon_r = 44$ and $a = 3.1$ cm we obtain a resonant frequency of 637 MHz.

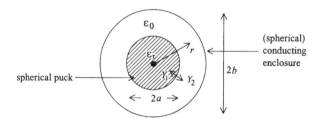

Figure 7.23 Spherical dielectric resonator

Dielectric resonator filters 293

A perfectly accurate expression for resonant frequency can be obtained by allowing the fields to leak out of the resonator into the air space and considering the whole structure. The resonator may then be considered as a cascade of two spherical transmission lines.

A spherical transmission line may be used as the single-mode equivalent circuit between two spherical surfaces at different radial points. Since the spherical resonator is perfectly symmetrical the spherical transmission line may be considered as a one-dimensional (radial) structure connecting modal voltages and currents which are related to the transverse fields.

The transfer matrix of a spherical transmission line is given by

$$\begin{bmatrix} V_1 \\ I_1 \end{bmatrix} = \begin{bmatrix} A & B \\ C & D \end{bmatrix} \begin{bmatrix} V_2 \\ I_2 \end{bmatrix} \tag{7.69}$$

where the relationship between transverse fields and voltages and currents is given by

$$E_t(r, \theta, \phi) = \frac{V(r)e(\theta, \phi)}{r} \tag{7.70}$$

$$H_t(r, \theta, \phi) = \frac{I(r)h(\theta, \phi)}{r} \tag{7.71}$$

From (7.64) and (7.65) the transverse functions e and h are identified for the TE_{01} mode, thus justifying a transmission line equivalent circuit. For TE modes the $ABCD$ parameters are [17]

$$A = \hat{J}_n(kr_1)\hat{N}'_n(kr_2) - \hat{N}_n(kr_1)\hat{J}'_n(kr_2) \tag{7.72}$$

$$B = jZ_0[\hat{N}_n(kr_2)\hat{J}_n(kr_1) - \hat{J}_n(kr_2)\hat{N}_n(kr_1)] \tag{7.73}$$

$$C = jY_0[\hat{N}'_n(kr_2)\hat{J}'_n(kr_1) - \hat{J}'_n(kr_2)\hat{N}'_n(kr_1)] \tag{7.74}$$

$$D = \hat{J}_n(kr_2)\hat{N}'_n(kr_1) - \hat{N}_n(kr_2)\hat{J}'_n(kr_1) \tag{7.75}$$

and for TM modes

$$A = \hat{J}_n(kr_2)\hat{N}'_n(kr_1) - \hat{N}_n(kr_2)\hat{J}'_n(kr_1) \tag{7.76}$$

$$B = jZ_0[\hat{N}'_n(kr_2)\hat{J}'_n(kr_1) - \hat{J}'_n(kr_2)\hat{N}'_n(kr_1)] \tag{7.77}$$

$$C = jY_0[\hat{N}_n(kr_2)\hat{J}_n(kr_1) - \hat{J}_n(kr_2)\hat{N}_n(kr_1)] \tag{7.78}$$

$$D = \hat{J}_n(kr_1)\hat{N}'_n(kr_2) - \hat{N}_n(kr_1)\hat{J}'_n(kr_2) \tag{7.79}$$

where Z_0 is the characteristic impedance of free space in the medium.

In regions which contain the origin the N functions cannot exist and we have, for TE modes,

$$V(r) = I\hat{J}_n(kr) \tag{7.80}$$

$$I(r) = jY_0 I\hat{J}'_n(kr) \tag{7.81}$$

and, for TM modes,

$$V(r) = jZ_0 I \hat{J}'_n(kr) \tag{7.82}$$

$$I(r) = I\hat{J}_n(kr) \tag{7.83}$$

The functions $\hat{J}_n(x)$ and $\hat{N}_n(x)$ are similar to the spherical Bessel functions, e.g.

$$\hat{J}_0(x) = \sin(x) \tag{7.84}$$

$$\hat{J}_1(x) = -\cos(x) + \frac{\sin(x)}{x} \tag{7.85}$$

$$\hat{N}_0(x) = -\cos(x) \tag{7.86}$$

$$\hat{N}_1(x) = -\sin(x) + \frac{\cos(x)}{x} \tag{7.87}$$

The equivalent circuit of the dielectric resonator consists of a cascade of two spherical waveguides (Figure 7.24). The resonance condition is thus

$$Y_1 + Y_2 = 0 \tag{7.88}$$

where Y_1 is given by the ratio of current to voltage from (7.80) and (7.81), or (7.82) and (7.83). Y_2 is the input impedance of a short circuited section of spherical waveguide of length $b - a$. Thus

$$Y_2 = \frac{D}{B}\bigg|_{r_1=a,\,r_2=b} \tag{7.89}$$

The resonance equations for TE and TM modes are then given by

$$\frac{\hat{J}'(ka)N(kb) - \hat{N}'(ka)\hat{J}(kb)}{\hat{N}(kb)\hat{J}(ka) - \hat{J}(kb)\hat{N}(ka)} + \frac{\sqrt{\varepsilon_r}\hat{J}'_n(k\sqrt{\varepsilon_r}\,a)}{\hat{J}_n(k\sqrt{\varepsilon_r}\,a)} \tag{7.90}$$

$$\frac{\hat{N}(ka)\hat{J}'(kb) - \hat{J}(ka)\hat{N}'(kb)}{\hat{N}'(kb)\hat{J}'(ka) - \hat{J}'(kb)\hat{N}'(ka)} - \frac{\sqrt{\varepsilon_r}\hat{J}_n(k\sqrt{\varepsilon_r}\,a)}{\hat{J}'_n(k\sqrt{\varepsilon_r}\,a)} \tag{7.91}$$

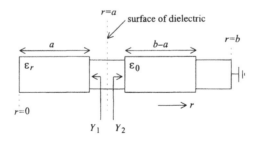

Figure 7.24 Equivalent circuit of a spherical dielectric resonator

respectively. Taking the previous numerical example with $\varepsilon_r = 44$, $a = 3.1$ cm and $b = 6.2$ cm we obtain resonant frequencies of 754.4 MHz and 1000 MHz for the lowest ordered TE and TM modes. The analysis is exact and agrees with experimental data and field simulations using finite element analysis. This is one of the few complex structures that is solvable by simple equations.

The variation of transverse field intensity along the (radial) direction of propagation may be obtained by analysing the equivalent circuit of the spherical dielectric resonator. We have for the TE modes, for $0 \leq r \leq a$,

$$V(r) = J_n(k\sqrt{\varepsilon_r} r) \tag{7.92}$$

$$I(r) = \frac{j\sqrt{\varepsilon_r}}{377} J_n'(k\sqrt{\varepsilon_r} r) \tag{7.93}$$

with

$$E_t = V(r)/r \tag{7.94}$$

$$H_t = I(r)/r \tag{7.95}$$

and, for $r = b$,

$$V(b) = 0 \tag{7.96}$$

For $a < r \leq b$

$$\begin{bmatrix} V(r) \\ I(r) \end{bmatrix} = \begin{bmatrix} A(r,b) & jB(r,b) \\ jC(r,b) & D(r,b) \end{bmatrix} \begin{bmatrix} 0 \\ I(b) \end{bmatrix} \tag{7.97}$$

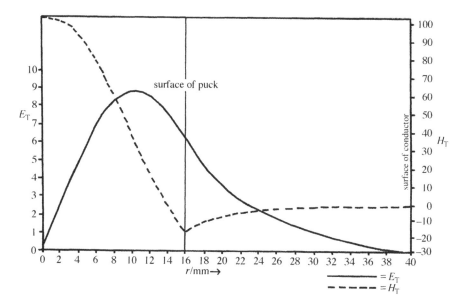

Figure 7.25 Transverse fields versus radial position for TE modes in a spherical dielectric resonator

Therefore
$$V(r) = jB(r, b)I(b) \tag{7.98}$$
and
$$V(a) = jB(a, b)I(b) = J_n(k\sqrt{\varepsilon_r}\, a) \tag{7.99}$$
Therefore
$$I(b) = \frac{J_n(k\sqrt{\varepsilon_r}\, a)}{jB(a, b)} \tag{7.100}$$
Hence
$$E_t = \frac{V(r)}{r} = \frac{B(r, b)J_n(k\sqrt{\varepsilon_r}\, a)}{rB(a, b)} \tag{7.101}$$
Similarly
$$H_t = \frac{I(r)}{r} = \frac{j\sqrt{\varepsilon_r}}{377}\frac{J'_n(k\sqrt{\varepsilon_r}\, a)D(r, b)}{D(a, b)} \tag{7.102}$$

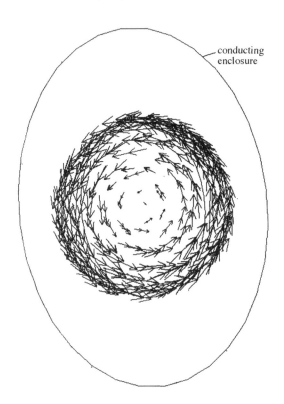

Figure 7.26   E field of a $\mathrm{TE}_{01\delta+1}$ spherical dielectric resonator

Equations (7.92)–(7.93) and (7.101)–(7.102) may be used to compute the transverse E fields as a function of radial position. This is shown for the lowest TE mode in Figure 7.25.

Scrutiny of Figure 7.25 shows a near magnetic wall at the surface of the dielectric. Furthermore, the variation in fields is similar to that for the $TE_{01\delta+1}$ mode in a cylindrical resonator. The correct name for this mode is thus a 'spherical-$TE_{01\delta+1}$' mode ($\theta = 1, \psi = 0, r = \delta + 1$). Field plots obtained from finite element analysis are shown in Figures 7.26–7.29. Here we see a remarkable similarity with the $TE_{01\delta}$ mode in a cylindrical resonator. The E field circles the equator of the puck and the H field forms loops in the meridian plane. The E field is zero in the centre of the puck where the H field is maximum. Imagine a cylinder aligned with the structure such that the E field rotates around the lateral surface of the cylinder. The components of magnetic field which are tangential to the axis of the cylinder would have maximum values on or near the

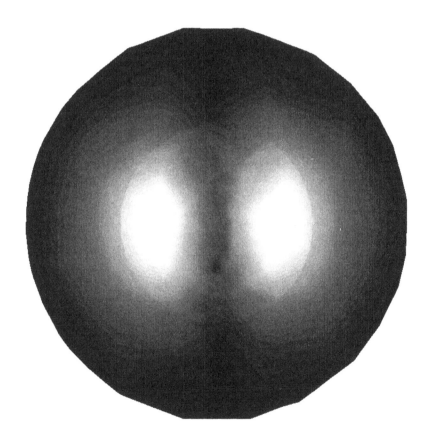

Figure 7.27 Intensity of the E field of a $TE_{01\delta+1}$ *spherical dielectric resonator*

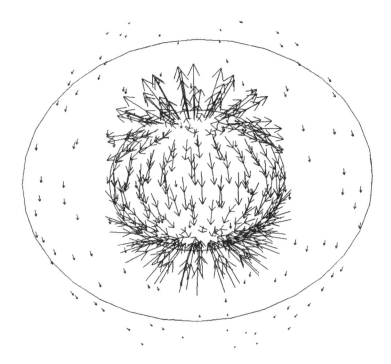

Figure 7.28   H field of a $TE_{01\delta+1}$ spherical dielectric resonator

flat surfaces of the cylinder. The difference in mode numbers thus arises from a difference in coordinate system rather than field pattern. It is also important to realise that we are only looking at one of the three degenerate modes; the other two have orthogonal polarisations.

The ratio of spurious frequency to fundamental in the spherical resonator is 1.32:1. This may be improved, as before, by introducing a hole into the centre of the puck. This will increase the ratio to 1.4:1. It is also interesting to note that the resonant frequency of such a structure can be evaluated by the transfer matrix procedure described here. The equivalent circuit would then consist of a cascade of three spherical transmission lines rather than two.

7.4.2 Cubic dielectric resonators

The spherical resonator has a nice simple structure with perfect symmetry, which is easy to analyse in terms of spherical waveguide modes. This yields exact expressions for resonant frequency and field patterns. Furthermore it has a reasonably good spurious-free bandwidth. Unfortunately it is difficult to manufacture in large volumes at low cost. This is because ceramic processing normally involves powder pressing which is easier to do on objects with flat surfaces. A cubic puck is a more practical shape.

Dielectric resonator filters 299

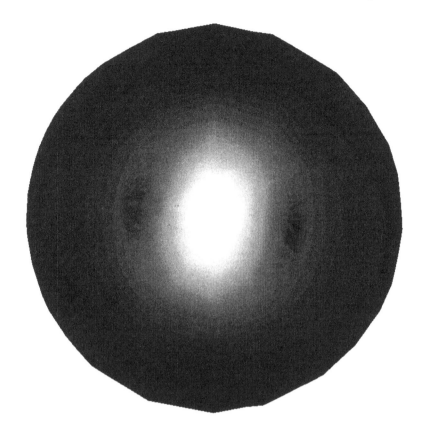

Figure 7.29 Intensity of the H field of a $TE_{01\delta+1}$ spherical dielectric resonator

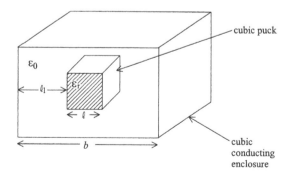

Figure 7.30 Cubic dielectric resonator

Figure 7.30 shows a cubic dielectric resonator consisting of a cubic puck suspended centrally in a cubic conducting enclosure. It is not possible to derive a simple exact expression for the resonant frequency; however, an approximate Cohn model can be developed. First because of symmetry we can assume an arbitrary direction for the z axis of the cube. The resonator will behave identically for the three degenerate modes along the x, y and z axes. Second, we assume as before that dielectric–air interfaces on surfaces parallel to the z axis are ideal magnetic conductors. Fields are allowed to leak out of the flat surfaces which are normal to z, and as before the magnetic conductors are extended to the top and bottom surfaces (Figure 7.31).

The equivalent circuit is thus similar to that for the cylindrical resonator except that the propagation constants and characteristic impedances are different. The Helmholtz equation for TE modes is

$$\nabla_T^2 H_z = -k_c^2 H_z \tag{7.103}$$

For magnetic wall boundary conditions

$$H_z = 0 \big|_{\substack{x=0, x=\ell \\ y=0, y=\ell}} \tag{7.104}$$

Hence

$$H_z = E \sin\left(\frac{m\pi x}{\ell}\right) \sin\left(\frac{n\pi y}{\ell}\right) \exp(\pm \gamma z) \tag{7.105}$$

The lowest order non-zero mode is the TE_{11} mode and

$$\gamma = k_c^2 - \omega^2 \mu \varepsilon \tag{7.106}$$

Thus for propagating waves in the dielectric

$$\gamma = j\beta = [\omega^2 \mu_0 \varepsilon_0 \varepsilon_r - 2(\pi/\ell)^2]^{1/2} \tag{7.107}$$

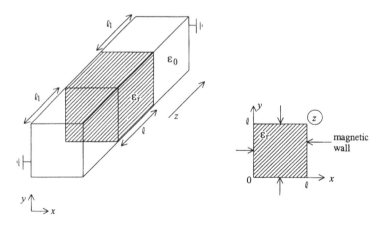

Figure 7.31 Cohn model for a cubic dielectric resonator

and for non-propagating waves in the air region

$$\gamma = \alpha = [2(\pi/\ell)^2 - \omega^2\mu_0\varepsilon_0]^{1/2} \tag{7.108}$$

The characteristic impedance is

$$Z_0 = \frac{E_y}{H_x} = \frac{-j\omega\mu}{\gamma} \tag{7.109}$$

The characteristic impedances in the propagating and non-propagating regions are

$$Z_p = \frac{-\omega\mu}{\beta} \tag{7.110}$$

$$Z_c = \frac{-j\omega\mu}{\alpha} \tag{7.111}$$

The resonance equation is thus given by

$$\frac{\beta}{\alpha}\tan\left(\frac{\beta\ell}{2}\right)\tanh(\alpha\ell_1) = 1 \tag{7.112}$$

This is identical to the resonance equation for the $TE_{01\delta}$ mode in cylindrical resonators, except with different values for β and α. The variation in transverse fields with axial position is the same as for the $TE_{01\delta}$ mode. Thus the mode is very similar to the $TE_{01\delta}$ cylindrical mode and $TE_{01\delta+1}$ spherical mode. Analysis of TM modes in the cube shows that the lowest order mode is the $TM_{11\delta+1}$ with a higher resonant frequency than the TE mode. Thus we can say that the TE mode, designated $TE_{11\delta}$ cartesian, is the fundamental resonant mode in a cubic dielectric resonator.

A resonator has been constructed using ZTS ceramic with $\varepsilon_r = 36$, $\ell = 2.5$ cm and $\ell_1 = 2.7$ cm. The measured resonant frequency of the fundamental mode was 1.67 GHz. Solution of (7.112) yields a frequency of 1.575 GHz, indicating a similar accuracy to the Cohn model for the cylindrical resonator. The measured Q factor of 23 000 is similar to that for cylindrical and spherical resonators of the same physical size. The temperature stability of the resonator was almost identical for each of the three degenerate modes with a value of 1.5 kHz/°C. This is very desirable for narrowband applications.

7.4.3 Design of triple-mode dielectric resonator reflection filters

A triple-mode resonator may be excited by an input probe which couples into one of three degenerate modes. If we assume that there are no non-adjacent couplings then the equivalent circuit of the resonator is a simple one-port ladder network. Alternatively there may indeed be non-adjacent couplings giving a more complex cross-coupled equivalent circuit. These situations are shown in Figure 7.32 where circles represent resonators and lines represent inverters.

Typical specifications for narrowband low loss filters usually require generalised Chebyshev transfer functions with arbitrary transmission zero locations. As an example a cross-coupled ladder network of degree 6 is shown in Figure 7.33. With this network it is possible to realise transfer

302 *Theory and design of microwave filters*

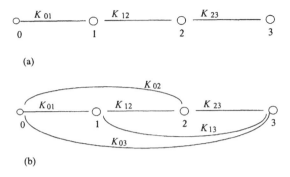

Figure 7.32 *One-port equivalent circuits for triple-mode resonators: (a) without non-adjacent couplings; (b) with non-adjacent couplings*

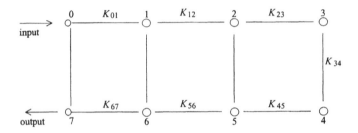

Figure 7.33 *Symmetrical cross-coupled prototype network*

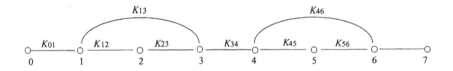

Figure 7.34 *Cross-coupled prototype with two finite real frequency transmission zeros*

functions with all transmission zeros at finite frequencies such as elliptic function filters.

A triple-mode realisation of this prototype is quite a challenge! Multiple couplings between modes in different cavities would be required and it would be hard to eliminate unwanted couplings. A less complicated realisation is possible using the network shown in Figure 7.34. In this case there is only one coupling between a pair of cavities and the realisation is feasible. However, the choice of transmission zero locations has been restricted.

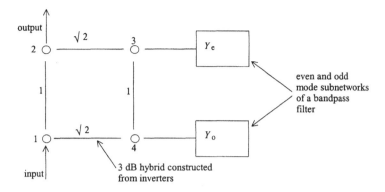

Figure 7.35 Hybrid reflection mode bandstop filter

An alternative approach is to use the hybrid reflection mode filter [18] shown in Figure 7.35. The hybrid reflection mode filter enables a simple realisation of all the transfer functions realisable by the network shown in Figure 7.33. It consists of a 3 dB quadrature hybrid with networks Y_e and Y_o connected to nodes 3 and 4. Y_e and Y_o are the even- and odd-mode subnetworks of a bandpass filter which is normally realised using the network shown in Figure 7.33. The complete device is a two-port network with input and output ports at nodes 1 and 2. The analysis of this circuit is relatively straightforward.

Consider an input signal of unity amplitude applied at port 1 of the hybrid. This will produce outputs $j/\sqrt{2}$ at port 3 and $1/\sqrt{2}$ at port 4. These signals will then reflect off the even- and odd-mode subnetworks producing an output at port 2 of

$$S_{12} = \frac{j}{2}(\Gamma_o + \Gamma_e) \tag{7.113}$$

and an output at port 1 of

$$S_{11} = \tfrac{1}{2}(\Gamma_o - \Gamma_e) \tag{7.114}$$

where Γ_e and Γ_o are the reflection coefficients of networks Y_e and Y_o, with

$$\Gamma_e = \frac{Y_e - 1}{Y_e + 1} \tag{7.115}$$

$$\Gamma_o = \frac{Y_o - 1}{Y_o + 1} \tag{7.116}$$

Hence

$$S_{12} = \frac{j(Y_e Y_o - 1)}{(1 + Y_e)(1 + Y_o)} \tag{7.117}$$

and

$$S_{12} = \frac{Y_e - Y_o}{(1 + Y_e)(1 + Y_o)} \tag{7.118}$$

Now a symmetrical two-port network such as Figure 7.33 with scattering parameters S'_{11} and S'_{12} can be described in terms of its even- and odd-mode admittances by

$$S'_{11} = \frac{Y_e Y_o - 1}{(1 + Y_e)(1 + Y_o)} \tag{7.119}$$

$$S'_{12} = \frac{Y_e - Y_o}{(1 + Y_e)(1 + Y_o)} \tag{7.120}$$

Thus

$$S_{11} = S'_{12} \tag{7.121}$$

$$S_{12} = jS'_{11} \tag{7.122}$$

Apart from a 90° phase shift the reflection and transmission functions of the original bandpass filter and the reflection filter are interchanged. In other words the transmission function of the hybrid filter is a bandstop filter exactly equal to the reflection function of the original bandpass filter.

Furthermore, we can also create a bandpass transmission function by inserting an inverter between port 3 of the hybrid and Y_e (Figure 7.36). Thus signals at port 3 of the hybrid experience an additional 180° of phase shift and the sign of Γ_e in equations (7.113)–(7.114) is reversed; hence

$$S_{11} = S'_{11} \tag{7.123}$$

$$S_{12} = jS'_{12} \tag{7.124}$$

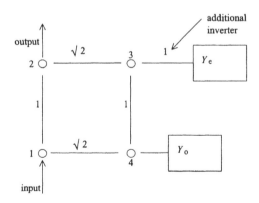

Figure 7.36 Hybrid reflection mode bandpass filter

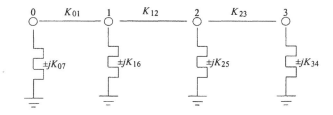

Figure 7.37 Even- and odd-mode subnetworks of the cross-coupled prototype

Thus provided that the even- and odd-mode subnetworks can be constructed then either a bandpass or bandstop filter may be constructed using the same physical hardware.

The even- and odd-mode subnetworks of the cross-coupled prototype network of Figure 7.33 are shown in Figure 7.37. These even- and odd-mode networks are simple third-order one-port ladder networks identical in form to the simplest equivalent circuit of the triple-mode resonator. The three shunt reactances at nodes 1–3 may be absorbed by tuning the resonant frequencies of the resonators. The even-mode resonators will be tuned up and the odd-mode resonators will be tuned down in frequency for positive Ks. Consequently a degree 6 cross-coupled filter may be constructed by realising the even- and odd-mode subnetworks as separate triple-mode resonators.

7.4.4 Design example

The design of a bandstop filter will now be described. The filter specification is based on a cellular radio base station application for separating A and B operators in the AMPS band. The filter specification was for a centre frequency of 845.75 MHz, a 1.5 dB passband of less than 1.5 MHz and a 20 dB stopband of greater than 1.1 MHz. The specification can be realised by a degree 6 elliptic function filter with

$$|S_{11}(j\omega)|^2 = \frac{F_N^2(\omega)}{1 + F_N^2(\omega)} \tag{7.125}$$

In this application it is convenient to choose the stopband insertion loss level to be equal to the passband return loss level. Thus

$$|S_{11}(j\omega)|^2 = |S_{12}(j/\omega)|^2 \tag{7.126}$$

Thus

$$F_N^2(\omega) = 1/F_N^2(1/\omega) \tag{7.127}$$

and

$$F_N^2 = \frac{(\omega^2 - \omega_1^2)(\omega^2 - \omega_2^2)(\omega^2 - \omega_3^2)}{(1 - \omega^2\omega_1^2)(1 - \omega^2\omega_2^2)(1 - \omega^2\omega_3^2)} \tag{7.128}$$

For a 22 dB return loss we have $\omega_1 = 0.36492$, $\omega_2 = 0.8155$ and $\omega_3 = 0.94835$.

Forming $S_{11}(p)$ and $S_{12}(p)$ by taking the left-hand plane zeros of $1 + F_N^2(\omega)$ we obtain

$$S_{11}(p) = \frac{N(p)}{D(p)} = \frac{0.9968p^6 + 1.6922p^4 + 0.8039p^2 + 0.0794}{p^6 + 1.992p^5 + 3.734p^4 + 4.025p^3 + 3.7347p^2 + 1.992p + 1}$$

(7.129)

$$S_{12}(p) = \frac{j(0.0794p^6 + 0.8039p^4 + 1.6922p^2 + 0.9968)}{D(p)}$$

(7.130)

The multiplication of $S_{12}(p)$ by j ensures that $Y_o = Y_e^*$. Now

$$S_{11}(p) = \frac{1 - Y_e Y_o}{(1 + Y_e)(1 + Y_o)}$$

(7.131)

and

$$S_{12}(p) = \frac{Y_e - Y_o}{(1 + Y_e)(1 + Y_o)}$$

(7.132)

Thus

$$S_{11} + S_{12} = \frac{(1 + Y_e)(1 - Y_o)}{(1 + Y_e)(1 + Y_o)} = \frac{1 - Y_o}{1 + Y_o}$$

(7.133)

Hence

$$Y_e = \frac{1 - S_{11} + S_{12}}{1 + S_{11} - S_{12}}$$

(7.134)

and

$$Y_e = \frac{25.1499jp^3 - 15.6939p^2 + 29.2789j - 18.490}{p^3 + 25.7124jp^2 - 7.0843p + 17.0766j}$$

(7.135)

Y_e can be synthesised using a continued fraction expansion into a ladder network composed of capacitors, invariant reactances and inverters. The network is shown in Figure 7.38 with element values

$B_0 = 25.1499$, $B_1 = 0.040229$

$B_2 = -580.807$, $B_3 = 0.00556$ (7.136)

$C_1 = 0.0015848$, $C_2 = 1034.60$, $C_3 = 0.006376$

The shunt frequency invariant reactance at the input to each subnetwork is not associated with a resonator. This may be realised as a shunt capacitor or capacitor or more elegantly it may be absorbed into a phase shifter. Consider the network shown in Figure 7.39. This shows the input part of the even- and odd-mode subnetworks. This may be equated to a phase shifter followed by an inverter, as follows.

Dielectric resonator filters

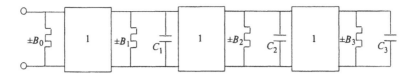

Figure 7.38 Synthesised even- and odd-mode subnetworks for a degree 6 elliptic function filter

The transfer matrix of the first network is given by

$$[T] = \begin{bmatrix} 1 & 0 \\ jB_0 & 1 \end{bmatrix} \begin{bmatrix} 0 & j \\ j & 0 \end{bmatrix} \begin{bmatrix} 1 & 0 \\ jB_1' & 1 \end{bmatrix}$$

$$= \begin{bmatrix} -B_1' & j \\ j(1 - B_0 B_1') & -B_0 \end{bmatrix} \qquad (7.137)$$

and for the second network

$$[T] = \begin{bmatrix} \cos(\psi) & j\sin(\psi) \\ j\sin(\psi) & \cos(\psi) \end{bmatrix} \begin{bmatrix} 0 & j/K \\ jK & 0 \end{bmatrix}$$

$$= \begin{bmatrix} -K\sin(\psi) & \dfrac{j\cos(\psi)}{K} \\ jK\cos(\psi) & \dfrac{-\sin(\psi)}{K} \end{bmatrix} \qquad (7.138)$$

Now equating the two matrices, from the B parameter, $\cos(\psi) = K$; thus from

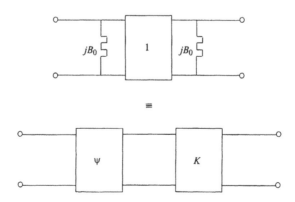

Figure 7.39 Removal of input shunt reactances

the D parameters

$$\frac{(1-K^2)^{1/2}}{K} = B_0 \tag{7.139}$$

Therefore

$$K = \frac{1}{(1+B_0^2)^{1/2}} \tag{7.140}$$

and

$$\begin{aligned}
\sin(\psi) &= [1 - \cos^2(\psi)]^{1/2} \\
&= (1-K^2)^{1/2} \\
&= \frac{B_0}{(1+B_0^2)^{1/2}}
\end{aligned} \tag{7.141}$$

Therefore

$$\psi = \sin^{-1}\left[\frac{B_0}{(1+B_0^2)^{1/2}}\right] \tag{7.142}$$

and from the A parameter

$$\begin{aligned}
B_1' &= K \sin(\psi) \\
&= \frac{B_0}{1+B_0^2}
\end{aligned} \tag{7.143}$$

B_0 has now been removed and B_1' can be absorbed into the first reactance B_1. The final network is as shown in Figure 7.40.

The value of ψ in this example is 87.72° for the even-mode network and $-87.82°$ for the odd-mode network. Only the difference between ψ_e and ψ_o is important and this is integrated into the hybrid circuit. The lowpass prototype even- and odd-mode networks are transferred into bandpass circuits using techniques described in Chapter 4.

The simulated frequency response of the complete filter is shown in Figure 7.41. The 3 dB hybrid may be constructed using TEM low loss transmission lines with a reasonable ground plane spacing. The construction of the cubic resonator assembly is shown in Figure 7.42. The measured performance of the prototype device using $\varepsilon_r = 44$ resonators with $Q_u = 28\,000$ is shown in Figure 7.43.

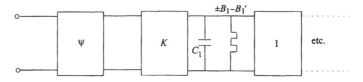

Figure 7.40   The final even- and odd-mode networks

Dielectric resonator filters 309

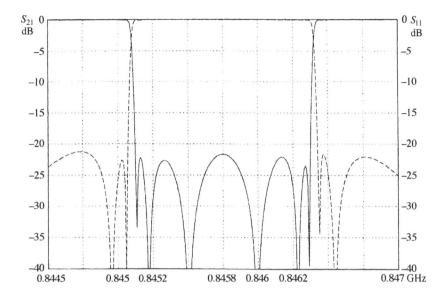

Figure 7.41 *Simulated frequency response of a hybrid reflection mode bandstop filter (assumed lossless)*

7.5 Dielectric-loaded filters

The application of high permittivity ceramics is not restricted to dielectric resonator filters. They may also be used to miniaturise conventional filters by partially or completely loading TEM and waveguide resonators with ceramic [19, 20]. For example, a TEM wave has a free space quarter wavelength at

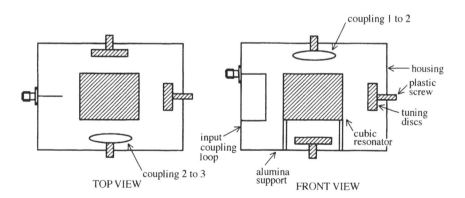

Figure 7.42 *Construction of a triple-mode resonator*

310 *Theory and design of microwave filters*

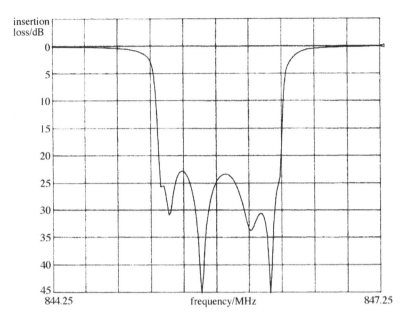

Figure 7.43 *Measured performance of a triple-mode reflection filter*

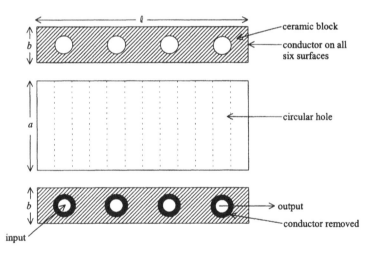

Figure 7.44 *Dielectric-loaded TEM coupled-line structure*

1 GHz of 7.5 cm. If the wave is propagating in a dielectric medium the wavelength is reduced by the square root of the permittivity and for $\varepsilon_r = 80$ the 1 GHz quarter wavelength reduces to 8.385 mm.

Consider the structure shown in Figure 7.44. This structure consists of a

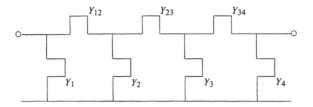

Figure 7.45 Equivalent circuit of the TEM coupled-line structure

rectangular block of high permittivity ceramic with circular holes introduced between the ground planes b in the direction a. The entire structure is metallised, apart from small regions round one end of the circular holes. The interiors of the holes are also metallised and so they form an array of coupled TEM transmission lines which are open circuited at one end and short circuited at the other end. The equivalent circuit is shown in Figure 7.45.

Since the dielectric is homogeneous all the phase velocities of the lines are equal and the circuit is an all-stop structure. One could introduce capacitors between the open circuited ends of the lines and ground as in the combline filter. However, this is difficult to do and it is preferable to have a single integrated structure. Alternatively one can alternate the shorts and open circuits between opposite ends of the lines so that the structure becomes an interdigital filter. In this case the device has a definite passband but the couplings are so strong that for narrow bandwidths the holes would have to be physically far apart. Alternatively we can introduce a discontinuity into the structure such that the dielectric loading is not homogeneous and the even- and odd-mode phase velocities are different. For example we can introduce a layer of lower dielectric constant on one of the flat surfaces as shown in Figure 7.46.

In this case the ε_{r2} layer could be a printed circuit which is used to interface with the outside world. ε_{r2} would then be much less then ε_{r1} and if we examine the even and odd modes of a pair of coupled lines we see that the even-mode phase velocity will be less than the odd-mode phase velocity. This is shown in Figure 7.47.

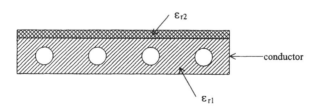

Figure 7.46 TEM coupled-line structure with inhomogeneous dielectric loading

312 *Theory and design of microwave filters*

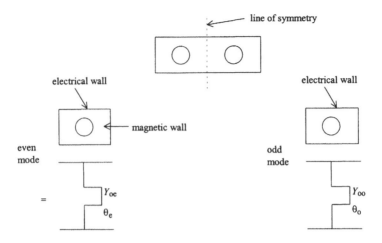

Figure 7.47 Even and odd modes of a coupled-line structure

In general an inhomogeneous dielectric structure cannot support a pure TEM mode. There will always be some longitudinal field components and therefore TEM equivalent circuits are not strictly valid. However, they are a reasonable approximation over narrow bandwidths and the pair of coupled inhomogeneous lines may be approximated by the pi equivalent circuit shown in Figure 7.48.

The values of Y_{oe}, Y_{oo}, θ_e and θ_o may be obtained for a particular set of dimensions and dielectric constants by electromagnetic field simulations.

Examining Figure 7.48 we see that the series branch is a parallel tuned circuit.

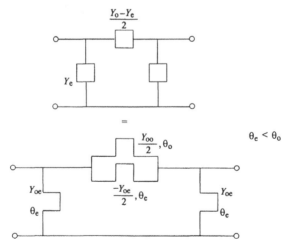

Figure 7.48 Approximate equivalent circuit of inhomogeneous coupled lines

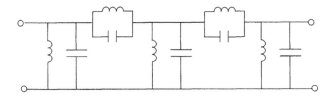

Figure 7.49 Equivalent circuit of a dielectric-loaded TEM filter of degree 3

For θ_e and θ_o less then 90° the values of $\tan(\theta_e)$ and $\tan(\theta_o)$ are both positive. Thus the element $Y_{oo}/2$ is inductive and the element $-Y_{oe}/2$ is capacitive. Consequently the series coupling branch will resonate below the resonant frequency of the shunt stubs.

The series coupling branches produce transmission zeros below the centre frequency of the passband at a frequency given by

$$Y_{oe} \tan(\theta_o) - Y_{oo} \tan(\theta_e) = 0 \tag{7.144}$$

The shunt elements resonate with $\theta_e = \pi/2$.

The lumped element equivalent circuit of a device with N holes is an nth-degree network with a single transmission zero at infinity and the remaining $N - 1$ transmission zeros at real frequencies on the low side of the passband (Figure 7.49).

This type of network may be designed using the asymmetric generalised Chebyshev lowpass prototype described in Chapter 3, as shown in Figure 7.50. The prototype network can be converted into a bandpass network by applying the conventional lumped bandpass transformation:

$$\omega \rightarrow \alpha \left(\frac{\omega}{\omega_0} - \frac{\omega_0}{\omega} \right) \tag{7.145}$$

After the transformation the bandpass network is as shown in Figure 7.51. This circuit may be equated to the purely lumped circuit of Figure 7.49 by equating the resonant frequencies and reactance slopes of the resonators.

Having designed a bandpass prototype we know the resonant frequencies and bandwidths of all the resonators. The series resonators are at arbitrary

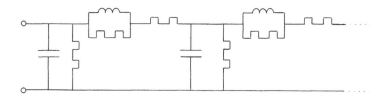

Figure 7.50 Asymmetric lowpass prototype

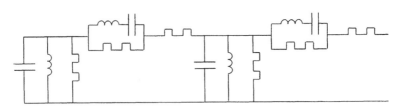

Figure 7.51 Bandpass transformed asymmetric filter

frequencies above or below the passband whereas the natural series resonances in the ceramic block all occur below the passband. We require a method of adjusting the transmission zero frequencies by adjusting these resonant frequencies. This may be achieved by introducing capacitive or inductive coupling between resonators as shown in Figure 7.52. The capacitive probe between two resonators is achieved by removing metallisation on the surface of the filter leaving a floating strip which couples across the resonators. Inductive coupling can be achieved by grounding the central part of this strip.

An exploded view of a typical ceramic TEM diplexer for the AMPS band is shown in Figure 7.53. Its measured performance is shown in Figure 7.54.

7.5.1 Dielectric-loaded waveguide filters

A dielectric-loaded waveguide resonator has an unloaded Q_u which is at least double the value for a TEM resonator of the same physical size and resonant

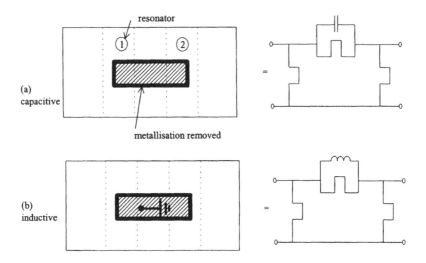

Figure 7.52 Capacitive and inductive couplings in the ceramic block

Dielectric resonator filters 315

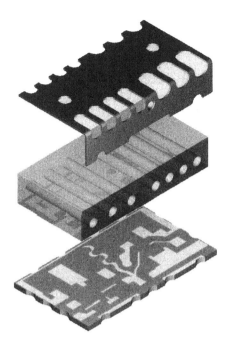

Figure 7.53 *A typical ceramic TEM diplexer*
(courtesy of Filtronic plc)

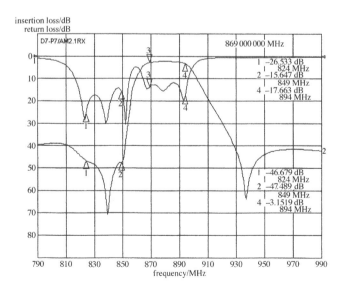

Figure 7.54 *Measured performance of a typical ceramic TEM diplexer*

316 *Theory and design of microwave filters*

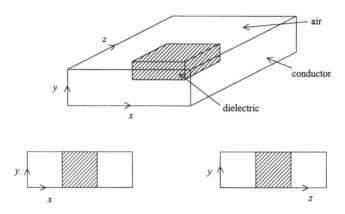

Figure 7.55   *Centrally dielectric-loaded waveguide cavity*

frequency. This is because the TEM resonator has current concentration in the centre conductor where most of the loss occurs. Although it is possible to construct dielectric waveguide filters from metallised slabs of dielectric [21] it is useful to limit the loading to *E*-plane dielectric slabs. By restricting the dielectric loading to the centre region of the *yz* plane most of the size reduction can be achieved while the amount of dielectric used is reduced dramatically.

As an example a silver-plated cavity with dimensions 20 mm × 20 mm × 9 mm, loaded with a slab of ceramic with $\varepsilon_r = 44$ and dimensions 7.7 mm × 7.7 mm × 9 mm, achieves a Q factor of 3000 at 2 GHz. The resonant mode is a slightly distorted version of the TE_{101} mode. Ninety-nine per cent of the E field and 27.26 per cent of the H field are stored in the dielectric; thus the magnetic fields must be used for coupling resonators in filters. A six-pole filter with

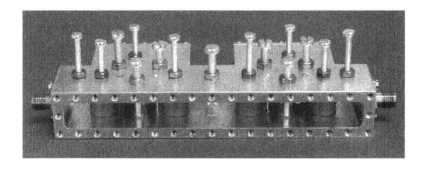

Figure 7.56   *Dielectric-loaded waveguide filter with six poles and two cross couplings*
(courtesy of Filtronic plc)

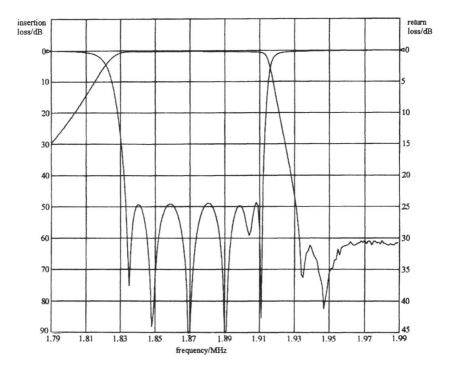

Figure 7.57 Measured performance of a six-pole filter with cross-couplings

cross-couplings has been constructed [22] (Figure 7.56). Its measured performance is shown in Figure 7.57.

7.6 Summary

This chapter has been concerned with the theory of the design of filters using resonators which are constructed using low loss high permittivity ceramics. Initially the fundamentals of modes in dielectric rod waveguides are discussed. This is followed by the derivation of the simple Cohn model for the $TE_{01\delta}$ mode dielectric resonator. An example of a filter using these resonators is presented. Next a discussion of dual-mode in-line dielectric filters is followed by details of the design of dual-mode conductor-loaded dielectric resonators. These find application in cellular radio base stations and a design example is presented. The use of triple-mode resonators enables significant size reduction compared with single-mode designs. The exact theoretical modelling of triple-mode spherical dielectric resonators is described. This is followed by the design theory and an example of a triple-mode reflection filter. Finally the use of dielectrics for extreme miniaturisation by loading TEM and waveguide structures is discussed.

7.7 References

1. RAYLEIGH: *Philosophical Magazine*, 1897, **43** (125)
2. RICHTMAYER, R.D.: 'Dielectric resonators', *Journal of Applied Physics*, 1938, **10**, pp. 391–98
3. COHN, S.B.: 'Microwave bandpass filters containing high-Q dielectric resonators', *IEEE Transactions on Microwave Theory and Techniques* 1968, **16** (4), pp. 218–72
4. WERSING, W.: 'Microwave ceramics for resonators and filters', *Current Opinion on Solid State and Materials Science*, 1996, **1**, pp. 715–29
5. KAJFEZ, D., and GUILLON, P.: 'Dielectric resonators' (Artech House, Norwood, MA, 1986) pp. 72–95
6. CLARRICOATS, P.J.B., and TAYLOR, B.C.: 'Evanescent and propagating modes of dielectric loaded circular waveguide', *Proceedings of the IEE*, 1964, **111** (12), pp. 1951–56
7. ITOH, T., and RUDOKAS, R.S.: 'New method for computing the resonant frequencies of dielectric resonators', *IEEE Transactions on Microwave Theory and Techniques*, 1977, **25** (1), pp. 52–54
8. JAWORSKI, M. and POSPIESZALSKI, M.W.: 'An accurate solution for the cylindrical dielectric resonator problem', *IEEE Transactions on Microwave Theory and Techniques*, 1979, **27** (7), pp. 639–43
9. CHEN, S.: 'Dielectric ring resonators loaded in waveguides and on substrate', *IEEE Transactions on Microwave Theory and Techniques*, **39** (12), pp. 2069–76
10. LIANG, X., and ZAKI, K.A.: 'Modelling of cylindrical dielectric resonators in rectangular waveguides and cavities', *IEEE Transactions on Microwave Theory and Techniques*, **41** (12), pp. 2174–81
11. KAJFEZ, D., and GUILLON, P.: 'Dielectric resonators' (Artech House, Norwood, MA, 1986), pp. 165–69
12. FIEDZIUSZKO, S.J.: 'Dual-mode dielectric resonator loaded cavity filters', *IEEE Transactions on Microwave Theory and Techniques*, 1982, **30** (9), pp. 1311–16
13. FIEDZIUSZKO, S.J.: 'Engine block dual-mode dielectric resonator cavity filter with monoadjacent cavity couplings', IEEE International Microwave Symposium Digest, San Francisco, CA, May–June 1984, pp. 285–87
14. KUDSIA, C., CAMERSON, R., and TANG, W.: 'Innovations in microwave filters and multiplexing networks for communications satellite systems", *IEEE Transactions on Microwave Theory and Techniques*, 1992, **40** (6), pp. 1133–49
15. HUNTER. I.C., RHODES, J.D., and DASSONVILLE, V.: 'Dual mode filters with conductor loaded dielectric resonators', *IEEE Transactions on Microwave Theory and Techniques*, 1999, **47** (12), pp. 2304–11
16. STRATTON, J.A.: 'Electromagnetic theory', (McGraw-Hill, New York, 1941) ch. 7
17. MARCUVITZ, N.: 'Waveguide handbook' (IEE Press, Stevenage, 1986), pp. 47–54
18. HUNTER. I.C., RHODES, J.D., and DASSONVILLE, V.: 'Triple mode

dielectric resonator hybrid reflection filters', *IEE Proceedings on Microwaves, Antennas and Propagation*, 1998, **145**, pp. 337–43

19 YAMASHITA, S., and MAKIMOTO, M.: 'Miniaturised coaxial resonator partially loaded with high-dielectric-constant microwave ceramics', *IEEE Transactions on Microwave Theory and Techniques*, 1983, **31** (9), pp. 697–703

20 KOBAYASHI, S., and SAITO, K.: 'A miniaturised ceramic bandpass filter for cordless phone systems', IEEE International Microwave Symposium Digest, 1995, **2**, pp. 391–94

21 WAKINO, K., NISHIKAWA, T., and ISHIKAWA, Y.: 'Miniaturisation technologies of dielectric resonator filters for mobile communications'. *IEEE Transactions on Microwave Theory and Techniques*, 1994, **42** (7), pp. 1295–300

22 DASSONVILLE, V., and HUNTER, I.C.: 'Dielectric loaded waveguide filter', Submitted to *IEE Proceedings on Microwaves, Antennas and Propagation*

Chapter 8
Miniaturisation techniques for microwave filters

8.1 Introduction

Among the most important specifications for microwave filters are selectivity, bandwidth, passband insertion loss and physical size. In fact as shown in Chapter 4 these are related for an all-pole bandpass filter by

$$L = \frac{4.343 f_0}{\Delta f Q_u} \sum_{r=1}^{N} g_r \qquad (8.1)$$

where f_0 is the centre frequency, Δf is the passband bandwidth, Q_u is the unloaded Q of the resonators and g_r is the element value of the rth element in the lowpass prototype. From this equation we can see that as we increase the selectivity of the filter then the number of elements, and hence the passband loss, increases. Furthermore, the roll-off of insertion loss across the passband also increases. Obviously we can use the optimum transfer function but the same relationship still holds. Also as we reduce the filter bandwidth we must increase the resonator Q if we are to maintain a fixed insertion loss. Now since Q_u is proportional to volume for a microwave resonator, a highly selective, narrowband, low loss filter will require a significant physical volume. The question is, are there any ways in which we can overcome this problem?

Several alternative approaches will be discussed. These are dielectric resonators, high temperature superconductivity, surface acoustic wave devices, active filters and finally, the use of new subsystem architectures combined with predistorted reflection mode filters.

8.2 Dielectric resonator filters

The loss of a dielectric resonator is largely determined by the dielectric loss tangent of the ceramic puck. This is because if the puck is physically remote from the walls of the conducting enclosure then the energy storage is largely confined to the interior of the puck. Furthermore, the high permittivity results in a dramatic reduction in wavelength compared with free space. Consequently unloaded Q factors of up to 50 000 may be obtained in a reasonable physical volume. This is not possible using TEM or metallic waveguide resonators. One would hope that by increasing ε_r *ad infinitum*, a very small resonator could be obtained. Unfortunately, as ε_r increases, the Q_u of available materials decreases and we rapidly reach a limit. We can improve this situation by using multiple degenerate modes to increase the efficiency in terms of volume per resonance for a given Q_u. However, we again reach a limit as increasing the number of degenerate modes beyond three results in a dramatic increase in physical complexity and poorer spurious performance, for a minimal improvement in volume per resonance. Thus we can conclude that dielectric resonators may improve the situation but in no way do they eliminate the basic problem.

8.3 Superconducting filters

It was discovered in 1911 [1] that the resistance of electrical conductors dropped to near zero at temperatures of a few kelvins. Furthermore, in 1986 [2] similar observations were made at 77 K. The most popular of these high temperature superconductors is $YBa_2Cu_3O_{7-x}$ (YBCO). As a comparison the surface resistance of YBCO at 10 GHz is 0.1 mΩ at 77 K compared with 8.7 mΩ for copper. The surface resistance of a superconductor increases more rapidly with frequency than that of a normal conductor, resulting in a cross-over frequency at which both have equal surface resistance. A typical value is 23 GHz, and since the resistivity of superconductors varies as the square of frequency then they work well in the 1–2 GHz band.

In principle high temperature superconductivity enables resonators with near infinite unloaded Q to be constructed. As an example a YBCO cavity resonator with a Q of 400 000 at 10 GHz and 77 K has now been used in a down-converter [3]. Coupled cavity resonator filters for mobile communications base stations have also been reported. As an example a B band notch filter for the American AMPS system with eight poles and a notch bandwidth of 1 MHz at 55 dB rejection has been reported. This used dielectric resonators within a superconducting cavity with resonator Q of 40 000 [4]. The resonators achieved higher Q per unit volume than a conventional $TE_{01\delta}$ resonator and were roughly comparable with triple-mode devices. On the other hand the cooling system required a power consumption of 400 W and weighed approximately 50 kg. Furthermore, the measured two-tone intermodulation performance was −85 dBm for an input power of −10 dBm. This corresponds to a third-order intercept point of +27.5 dBm. If the input power were increased to +10 dBm the intermodulation products would then be at −25 dBm, only 35 dB below the carrier, and there would be little point in having a 55 dB notch. Improvements in the non-linear performance of these devices would probably make them acceptable for certain low power receive applications. They would not be suitable for a GSM transmit filter where a typical requirement is for third-order products at −115 dBm with input powers of 30 W, corresponding to an intercept point of +122 dBm.

8.4 Surface acoustic wave filters

Surface acoustic wave (SAW) devices have been used for IF filtering and other low frequency applications since the 1970s [5]. However, in recent years new device architectures have been developed for frequencies up to 3 GHz. Their main advantage over other technologies is their very small size (typically 3 mm × 3 mm × 1 mm) in applications such as cellular handsets where their insertion loss and power handling are tolerable. Typically they have 3 dB loss and 2 W power handling.

SAW devices operate by manipulating acoustic waves propagating near the surface of piezoelectric crystals. Typically the speed of propagation of these waves is 10 000 times slower than the speed of light. Hence structures many acoustic wavelengths long can be made on surfaces only a few millimetres long. Conventional IF filters use Rayleigh waves where the molecules on the surface of the crystal move in an elliptical path. These waves are generated by applying an RF electrical field to the surface of the crystal via an interdigital transducer. A typical IF SAW filter design is shown in Figure 8.1.

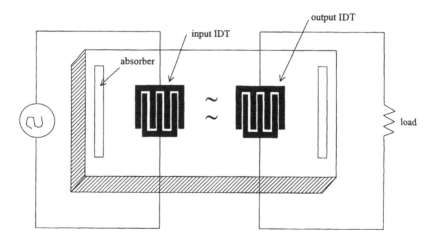

Figure 8.1 Conventional IF SAW filter

The separation between adjacent fingers of the transducer is one-half of the acoustic wavelength. The amplitude of the SAW wave generated by the nth finger on the transducer is related to the length of the finger. A particular pattern of overlaps along the transducer, or apodisation pattern, determines the impulse response of the transducer. Its frequency response is the Fourier transform of the impulse response. In a conventional SAW filter the input wave propagates a signal through the crystal and it is received by the output transducer. The transfer characteristic of the filter is then determined by the product of the transfer functions of the individual transducers. Unfortunately the insertion loss of these devices is high as the transducers are bidirectional. Half the power from each transducer propagates in the wrong direction and must be absorbed. Thus the minimum insertion loss is 6 dB. In addition multiple reflections, known as triple transits, give rise to significant amplitude and phase ripple across the passband.

Recently relatively low loss SAW filters have been developed using SAW resonators which are formed between acoustically reflective gratings on the surface of the SAW crystal [6] (Figure 8.2). Energy is coupled in and out of the SAW resonator by placing a transducer between the gratings. The transducer has only a few fingers and is relatively broad band. Furthermore, as the transducer couples into waves in both directions the insertion loss problems of bidirectional transducers are avoided. In addition to a different architecture most RF SAW filters use a leaky SAW which has three main advantages over Rayleigh waves at RF frequencies. First, the speed of propagation is approximately 1.5 times faster than a Rayleigh wave with a corresponding reduction in acoustic wavelength, enabling transducers and gratings to be fabricated at higher frequencies. Second, higher values of electromechanical coupling also enable relatively broadband filters to be constructed. Finally leaky SAWs penetrate deep into the crystal enabling higher power handling capability.

The equivalent circuit of a SAW resonator shows that it has a pole and a zero. Typically RF SAW resonator filters are made by cascading resonators in a ladder network, Figure 8.3.

Careful resonator design enables the poles in the series resonators and the

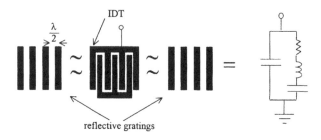

Figure 8.2 One-pole SAW resonator and its equivalent circuit

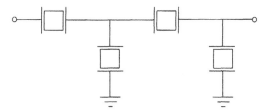

Figure 8.3 SAW resonator ladder filter

zeros in the shunt resonators to produce real frequency transmission zeros. Typical performance of these devices is as one would expect from low degree filters with resonator Q factors of 300–400. As an example a GSM receive filter would have an insertion loss of 3.5 dB from 925 to 960 MHz and 25 dB rejection at 915 MHz. These devices offer a size reduction compared with ceramic-loaded TEM filters provided the specification is fairly modest. SAW filters have poorer power handling and temperature stability than ceramic filters. It seems unlikely that they will perform as well as ceramic filters for high performance applications such as handsets for third generation cellular systems where the transmitter and receiver operate simultaneously. However, acoustic wave technology is developing rapidly and improvements are to be expected. Indeed, bulk acoustic resonator devices with impressive performance have recently been demonstrated [7].

8.5 Active microwave filters

Miniaturisation of filters normally results in an increase of insertion loss. Consequently it is worth investigating whether there is any merit in integrating active devices into filters to compensate for losses. In this way it is possible, at least in principle, to make very small resonators. One of the most interesting

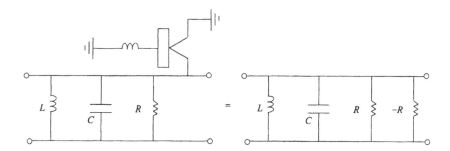

Figure 8.4 Active resonator with negative resistance circuit

techniques which has been reported [8] is to use the negative real part of the input impedance of an inductively loaded common base bipolar transistor circuit in order to cancel the losses in a low Q resonator (Figure 8.4).

In this way both active fixed tuned and varactor tuned filters have been constructed where the active circuit compensates not only for the resonator losses but also for the much larger losses associated with varactor diodes. Experimental results show that the small signal performance of the devices is as would be expected from devices with infinite unloaded Q. This was particularly apparent in a notch filter where switching the active device off resulted in a reduction in stopband attenuation from 20 dB to 3 dB. However, these results were somewhat misleading as they were single frequency low power measurements. In reality a filter has multiple inputs and its purpose is to pass the desired passband signal undistorted while rejecting one or more, possibly high power, unwanted stopband signals. Under these conditions the non-linear characteristics of the active devices used will cause various distortion effects. These include a reduction in the unloaded Q of the resonators as the devices saturate, and the generation of intermodulation products.

Measurements of a two-pole 75 MHz bandwidth 1.8 GHz bandpass filter showed a gain compression of 4 dB at +7.5 dBm input power, at a device collector current of 5 mA [9]. Power saturation effects on a three-pole 75 MHz bandstop filter showed a reduction in stopband attenuation from 35 dB to only 12 dB as the input power was increased from −20 dBm to 0 dBm. The saturation effects can be overcome to a certain degree by changing the device bias when large input signals are present, although this would not be trivial. A more significant problem is third-order intermodulation distortion.

Consider the parallel tuned circuit shown in Figure 8.5. Analysis of this circuit yields

$$I = V(G + j\omega C - j/\omega L) \qquad (8.2)$$

and the Q factor is given by

$$Q = \frac{\omega_0 C}{G} \qquad (8.3)$$

where

$$\omega_0 = \frac{1}{(LC)^{1/2}} \qquad (8.4)$$

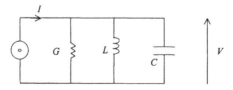

Figure 8.5   Parallel tuned circuit

At resonance

$$V = \frac{I}{G} \tag{8.5}$$

and at resonance the current flowing through the capacitor is

$$I_C = j\omega_0 CV = \frac{j\omega_0 CI}{G} \tag{8.6}$$

Therefore

$$|I_C| = QI \tag{8.7}$$

Thus at resonance the current flowing through the capacitor can be much higher than the applied current.

We can consider the non-linearity in the active resonator in terms of a non-linear current transfer characteristic given by

$$I = aI_{in} + bI_{in}^2 + cI_{in}^3 \ldots \tag{8.8}$$

This generates third-order intermodulation products at the output of the device of a power level P_{IM} where

$$P_{IM} = 3P_{in} - 2IP_3 \tag{8.9}$$

P_{in} is the input power and IP_3 is the two-tone third-order intercept point.

These intermodulation products are generated by the third-order term in (8.8). However, from (8.7) the current in this expression is proportional to the loaded Q and inversely proportional to the percentage bandwidth of the filter. Thus the third-order intercept point of an active filter will be reduced by 6 dB each time the filter bandwidth is halved. Typical results for a 1 per cent bandwidth filter gave an IP_3 at 8 dBm. Obviously this depends on the exact circuit and the type of device used but such low intercept points are very difficult to work with. With an IP_3 of 8 dBm and input signals of 0 dBm the third-order products would be at −16 dBm; thus there is no point in designing the filter to have a stopband rejection of more than 16 dB. An active filter would only be used when the bandwidth was narrow, thus requiring high Q resonators. This is exactly the situation when the current magnification in the resonators causes a reduction in IP_3. Furthermore, an active filter has a noise figure which is equal to the insertion loss of the passive part of the filter plus a contribution from the active devices. The conclusion is thus that active filters are not a useful solution for miniaturisation.

8.6 Lossy filters

As we have seen, integrating active devices into resonators is not a good solution. Although this restores the small signal shape factor of the filter there are other problems of non-linearity. However, it is possible to design a

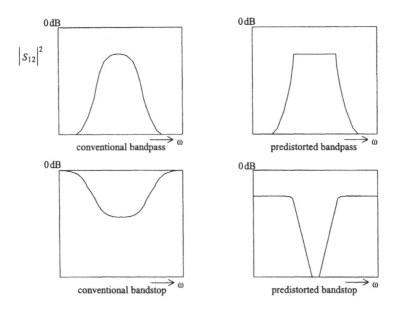

Figure 8.6 *Effect of finite Q_u on conventional and predistorted bandpass and bandstop filters*

purely passive filter to have a sharp, selective response even with low Q resonators. This is shown in Figure 8.6. Here we see that the effect of losses on a bandpass filter is to round the passband. Alternatively it is possible to retain a sharp characteristic by a technique known as predistortion. In the case of a conventional bandstop filter the losses cause a rounding of passband and reduce the stopband attenuation. Alternatively a predistorted bandstop filter retains a good characteristic with a sharp response and high stopband attenuation. Thus we can say that it is possible to design filters with low Q resonators provided we can tolerate a certain level of passband insertion loss. The real question is whether there is any real application for a filter with significant passband insertion loss.

Consider the situation of the transmitter shown in Figure 8.7 where a power amplifier is followed by a lossy bandpass filter. In this case the filter would be required to remove out-of-band noise from the spectrum of the (non-linear) power amplifier. Normally in this situation the filter would be highly selective requiring high resonator Qs. A lossy filter could still be selective, and indeed remove the noise, but it would reduce the output power of the amplifier by its insertion loss. This would require the amplifier to be driven harder, producing more noise, and would be self-defeating. It would also reduce the efficiency of the amplifier.

Now consider the situation of the receiver shown in Figure 8.8. This shows a low noise receiver front end, possibly for a cellular radio base station. The front

Miniaturisation techniques 329

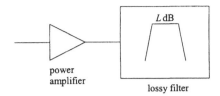

Figure 8.7 Power amplifier followed by a lossy filter

end must have good sensitivity and a wide dynamic range. Thus the noise figure must be low and the intercept point should be high. In a situation where known frequency, high power interfering signals are present, we can use a bandstop filter prior to the amplifier to stop the interferers causing intermodulation distortion in the amplifier. The passband insertion loss of the bandstop filter must be low to preserve the system noise figure. Normally the stopband bandwidth of the filter would be narrow and it would be highly selective. Thus high Q resonators would be necessary and the filter would be large. Alternatively we can use a lossy filter prior to the amplifier. In this case, in order to preserve the system noise figure this must be preceded by a further low noise amplifier. This would have just enough gain to minimise the effect of the passband loss L_1 on the noise figure. If the noise figure of the first amplifier is F then the noise figure contribution from the first two stages is

$$\text{NF} = F + \frac{L_1 - 1}{G_1} \tag{8.10}$$

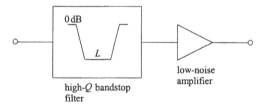

(a) conventional design

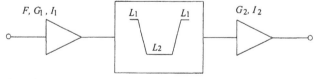

(b) alternative design

Figure 8.8 Architectures for low noise front end with receiver protection

Thus if F is 1 dB and L_1 is 6 dB, G_1 needs to be around 10–20 dB to minimise the effect of the filter. The gain should be just enough to achieve this and no more, so that the first amplifier does not contribute too much intermodulation distortion. This can be checked by analysing the circuit. We assume the filter has a passband loss L_1 and a stopband attenuation L_2, and the amplifiers have gains G_1 and G_2 and intercept points I_1 and I_2.

The system gain is given by

$$G = G_1 - L_1 + G_2 \tag{8.11}$$

Now we assume that two interfering signals are present at the system input with power levels of X dBm. The frequencies of these signals lie in the stopband of the bandstop filter. First we calculate the power levels of intermodulation products at the output of the system which are generated by the first amplifier. We assume that the frequencies of these products occur within the passband of the filter. The power levels are

$$P_1 = 3X - 2I_1 + G_1 - L_1 + G_2 \tag{8.12}$$

Intermodulation products generated by the second amplifier have system output levels of

$$P_1 = 3(X + G_1 - L_2) - 2I_2 + G_2 \tag{8.13}$$

The bandstop filter obviously protects the second amplifier from the interferers but the first amplifier is not protected. Thus the optimum value of L_2, the stopband attenuation, is when $P_1 = P_2$. There is no point in reducing intermodulation from the second amplifier below the level produced by the first amplifier. Thus

$$3X - 2I_1 + G_1 - L_1 + G_2 = 3X + 3G_1 - 3L_2 - 2I_2 + G_2 \tag{8.14}$$

and

$$L_2 = \frac{2G_1 + L_1 + 2(I_1 - I_2)}{3} \tag{8.15}$$

Now if we assume $I_1 = I_2$ then

$$L_2 = \frac{2G_1 + L_1}{3} \tag{8.16}$$

As an example if $G_1 = 21$ dB and $L_1 = 6$ dB then $L_2 = 16$ dB, and in this situation there would be no practical reason to make L_2 greater than 16 dB. However, a 16 dB reduction in interferers at the input to the second amplifier is equivalent to increasing the intercept point of this amplifier by 24 dB.

With an interfering signal level X of $+10$ dBm the output products would be at $+6$ dBm compared with $+54$ dBm for an unfiltered amplifier with the same gain and intercept point. The system works equally well as having a low loss filter at the input of the first amplifier with 18 dB stopband rejection. In this case intermodulation from the first amplifier would be reduced but the second amplifier would still produce products at $+6$ dBm.

8.6.1 Design of lossy filters – classical predistortion

First we will review classical predistortion techniques as in Reference 10. Predistortion is a method by which the correct transmission characterisation of a bandpass filter may be preserved when the resonators have finite Q factor. This is achieved as follows.

The transfer function of an all-pole (ladder) lowpass prototype is given by

$$S_{12}(p) = \frac{1}{D(p)} \tag{8.17}$$

In the presence of dissipation loss the lowpass prototype ladder network is as shown in Figure 8.9. If this dissipation loss is uniform then the transfer function of the device is given by

$$S'_{12}(p) = S_{12}(p + \alpha)$$

$$= \frac{1}{D(p + \alpha)} \tag{8.18}$$

where α is related to the Q of the structure and

$$G_r = \alpha C_r \tag{8.19}$$

where C_r and G_r are the rth elements in the lossy filter.

We have already established that introducing dissipation into the filter causes a rounding of the passband. However, this can be avoided by shifting p to $p - \alpha$ and synthesising a network with a transfer function $S_{12}(p - \alpha)$. This results in a lossless ladder network which when we add loss and p shifts to $p + \alpha$ has the correct transfer function other than a constant offset in insertion loss. This is best illustrated by an example, in this case a degree 2 Butterworth filter where

$$S_{12}(p) = \frac{1}{p^2 + \sqrt{2}p + 1} \tag{8.20}$$

We now shift p to $p - \alpha$ and multiply by a constant to obtain

$$S_{12}(p) = \frac{K}{(p-\alpha)^2 + \sqrt{2}(p-\alpha) + 1}$$

$$= \frac{K}{p^2 + p(\sqrt{2} - 2\alpha) + 1 - \sqrt{2}\alpha + \alpha^2} \tag{8.21}$$

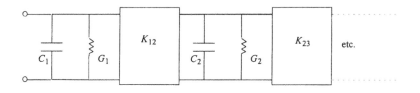

Figure 8.9   Lowpass ladder network with finite uniform dissipation

and

$$|S_{12}(jw)|^2 = \frac{K^2}{[(1-\sqrt{2\alpha}+\alpha^2)-w^2]^2 + w^2(\sqrt{2}-2\alpha)^2}$$

$$= \frac{K^2}{X^2 - 2\sqrt{2\alpha}X + 1 - 2\sqrt{2} + 4\alpha^2} \qquad (8.22)$$

where

$$X = w^2 + \alpha^2 \qquad (8.23)$$

The act of shifting p to $p - \alpha$ makes $|S_{12}|^2$ peak near the band-edges. The value of K should be chosen so that $|S_{12}|^2$ peaks at exactly unity. This ensures minimum insertion loss and a passive realisation.

Now for $|S_{12}(jw)|^2$ to have unity magnitude at a turning point X_0 then

$$X_0^2 - 2\sqrt{2\alpha}X_0 + 1 - 2\sqrt{2\alpha} + 4\alpha^2 = K^2 \qquad (8.24)$$

and

$$\frac{\partial}{\partial X}(X^2 - 2\sqrt{2\alpha}X + 1 - 2\sqrt{2\alpha} + 4\alpha^2) = 0\Big|_{X_0} \qquad (8.25)$$

or

$$X_0 = \sqrt{2\alpha} \qquad (8.26)$$

From (8.24) and (8.26)

$$K^2 = 1 - 2\sqrt{2\alpha} + 2\alpha^2 \qquad (8.27)$$

Hence

$$K = 1 - \sqrt{2\alpha} \qquad (8.28)$$

and

$$S_{12}(p) = \frac{1-\sqrt{2\alpha}}{p^2 + p(\sqrt{2}-2\alpha) + 1 - \sqrt{2\alpha}+\alpha^2} \qquad (8.29)$$

Now

$$|S_{11}(jw)|^2 = 1 - |S_{12}(jw)|^2$$

$$= \frac{X^2 - 2\sqrt{2\alpha}X + 2\alpha^2}{X^2 - 2\sqrt{2\alpha}X + 1 - 2\sqrt{2\alpha}+4\alpha^2}$$

$$= \frac{(X-\sqrt{2\alpha})^2}{X^2 - 2\sqrt{2\alpha}X + 1 - 2\sqrt{2\alpha}+4\alpha^2} \qquad (8.30)$$

Thus

$$S_{11}(p) = \frac{p^2 + \sqrt{2\alpha} - \alpha^2}{p^2 + p(\sqrt{2}-2\alpha) + 1 - \sqrt{2\alpha}+\alpha^2} \qquad (8.31)$$

Now forming the input admittance $Y(p)$

$$Y(p) = \frac{1 + S_{11}(p)}{1 - S_{11}(p)}$$

$$= \frac{2p^2 + p(\sqrt{2} - 2\alpha) + 1}{p(\sqrt{2} - 2\alpha) + 1 - 2\sqrt{2}\alpha + 2\alpha^2} \quad (8.32)$$

Synthesising $Y(p)$ as a ladder network we first extract a capacitor of value

$$C_1 = \frac{\sqrt{2}}{1 - \sqrt{2}\alpha} \quad (8.33)$$

leaving a remaining admittance

$$Y_1(p) = \frac{p\left[\sqrt{2} - 2\alpha - \frac{\sqrt{2}p}{1 - \sqrt{2}\alpha}(1 - 2\sqrt{2}\alpha + 2\alpha^2)\right] + 1}{p(\sqrt{2} - 2\alpha) + 1 - 2\sqrt{2}\alpha + 2\alpha^2}$$

$$= \frac{1}{p(\sqrt{2} - 2\alpha) + 1 - 2\sqrt{2}\alpha + 2\alpha^2} \quad (8.34)$$

Now inverting with an inverter of value

$$K_{12} = \frac{1}{1 - \sqrt{2}\alpha} \quad (8.35)$$

we obtain

$$Y_2(p) = \frac{\sqrt{2}p}{1 - \sqrt{2}\alpha} + 1 \quad (8.36)$$

which is a capacitor of value

$$C_2 = \frac{\sqrt{2}}{1 - \sqrt{2}\alpha} = C_1 \quad (8.37)$$

in parallel with a 1 Ω load.

Now we add loss to the network by letting $p = p + \alpha$. The final network is shown in Figure 8.10. The transfer function of the filter is

$$|S_{12}(j\omega)|^2 = \frac{(1 - \sqrt{2}\alpha)^2}{1 + \omega^4} \quad (8.38)$$

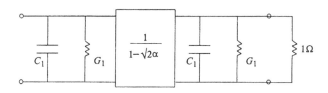

Figure 8.10   Predistorted maximally flat prototype

with a realisability condition

$$\alpha \leq \frac{1}{\sqrt{2}} \tag{8.39}$$

Thus the transfer function is preserved apart from a constant offset in insertion loss.

The input reflection coefficient can be found from (8.31) with p replaced by $p + \alpha$.

$$S_{11}(p) = \frac{(p+\alpha)^2 + \sqrt{2}\alpha - \alpha^2}{p^2 + \sqrt{2}p + 1} \tag{8.40}$$

and

$$|S_{11}(j\omega)|^2 = \frac{\omega^4 + \omega^2(4\alpha^2 - 2\sqrt{2}\alpha) + 2\alpha^2}{1 + \omega^4} \tag{8.41}$$

Thus the predistorted insertion loss characteristic has been obtained by modifying the numerator of the return loss function. The insertion loss of a conventional Butterworth filter rolls off near the edge of the passband. However, by reflecting energy in the middle of the passband the predistorted filter recovers the original transfer function. As an example, if $\alpha = 1/2\sqrt{2}$ then the insertion loss at d.c. is 6 dB and the return loss is also 6 dB. The output return loss is equal to the input return loss, but higher degree solutions result in asymmetric networks with different values for S_{11} and S_{22}. Furthermore, it can be shown that for higher degree solutions a significant price is paid in terms of extra insertion loss above the band-edge loss of the original non-predistorted network. In addition it is difficult to obtain simple solutions for predistorted highpass networks with finite real frequency transmission zeros. For example consider the highpass prototype ladder network with finite dissipation loss shown in Figure 8.11.

The transmission zeros in this network occur when the impedance of the rth shunt branch is zero, i.e.

$$L_r p_r + R_r = 0 \tag{8.42}$$

That is,

$$p_r = -\frac{R_r}{L_r} \tag{8.43}$$

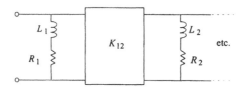

Figure 8.11   Highpass prototype network with finite dissipation loss

The effect of uniform loss is to move the transmission zeros from d.c. onto the left-hand real axis in the complex plane, hence limiting the maximum stopband insertion loss. This is different from the lowpass case shown in Figure 8.10 where the transmission zeros are still at infinity. Predistortion in the lowpass case preserves the passband shape by modifying the capacitors whereas modifying the inductors will not move the transmission zeros of the highpass filter back to the origin. There are various techniques for predistorting highpass prototypes by adding additional loss (see, for example, Reference 11) but they are of little value. For these reasons predistortion of the transmission response is not the best solution for lossy filters.

8.6.2 Design of lossy filters – reflection mode type

Consider the network shown in Figure 8.12 where a resonant circuit with finite loss is coupled to one of the ports of an ideal circulator. (For a discussion on circulators see Reference 12.) The transmission characteristic from ports 1 to 3 is the reflection coefficient of the resonator. Assume that we adjust the input coupling to the resonant circuit so that the real part of its input impedance is matched to the circulator. Thus in a 1 Ω system we have

$$\operatorname{Re} Z(j\omega) = 1 \qquad (8.44)$$

Thus

$$\operatorname{Re} \frac{G + j(\omega C - 1/\omega L)}{K^2} = 1 \qquad (8.45)$$

That is,

$$K = \sqrt{G} \qquad (8.46)$$

In this case at the resonant frequency all the power incident at port 1 will be absorbed in the resistive part of the resonator. Hence the transmission

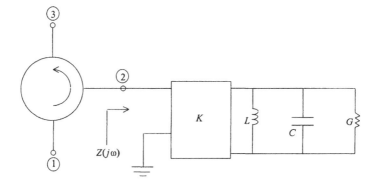

Figure 8.12   Reflection mode bandstop resonator

characteristic from port 1 to port 3 is that of a bandstop resonator with infinite unloaded Q factor at its resonant frequency. Thus we have created a resonant circuit with loss which has a finite real frequency transmission zero, overcoming the problem described in the previous section.

The basic objective is thus to synthesise a multi-element version of this network, in other words to synthesise lossy networks with prescribed reflection functions. This can be achieved by predistorting the reflection function of a lossless prototype network as follows. Given a reflection function $S_{11}(p)$ we let

$$S_{11}(p) \to KS_{11}(p - \alpha) \tag{8.47}$$

The choice of K is made on a similar basis to that for predistortion of $S_{12}(p)$. We evaluate the maximum value of K such that the resultant network is passive. This is achieved by choosing K so that $|S_{11}(j\omega)|^2$ peaks at a value of unity.

Thus we determine the frequency ω_0 and value of K such that

$$K^2 |S_{11}(j\omega_0 - \alpha)|^2 = 1 \tag{8.48}$$

and

$$\frac{d}{d\omega}|S_{11}(j\omega_0 - \alpha)|^2 = 1 \tag{8.49}$$

are simultaneously satisfied. There will, in general, be more than one value of K which satisfies these equations so the minimum value must be chosen.

Having found the values K and ω_0 for a given value of α we then formulate the input impedance and synthesise the network. Now the input impedance (or admittance) is given by

$$Z_{in}(p) = \frac{1 \pm KS_{11}(p - \alpha)}{1 \mp KS_{11}(p - \alpha)} = \frac{N(p)}{D(p)} \tag{8.50}$$

We have chosen K so that S_{11} is completely reflective at $p = j\omega_0$; thus $Z_{in}(p)$ has a pair of transmission zeros at $p = \pm j\omega_0$. These may be extracted by removing a Brune section or its equivalent cross-coupled section shown in Figure 8.13. This

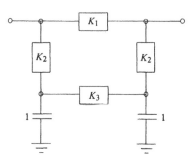

Figure 8.13   Cross-coupled Brune section

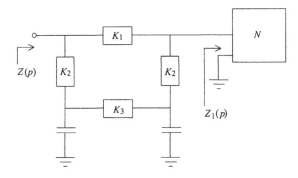

Figure 8.14   Extraction of a cross-coupled Brune section from $Z(p)$

is done by cascade synthesis. $Z(p)$ may be represented by the cascade of this Brune section followed by a remaining impedance (Figure 8.14).

Now the $ABCD$ parameters of the cross-coupled Brune section are given by

$$\begin{bmatrix} A & B \\ C & D \end{bmatrix} = \frac{1}{Y_o - Y_e} \begin{bmatrix} Y_o + Y_e & 2 \\ 2Y_e Y_o & Y_o + Y_e \end{bmatrix} \tag{8.51}$$

where

$$Y_e = jK_1 + \frac{K_2^2}{p + jK_3} \tag{8.52}$$

$$Y_o = Y_e^* \tag{8.53}$$

Now

$$Z(p) = \frac{AZ_1(p) + B}{CZ_1(p) + A} = \frac{N(p)}{D(p)} \tag{8.54}$$

Hence the remaining $Z_1(p)$ impedance is given by

$$Z_1(p) = \frac{BD(p) - AN(p)}{CN(p) - AD(p)} \tag{8.55}$$

and substituting for A, B and C in (8.55) we obtain

$$Z_1(p) = \frac{(p^2 + K_3^2)D(p) - K_2^2 p N(p)}{(K_2^4 + K_2^2 K_3^2 - 2K_1 K_2^2 K_3 + K_1^2 p^2)N(p) - K_2^2 p D(p)} \tag{8.56}$$

Now (8.56) is of degree 2 higher than $Z(p)$. For the transmission zeros to be successfully extracted it should be of degree 2 lower. Thus $Z_1(p)$ must contain the factor $(p^2 + \omega_0^2)^2$ in both numerator and denominator. Thus both the numerator and denominator, and their derivatives, should be zero at $p = \pm j\omega_0$.

From the numerator of $Z_1(p)$ we obtain

$$(p^2 + K_3^2)D(p) - K_2^2 p N(p) = 0|_{p=\pm j\omega_0} \tag{8.57}$$

and

$$\frac{d}{dp}[(p^2 + K_3^2)D(p) - K_2^2 p N(p)] = 0|_{p=\pm j\omega_0} \tag{8.58}$$

Hence

$$(K_3^2 - \omega_0^2)D(j\omega_0) - j\omega_0 K_2^2 N(j\omega_0) = 0 \tag{8.59}$$

$$(K_3^2 + 2j\omega_0)D(j\omega_0) + (K_3^2 - \omega_0^2)D'(j\omega_0)$$
$$- j\omega_0 K_2^2 N'(j\omega_0) - K_2^2 N(j\omega_0) = 0 \tag{8.60}$$

Solving (8.59) and (8.60) simultaneously we obtain

$$K_2 = \frac{2j\omega_0 D^2(j\omega_0)}{N(j\omega_0)D(j\omega_0) + j\omega_0[N'(j\omega_0)D(j\omega_0) - D'(j\omega_0)N(j\omega_0)]} \tag{8.61}$$

$$K_3 = \frac{j\omega_0 K_2^2 N(j\omega_0) + \omega_0 D(j\omega_0)}{D(j\omega_0)} \tag{8.62}$$

and similarly from the denominator of $Z_1(p)$

$$K_1 = K_2 \left[\frac{D'(j\omega_0)N(j\omega_0) - D(j\omega_0)N'(j\omega_0)}{2N^2(j\omega_0)} + \frac{D(j\omega_0)}{2j\omega_0 N(j\omega_0)} \right] \tag{8.63}$$

With these values of K_1, K_2 and K_3 the factor $(p^2 + \omega_0^2)^2$ will appear in the numerator and denominator of $Z_1(p)$ and it may be cancelled.

Synthesis of the remaining impedance $Z_1(p)$ now follows. In general it will have no poles or zeros on the imaginary axis or at infinity. Writing $Z_1(p)$ in terms of its even and odd parts we have

$$Z_1(p) = \frac{m_1(p) + n_1(p)}{m_2(p) + n_2(p)} \tag{8.64}$$

where $m_{1,2}$ and $n_{1,2}$ are even and odd polynomials in p. The real part of $Z_1(p)$ is then obtained from its even part, i.e.

$$\text{Re } Z(j\omega) = \text{Ev } Z(p)|_{p=j\omega_1}$$
$$= \frac{m_1(p)m_2(p) - n_1(p)n_2(p)}{m_2^2(p) - n_2^2(p)} \Big|_{p=j\omega_1} \tag{8.65}$$

At some frequency ω_1, $\text{Re } Z(j\omega)$ obtains its minimum value R_1; this is shown in Figure 8.15. We can extract a resistor R_1, leaving a positive real remainder, where ω_1 and R_1 are given by

$$R_1 = \text{Ev } Z(j\omega_1) \tag{8.66}$$

$$\frac{d}{dp} \text{Ev } Z(p) = 0|_{p=j\omega_1} \tag{8.67}$$

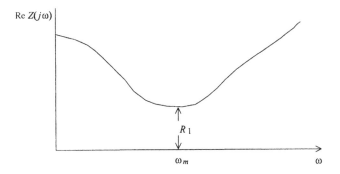

Figure 8.15 *The real part of the remaining impedance*

The remaining impedance is now $Z_2(p)$ where

$$Z_2(p) = Z_1(p) - R_1 \tag{8.68}$$

$Z(p)$ is purely reactive at ω_1 and thus has a transmission zero at ω_1. If ω_1 is finite we must extract a second cross-coupled Brune section. Alternatively if ω_2 is infinite then the remaining impedance may be synthesised as a lossy ladder network by a continued fraction expansion of parallel RC networks and inverters.

Finally having synthesised the network we add uniform dissipation by letting $p \to p + \alpha$. The result of a complete synthesis cycle for a degree 4 case is shown in Figure 8.16. One of the interesting features of this synthesis technique is that for most lowpass prototypes the network is of type (a). However, highpass prototypes may also be synthesised using this method and usually the network is of type (b). Further discussion of this and the synthesis of asymmetric prototypes is given in References 13 and 14.

8.6.3 Design example

The procedure has been applied to the design of a bandstop filter for a cellular radio base station application. The specification was

Centre frequency	$F_c = 845.75$ MHz
Stopband rejection	>20 dBc at $F_c \pm 550$ kHz
Passband loss	<1.5 dBc at $F_c \pm 750$ kHz

In this case dBc refers to attenuation with respect to the passband loss. This is the same specification as was previously described for triple-mode dielectric resonator filters. However, by using the approach described here a Q factor of 30 000 is no longer required. Instead we will use a resonator Q factor of 7000, yielding a loss of 6.8 dB. This may be tolerated by using a high intercept point low noise amplifier before the filter, as described previously. A Q factor of 7000 enables a smaller physical realisation than using dielectric resonators;

340 Theory and design of microwave filters

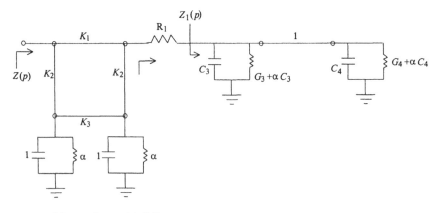

(a) case where ω_1 is infinite

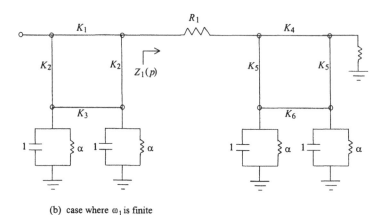

(b) case where ω_1 is finite

Figure 8.16 Synthesis of a lossy reflection prototype filter, $N = 4$

either a coaxial or a dual-mode conductor-loaded ceramic realisation may be considered.

The prototype network used was a degree 4 elliptic function filter with 22 dB passband return loss ripple and an equal stopband ripple with a reflection coefficient.

$$|S_{11}(j\omega)|^2 = \frac{F_N^2(\omega)}{1 + F_N^2\omega} \tag{8.69}$$

and

$$F_N(\omega) = \frac{(\omega^2 - \omega_1^2)(\omega^2 - \omega_2^2)}{(1 - \omega^2\omega_1^2)(1 - \omega^2\omega_2^2)} \tag{8.70}$$

$S_{11}(p)$ can then be found by factorisation giving

$$S_{11}(p) = \frac{(p^2 + 0.6124938)(p^2 + 0.1306136)}{(p^2 + 1.5189939p + 1)(p^2 + 0.31265488p + 1)} \tag{8.71}$$

In order to complete the synthesis the value of α must be computed from the ratio of loaded filter Q to unloaded resonator Q:

$$\alpha = \frac{f_0}{\Delta f Q_u} \tag{8.72}$$

In this case, because of the prototype used, Δf is the 3 dB bandwidth of the filter which is 1.299 MHz. In this example choosing resonator Q of 7000 yields α equal to 0.093. Then the value of the gain constant K and the resonant frequency of the Brune section are given from (8.48) and (8.49):

$$K = 0.457796 \qquad \omega_0 = 1.009825 \tag{8.73}$$

giving a passband loss of 6.78 dB.

The Brune section is extracted using the method described with

$$K_1 = -0.1157, \ K_2 = 0.05567, \ K_3 = 0.99651 \tag{8.74}$$

After extracting the Brune section and a unity inverter the remaining impedance $Z_1(p)$ is

$$Z_1(p) = \frac{0.0119251p^2 + 0.0284371p + 0.0179403}{2.396808p^2 + 2.117032p + 1.396287} \tag{8.75}$$

The minimum value of the real part of $Z_1(p)$ occurs at $\omega = \infty$. A resistor is then extracted of value

$$R_1 = Z_1(\infty) = 0.004975 \ \Omega \tag{8.76}$$

and the remaining impedance is

$$Z_2(p) = Z_1(p) - R_1 \tag{8.77}$$

Its admittance is

$$Y_2(p) = \frac{1}{Z_2(p)} = \frac{2.396808p^2 + 2.117032p + 1.396267}{0.179048p + 0.010993} \tag{8.78}$$

We then extract a capacitor C_3 of value

$$C_3 = \left.\frac{Y_2(p)}{p}\right|_{p=\infty} = 133.864 \text{ F} \tag{8.79}$$

The remaining admittance is then

$$Y_3(p) = Y_2(p) - C_3 p = \frac{0.645465p + 1.396267}{0.0179048p + 0.010993} \tag{8.80}$$

We then extract a shunt conductor of value

$$G_3 = \left.Y_3(p)\right|_{p=\infty} = 36.0498 \tag{8.81}$$

and the remaining impedance is

$$Y_4(p) = Y_3(p) - G_3 \tag{8.82}$$

Inverting Y_4 we obtain

$$Y_5(p) = \frac{1}{Y_4(p)} = 0.0179048p + 0.010993 \tag{8.83}$$

which is a capacitor C_4 in parallel with a conductance G_4 where

$$C_4 = 0.0179 \text{ F} \qquad G_4 = 0.01099 \tag{8.84}$$

After transforming p to $p + \alpha$ we obtain the final circuit of Figure 8.17.

After applying the appropriate bandpass transformation we obtain the simulated frequency response shown in Figure 8.18 where the return loss corresponds to the required bandstop response.

It is interesting to note that the first two resonators in the circuit in the cross-coupled Brune section have an unloaded Q factor of 7000 but the remaining elements have a lower Q. For example the first RC element in the ladder network has an admittance

$$C_3(p + \alpha) + G_3 = C_3\left(p + \alpha + \frac{G_3}{C_3}\right) \tag{8.85}$$

Thus new values of α for the ladder part of the filter are

$$\alpha_3 = \alpha + \frac{G_3}{C_3} = 0.3623 \tag{8.86}$$

$$\alpha_4 = \alpha + \frac{G_4}{C_4} = 0.7069 \tag{8.87}$$

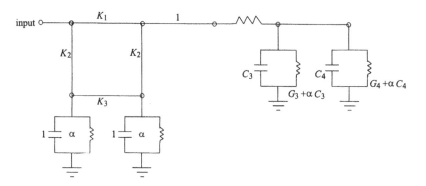

Figure 8.17 Network realisation of a degree 4 elliptic function lossy reflection mode lowpass prototype filter

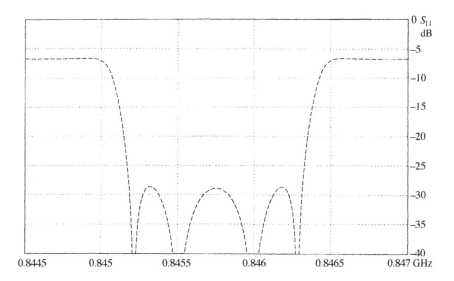

Figure 8.18   Simulated frequency response of a notch filter

Now from (8.77)

$$Q_u = \frac{f_0}{\Delta f \alpha} \qquad (8.88)$$

giving $Q_3 = 1797$ and $Q_4 = 921$. This remarkable result is explained by the fact that only the first two resonators contribute significantly to the sharpness of the response near band-edge; the other resonators contribute to the broader response.

8.7 Summary

Various techniques for the miniaturisation of filters are reviewed. A brief discussion on dielectric resonators points out the limitations in terms of increasing the dielectric constant or the number of degenerate modes. Superconducting filters offer near infinite Q in small physical size, but at the expense of complex cooling systems and poor intermodulation performance. SAW filters offer high levels of miniaturisation with relatively modest performance. Active filters exhibit near infinite Q when considered as small signal devices but suffer from poor large signal performance and have an associated noise figure. An alternative approach for receiver filters is to use predistorted lossy filters. High selectivity can be achieved in a small physical size provided the filter is preceded by a high intercept low noise amplifier. An intermodulation analysis is used to justify this approach and the details of the required filter synthesis procedure are presented.

It is to be expected that miniaturisation will remain an active area for future research and development efforts.

8.8 References

1. ONNES, H.K.: Communications of the Physics Laboratory, University of Leiden, 1911, **120b**, No. 3
2. BEDNORZ, J.G., and MÜLLER, K.A.: 'Possible high Tc superconductivity in the Ba–La–Cu–O system', *Zeitschrift für Physik*, 1986, **B64**, p. 189
3. PHILLIPS, W.A. et al.: 'An integrated 11 GHz cryogenic down converter', *IEEE Transactions on Applied Superconductivity*, 1995, **5** (2), pp. 2283–89
4. LANCASTER, M.J.: 'Passive microwave device applications of high-temperature superconductors' (Cambridge University Press, Cambridge, 1997), pp. 159–66
5. CAMPBELL, C.K.: 'Surface acoustic wave devices' (Academic Press, London, 1998)
6. TAGAMI, T., EHARA, H., NOGUCHI, K., and KOMASKI, T.: 'Resonator type SAW filter', *Oki Technical Review*, 1997, **63**, p. 59
7. LARSON, J.D., RUBY, R.C., BRADLEY, P., WEN, J., KOK, S., and CHIEN, A.: 'Power handling and temperature coefficient studies in FBAR duplexers for 1900 MHz PCS band', IEEE Ultrasonics Symposium, 2000, Paper no. 3H-6
8. CHANG, C.Y., and ITOH, T.: 'Microwave active filters base on a coupled negative resistance method', *IEEE Transactions on Microwave Theory and Techniques*, 1990, **30** (12), pp. 1879–84
9. HUNTER, I.C., and CHANDLER, S.R.: 'Intermodulation in active microwave filters', *IEE Proceedings on Microwaves, Antennas and Propagation*, 1998, **145** (1), pp. 7–12
10. WILLIAMS, A.E., BUSH, W.G., and BONETTI, R.R.: 'Predistortion techniques for multi-coupled resonator filters', *IEEE Transactions on Microwave Theory and Techniques*, 1985, **33** (5), pp. 402–7
11. FATHELBAB, W.M., and HUNTER, I.C.: 'A prototype for predistorted highpass networks', *International Journal of RF and Microwave Computer-aided Engineering*, 1998, **8** (2), pp. 156–60
12. POZAR, D.M.: 'Microwave engineering' (Addison-Wesley, Reading, MA, 1990), pp. 571–77
13. RHODES, J.D., and HUNTER, I.C.: 'Synthesis of reflection-mode prototype networks with dissipative circuit elements', *IEE Proceedings on Microwaves, Antennas and Propagation*, **144** (6), pp. 437–42
14. FATHELBAB, W.M., HUNTER, I.C., and RHODES, J.D.: 'Synthesis of lossy reflection-mode prototype networks with symmetrical and asymmetrical characteristics', *IEE Proceedings on Microwaves, Antennas and Propagation*, 1999, **146** (2), pp. 97–104

Index

ABCD matrix *see* transfer matrix
active microwave filters 325–7
  features 325–6
  limitations 326–7
  noise considerations 327
  third-order intermodulation distortion 326–7
admittance/impedance inverters 55–6
air-filled waveguide resonators 12
all-pole networks, Darlington synthesis 29
amplitude approximation to ideal lowpass filters 10–11
AMPS system 2–4
  notch filter requirements 2–4
  spectrum allocations 3
  transmit filter frequency response 3
applications 1–4
attenuation constant
  distributed circuits 46

bandpass filter loss effects 125–31
  finite dissipation effects 129–31
  lossy circuit element 130–1
  midband insertion loss 128–9
  Q factor approach 125–31
  stored energy 126
bandpass from lowpass transformation 107–18
bandpass resonators
  coupling bandwidth 134
  input impedance 133–4
  reflected group delay 132–3
  two resonator coupling measurement 132–4

bandpass transformation of a capacitor 108
bandpass transformation of an inductor 108
bandpass waveguide filter design
  see waveguide bandpass filter design
bandstop from lowpass transformations 118–25
bandwidth of filters
  and insertion loss 2
bandwidth scaling factor 108
bounded real condition 17
broadband TEM filters with generalised Chebyshev characteristics 151–65
  distributed lowpass filter 155–8
  highpass filters 162–5
  Richards transformation, applications 154–6, 162–3
  suspended substrate striplines 158–62, 164–5
Brune section
  Darlington synthesis 25, 27
  lossy predistortion filters 336–7, 341–2
Butterworth (maximally flat) prototype lowpass network approximation 49–56
  definition 49–50
  impedance/admittance inverters 55–6
  insertion loss characteristics/derivatives 50–1
  inverter coupled prototype filters 55–7
  ladder realisation 53, 55–6
  selectivity 51–2
  synthesis 52–4

Index

capacitors
  bandpass transformation 108
  impedance transformation 101–2
cascaded network analysis 30–1
cascaded network synthesis 25–9
cavity generalised direct-coupled
    waveguide filters *see* waveguide
    generalised direct-coupled cavity
    filters
cellular radio *see* GSM cellular radio
ceramic processing 298
Chebyshev bandstop filters 119
Chebyshev generalised prototype
    approximation 68–71
  synthesis with asymmetrically located
    transmission zeros 90–9
    degree 3 network example 90–9
    simulated response 94, 96
  synthesis with ladder-type networks
    85–90
    degree 3 filter example 86–9
    degree 7 filter example 89–90
  synthesis with symmetrically located
    transmission zeros 81–5
    degree 4 cross-coupled filter 85–6
  transmission and reflection functions
    71
  transmission zeros manipulation 70–1
  see also broadband TEM filters with
    generalised Chebyshev
    characteristics
Chebyshev polynomial 59
Chebyshev prototype lowpass network
    approximation 56–64
  Chebyshev polynomial 59
  for commensurate distributed
    networks 140
  equiripple response 58–9
  insertion loss ripple 58, 60
  lowpass ladder network synthesis
    62–3
  passband ripple 58, 60
  return loss 60
  third-order filter evaluation 59–62
Chebyshev response, stepped
    impedance unit element prototypes
    145–8
circular waveguide basic theory
  see waveguide basic theory

coaxial lines, Q factor 150–1
Cohn model
  cubic dielectric resonators 300
  resonators in the TE_{01} mode 276–7
combline filter 182–94
  admittance matrix 190
  design example, degree 4 Chebyshev
    filter 192–4
  equivalent circuit 183–7
  frequency transformation from
    lowpass prototype network 187–91
  input transformers 190–1
  inverter formulation 184–6
  narrow bandwidths 188
  phase shifter at input and output
    188–90
  principle of operation 182–3
  tunable 194
commensurate distributed networks
    137–44
  Chebyshev lowpass prototype 140
  design example 143–4
  distributed quasi-highpass filter 142–3
  distributed quasi-lowpass filter 140–1
  Richards' transformation 139–42
  short circuited stubs 139
  unit element (UE) 138
complex conjugate symmetry 239–40
complex frequency plane, Darlington
    synthesis 24
conservation of energy, for passive
    networks 37–9
continued fraction expansion, ladder
    networks 21
coupling bandwidth 134
coupling between resonators,
    measurement 132–4
coupling capacitance of coupled
    rectangular bars 180–2
cross-coupled array synthesis 244–8
cubic dielectric resonators 298–301
  ceramic processing 298
  characteristic impedance 301
  Cohn model 300
  Helmholtz equation for TE modes
    300

Darlington synthesis, two-port networks
    22–9

all-pole networks 29
Brune section 25, 27
cascade synthesis 25–9
Darlington C section 25, 27
Darlington D section 25, 27
input impedance 24
linear phase filter 28–9
transmission zeros 24–5
dielectric dual-mode resonator filters 283–90
  conductor loaded 285–90
    E field magnitude of fundamental mode 286
    example for GSM base station 289–90
    GSM cellular radio application 285
    Q factors 285
    unloaded Q factor and spurious performance 288–9
  principle of operation 283–5
dielectric material properties 271–2
dielectric resonator filters 150, 271–319
  features and principle of operation 271–2
  history and early designs 271
  miniaturisation 322
dielectric resonators, pucks 12, 271
  cylindrical 272–3
dielectric rod waveguides and the $TE_{01\delta}$ mode 272–83
  Cohn model 276–7
  couplings between dielectric resonators 282
  cross-coupled filters 282–3
  dielectric-air interface 275–6
  energy stored within dielectric and air cavity 282
  example analysis using resonance equation 281–3
  hybrid (HE) mode 273, 274
    transverse field components 275
  metallic enclosure 273–4
  resonant frequencies in resonators 281
  transverse electric mode (TE) 273, 274, 278–84
    dielectric-air interface 275–6
    ideal magnetic wall concept 276
    input impedance 279
    propagating/non-propagating modes 278
    resonance 279
    $TE_{01\delta}$ mode 279–84
    TE_{01} simplified model 275–84
    terminations 278–9
    transverse field components 275
    transverse field as a function of axial position 279–80
  transverse magnetic mode (TM) 273, 274
dielectric triple-mode resonator filters design 301–10
  cross-coupled prototype networks 305
  design example, bandstop filter 305–10
  simulated frequency response of complete filter 308–10
  hybrid reflection mode filter approach 303–5
  see also cubic dielectric resonators; spherical dielectric resonator filters
dielectric-loaded filters 309–17
  even and odd modes of a coupled-line structure 311–12
  exploded view of typical ceramic TEM diplexer 314–15
  inhomogeneous dielectric loading 311–13
  lumped equivalent circuit 313
  measured performance of a six-pole filter with cross-couplings 316–17
  measured performance of typical TEM diplexer 314–15
  principle of operation 309–14
  TEM coupled-line structure 310–11
  wave length in ceramic material 309–10
  waveguide filter construction 315–16
  waveguide filters performance 314–16
direct-coupled generalised cavity waveguide filters *see* waveguide generalised direct-coupled cavity filters
distributed circuits
  analysis 46–7
  attenuation constant 46
  input impedance 47
  phase constant 46
  transfer matrix 46–7

dual-mode filters *see* dielectric dual-mode resonator filters; waveguide dual-mode filters

elliptic function prototype network approximation 64–8
    characteristics 64
    disadvantages 68
    element value explicit formulae 67–8
    highpass filter 65–7
    passband/stopband turning points 65–7
    synthesis methods 67
    transmission zeros location 64
equiripple filter characteristics/transfer functions 11, 58–9
even- and odd-mode analysis of symmetrical networks 41–3
    image parameter analysis 45
    see also waveguide dual-mode filters; waveguide extracted pole filters
extracted pole synthesis 240–4
extracted pole waveguide filters *see* waveguide extracted pole filters

finite element analysis 137
forward transmission coefficient 37
Foster synthesis/realisation 20–1
Fourier transforms, ideal lowpass filter 5–6
fringing capacitance of coupled rectangular bars 180–2

generalised Chebyshev filters
    see Chebyshev generalised prototype approximation
generalised direct-coupled cavity waveguide filters *see* waveguide generalised direct-coupled cavity filters
group delay
    ideal lowpass filter 5
    importance of 71
    for ladder networks 71–3
    maximally flat group delay lowpass prototype 73–5
    minimum phase networks 9–10
    see also phase delay

GSM cellular radio 1–5
    base station filter requirements 2–5, 151, 285
    dielectric resonant filter for 289–90

Helmholtz equations
    cubic dielectric resonators 300
    spherical resonators 291
    waveguides 202–3, 203–4, 213, 215–16
highpass filters *see* broadband TEM filters with generalised Chebyshev characteristics; elliptic function prototype network approximation; interdigital filters; lossy predistortion filters
highpass from lowpass transformation 105–7
Hilbert transforms 8
Hurwitz polynomials
    minimum phase networks 7
    stepped impedance unit element prototypes 145

ideal lowpass filters 4–7
    definition 4
    frequency response 5
    group delay 5
    impulse response 6
    passband group delay 6–7
    phase response 5
ideal magnetic wall concept, dielectric waveguides 276
image parameters, analysis by 44–6
    image propagation function 44
impedance transformations
    see transformations on lumped prototype networks
impedance/admittance inverters 55–6
impulse response of filters 78–81
    ideal lowpass filter 5–7
    time domain sidelobe damping 79–81
inductors
    bandpass transformation 108
    impedance transformation 101–2
input coupling measurement 131
input impedance
    Darlington synthesis 24
    distributed circuits 47

lossless networks 18-19
one-port networks 17
and reflection coefficients 17
transfer matrix 31-3
insertion loss
and filter bandwidth 2
and the scattering matrix 39
interdigital filters 167-82
  design examples, degree 4 Chebychev
    prototype filter 173-4, 176-8
  narrowband interdigital filters 174-6
  physical design 177-82
  Richards' highpass transformations
    171-3
  three-wire interdigital line 168-70
  unit elements (UEs) 170-1, 174-6
inverters
  characteristic impedance 110-11
  impedance/admittance 55-6
  narrowband approximation 110-11
  pi network of reactances 109-10

ladder networks 20-3
  continued fraction expansion 21
  residues 22
linear passive time-invariant networks
  15-17
  see also one-port network
linear phase filter
  Darlington synthesis 28-9
linearity, one-port network 16
lossless networks 18-20
  reactance function 19
lossy predistortion filters 327-43
  classical predistortion design 331-5
    disadvantages 335
    highpass prototype network with
      finite dissipation loss 334
    insertion loss 334
    predistorted maximally flat
      prototype 333
    principle of operation 331
  intermodulation products 330
  low-noise front end 329-30
  reflection mode type design 335-43
    cross-coupled Brune section
      extraction 336-7, 341-2
    design example, bandstop filter for a
      cellular radio base station 339-43

network realisation of a degree 4
  filter 342
principle of operation 335-6
reflection mode bandstop resonator
  335
simulated frequency response for a
  notch filter 343
synthesis (for $N = 4$) 339-40
lowpass filters *see* broadband TEM
  filters with generalised Chebyshev
  characteristics; combline filter;
  commensurate distributed networks;
  group delay; lumped lowpass
  prototype networks; transformations
  on lumped prototype networks;
  waveguide dual-mode filters;
  waveguide generalised direct-
  coupled cavity filter
lowpass from lowpass transformation
  103-5
lowpass ladder network 10-11
lowpass to bandpass transformation
  107-18
  bandwidth scaling factor 108
  characteristic admittance 110-11
  design example 114-16
    simulated response 124, 125
  impedance scaling 112
  narrow bandwidth problems 111-13
  nodal admittance matrix scaling
    116-18
lowpass to bandstop transformations
  118-25
  Chebyshev filters 119
  design example 121-5
  inverter realisation 119
  narrowband 119-21
lowpass to highpass transformation
  105-7
lowpass to lowpass transformation
  103-5
lumped lowpass prototype networks *see*
  Butterworth (maximally flat)
  prototype lowpass network
  approximation; Chebyshev
  generalised prototype
  approximation; Chebyshev
  prototype lowpass network
  approximation; elliptic function

prototype network approximation; group delay; minimum phase networks; phase delay

Maxwell's equations
  spherical dielectric resonator filters 291
  waveguides 202, 203–4, 213
microstrip circuit pattern 149–50
microwave filters, practical realisation 12
miniaturisation techniques *see* active microwave filters; dielectric resonator filters, miniaturisation; lossy predistortion filters; superconducting filters; surface acoustic wave (SAW) resonators and filters
minimum phase networks 7–10
  group delay 9–10
multiple degenerate modes in waveguide cavities 255

N-wire lines *see* interdigital filters
narrowband bandstop filter 121–4
narrowband coaxial resonator filters 197–8
narrowband impedance transformer 112–13
Newton-Raphson technique, waveguide bandpass filters 226
nodal admittance matrix scaling 116–18

odd-mode analysis *see* even- and odd-mode analysis of symmetrical networks
one-port network
  bounded real condition 17
  Laplace transform for 15–16
  linearity 16
  passivity 17
  time invariance 16

parallel coupled transmission lines 165–7
  parallel coupled N wire line 165–7
  static capacitances of an N wire line 167
parallel coupled-line filter 194–7
  application 195

design using Richards' highpass transformation 196
  principle 194–5
parallel tuned circuit, impedance of 19–20
passivity 17
phase constant, distributed circuits 46
phase delay
  combined phase and amplitude approximation 77–8
  synthesis of 78
  equidistant linear phase approximation 75–6
  see also group delay
phase response, ideal lowpass filter 5
pi network transfer matrix 109–11
PIN diode switches 4
predistortion *see* lossy predistortion filters
pucks *see* dielectric resonators, pucks

Q factor
  for coaxial lines 150–1
  dielectric resonator filters 271–2
    dual-mode conductor-loaded 285
  GSM base station requirements 2
  interdigital filters 182
  and passband loss 12, 125–31

radar warning receivers 151
reactance function 19
reciprocal networks 41
reciprocity, odd- and even mode analysis 43
rectangular waveguide basic theory *see* waveguide basic theory
reflected group delay
  use of 131–3
reflection coefficients
  and input impedance 17
  scattering matrix 37
residues, ladder networks 22
reverse transmission coefficient 37
Richards' transformation
  application to interdigital filters 171–3
  applications to broadband TEM filters 154–6, 162–3
  with commensurate distributed networks 139–42
rotational transformations 260–1

SAW *see* surface acoustic wave (SAW) resonators and filters
scattering matrix/parameters 34–41
  conservation of energy 37–9
  Darlington technique as a ladder network 40
  degree 3 Butterworth filter 39
  input impedance 34–7
  input/output reflection coefficients 37
  insertion loss 39, 41
  network transducer power gain 35–7
  reciprocal networks 41
  and the transfer matrix 40–1
  unitary condition 38
short circuited stubs 139
spherical dielectric resonator filters 290–8
  E and H fields 297–9
  equivalent circuit 294
  field solutions with Maxwell's equations 291
  field solutions for symmetric TE and TM modes 291–2
  Helmholtz equation for TE modes 291
  resonance equations, TE and TM modes 294
  resonant frequency 292
  spurious to fundamental frequency ratio 298
  transmission line equivalent circuit 293
  transverse field intensity 295–7
stepped impedance unit element prototypes 144–51
  Chebyshev response 145–8
  coaxial lines 150–1
  dielectric resonator filters 150
  Hurwitz polynomial 145
  microstrip circuit pattern 149–50
  physical realisation 149–51
  unit elements (UEs) in 144–5, 148–9
superconducting filters 322–3
  YBCO high temperature superconductor 322
surface acoustic wave (SAW) resonators and filters 323–5
  applications 323
  performance 325
  principle of operation 323–4

suspended substrate striplines 158–62, 164–5
symmetry, odd- and even mode analysis 43
synthesis
  continued fraction expansion 21–2
  Foster synthesis 20
  ladder networks 20–3
  see also Butterworth (maximally flat) prototype lowpass network approximation; Chebyshev generalised prototype approximation; Darlington synthesis

TDMA *see* time division multiplex access (TDMA) technique
$TE_{01\delta}$ mode *see* dielectric rod waveguides and the $TE_{01\delta}$ mode
TEM transmission line filters *see* broadband TEM filters with generalised Chebyshev characteristics; combline filter; commensurate distributed networks; interdigital filters; narrowband coaxial resonator filters; parallel coupled transmission lines; parallel coupled-line filter; stepped impedance unit element prototypes
time division multiplex access (TDMA) technique 1–2
  see also GSM cellular radio
time domain characteristics of filters 78–81
  see also impulse response of filters
time invariance, one-port network 16
transfer matrix 29–34
  cascaded network analysis 30–1
  distributed circuits 46
  input impedance calculations 31–2, 33–4
  and the scattering matrix 40–1
  for series connected elements 32
  for shunt connected elements 32–3
transformations on lumped prototype networks 101–36
  bandpass of an inductor 108
  bandpass of a capacitor 108–9
  bandwidth scaling factor 108
  impedance transformations 101–3

lowpass to bandpass 107–18
lowpass to bandstop 118–25
lowpass to highpass 105–7
lowpass to lowpass 103–5
transmission coefficients, forward and reverse 37
transmission zeros, Darlington synthesis 24–5
triple-mode dielectric resonator filters *see* cubic dielectric resonators; dielectric triple-mode resonator filters design; spherical dielectric resonator filters
two-port network analysis/synthesis *see* Darlington synthesis; scattering matrix/parameters; transfer matrix

unit elements (UEs) 138
  with interdigital filters 170–1, 174–6
  in stepped impedance unit element prototypes 144–5
unitary condition
  for lossless network 38

waveguide bandpass filter design 220–30
  all-pole type design procedure 225–8
  design example 228–30
  inductive iris section 222–5
  Newton-Raphson technique 226
  normalisation 220–1
  principle of operation 220
  shunt inductive discontinuities 221–2
waveguide basic theory
  air filled resonators 12
  circular waveguides, TE modes 213–15
    boundary conditions 213–14
    Helmholtz equation 213
    Maxwell equations 213
    mode numbers 214–15
  circular waveguides, TM modes 215–17
    boundary conditions 216
    circular resonators 217
    Helmholtz equation 215–16
    numerical example 217–19
    resonators mode chart 218
  Helmholtz equations 202–3

main features 201–2
Maxwell's equations 202
rectangular waveguides, relative cut-off frequency, any mode 209
rectangular waveguides, spurious resonances 212
rectangular waveguides, TE modes 203–8
  E and H fields 206–8
  Helmholtz equation 203–4
  Maxwell's equations 203–4
  TE_{10} mode 206–8
  TE_{101} mode 210
rectangular waveguides, TM modes 208–9
rectangular waveguides, waveguide resonators 209–10
rectangular waveguides, waveguide unloaded Q, example 210–12
see also dielectric rod waveguides
waveguide dual-mode filters 255–67
  asymmetric realisations 265–7
    pivotal positions and rotational angles for in-line prototype networks 266–7
  equivalent circuit 255–6
  even-mode case representation 260
  even-mode network for the cross-coupled array prototype 258
  example, degree 6 linear phase filter with four transmission zeroes at infinity 263–6
  general cross-coupled array prototype network suitable for dual-mode in-line realisation 257
  lowpass prototype for a sixth-degree in-line filter 256
  nodal matrix transformations 259
  realisability conditions 263
  rotational transformations 260–1
waveguide extracted pole filters 239–54
  complex conjugate symmetry synthesis procedure 239–40
  cross-coupled array synthesis 244–8
  design example, sixth-degree prototype 252–4
  even-mode network for the complex-conjugate symmetric array 249–51

extracted pole synthesis 240–4
pole cavity pair synthesis 250–1
realisation 249–51
realisation in T_{011} mode cavities 254
rectangular waveguide arrangement 248
simulated response 248
waveguide generalised direct-coupled cavity filters 230–9
cross-coupled lowpass prototype 231–2
equivalent circuits 232–3, 235–6
limitations 238–9
midband susceptances and electrical lengths 231
transfer matrix 233–5
transformed (dual) lowpass prototype even-mode circuit with equal value conductors 236–7
waveguides
dielectric rod *see* dielectric rod waveguides
wideband TEM filters *see* broadband TEM filters
YBCO high temperature superconductor material 322

zero-bandwidth approximation *see* Butterworth (maximally flat) prototype lowpass network approximation